P9-CBI-333

Pitman Research Notes in Mathematics Series

Main Editors
H. Brezis, Université de Paris
R.G. Douglas, State University of New York at Stony Brook
A. Jeffrey, University of Newcastle upon Tyne *(Founding Editor)*

Editorial Board
R. Aris, University of Minnesota
A. Bensoussan, INRIA, France
S. Bloch, University of Chicago
B. Bollobás, University of Cambridge
W. Bürger, Universität Karlsruhe
S. Donaldson, University of Oxford
J. Douglas Jr, Purdue University
R.J. Elliott, University of Alberta
G. Fichera, Università di Roma
R.P. Gilbert, University of Delaware
R. Glowinski, Université de Paris
K.P. Hadeler, Universität Tübingen
K. Kirchgässner, Universität Stuttgart
B. Lawson, State University of New York at Stony Brook
W.F. Lucas, Claremont Graduate School
R.E. Meyer, University of Wisconsin-Madison
S. Mori, Nagoya University
L.E. Payne, Cornell University
G.F. Roach, University of Strathclyde
J.H. Seinfeld, California Institute of Technology
B. Simon, California Institute of Technology
S.J. Taylor, University of Virginia

Submission of proposals for consideration

Suggestions for publication, in the form of outlines and representative samples, are invited by the Editorial Board for assessment. Intending authors should approach one of the main editors or another member of the Editorial Board, citing the relevant AMS subject classifications. Alternatively, outlines may be sent directly to the publisher's offices. Refereeing is by members of the board and other mathematical authorities in the topic concerned, throughout the world.

Preparation of accepted manuscripts

On acceptance of a proposal, the publisher will supply full instructions for the preparation of manuscripts in a form suitable for direct photo-lithographic reproduction. Specially printed grid sheets can be provided and a contribution is offered by the publisher towards the cost of typing. Word processor output, subject to the publisher's approval, is also acceptable.

Illustrations should be prepared by the authors, ready for direct reproduction without further improvement. The use of hand-drawn symbols should be avoided wherever possible, in order to maintain maximum clarity of the text.

The publisher will be pleased to give any guidance necessary during the preparation of a typescript, and will be happy to answer any queries.

Important note

In order to avoid later retyping, intending authors are strongly urged not to begin final preparation of a typescript before receiving the publisher's guidelines. In this way it is hoped to preserve the uniform appearance of the series.

Longman Scientific & Technical
Longman House
Burnt Mill
Harlow, Essex, CM20 2JE
UK
(Telephone (0279) 426721)

Titles in this series. A full list is available on request from the publisher.

51 Subnormal operators
J B Conway

52 Wave propagation in viscoelastic media
F Mainardi

53 Nonlinear partial differential equations and their applications: Collège de France Seminar. Volume I
H Brezis and J L Lions

54 Geometry of Coxeter groups
H Hiller

55 Cusps of Gauss mappings
T Banchoff, T Gaffney and C McCrory

56 An approach to algebraic K-theory
A J Berrick

57 Convex analysis and optimization
J-P Aubin and R B Vintner

58 Convex analysis with applications in the differentiation of convex functions
J R Giles

59 Weak and variational methods for moving boundary problems
C M Elliott and J R Ockendon

60 Nonlinear partial differential equations and their applications: Collège de France Seminar. Volume II
H Brezis and J L Lions

61 Singular Systems of differential equations II
S L Campbell

62 Rates of convergence in the central limit theorem
Peter Hall

63 Solution of differential equations by means of one-parameter groups
J M Hill

64 Hankel operators on Hilbert Space
S C Power

65 Schrödinger-type operators with continuous spectra
M S P Eastham and H Kalf

66 Recent applications of generalized inverses
S L Campbell

67 Riesz and Fredholm theory in Banach algebra
B A Barnes, G J Murphy, M R F Smyth and T T West

68 Evolution equations and their applications
K Kappel and W Schappacher

69 Generalized solutions of Hamilton–Jacobi equations
P L Lions

70 Nonlinear partial differential equations and their applications: Collège de France Seminar. Volume III
H Brezis and J L Lions

71 Spectral theory and wave operators for the Schrödinger equation
A M Berthier

72 Approximation of Hilbert space operators I
D A Herrero

73 Vector valued Nevanlinna theory
H J W Ziegler

74 Instability, nonexistence and weighted energy methods in fluid dynamics and related theories
B Straughan

75 Local bifurcation and symmetry
A Vanderbauwhede

76 Clifford analysis
F Brackx, R Delanghe and F Sommen

77 Nonlinear equivalence, reduction of PDEs to ODEs and fast convergent numerical methods
E E Rosinger

78 Free boundary problems, theory and applications. Volume I
A Fasano and M Primicerio

79 Free boundary problems, theory and applications. Volume II
A Fasano and M Primicerio

80 Symplectic geometry
A Crumeyrolle and J Grifone

81 An algorithmic analysis of a communication model with retransmission of flawed messages
D M Lucantoni

82 Geometric games and their applications
W H Ruckle

83 Additive groups of rings
S Feigelstock

84 Nonlinear partial differential equations and their applications: Collège de France Seminar. Volume IV
H Brezis and J L Lions

85 Multiplicative functionals on topological algebras
T Husain

86 Hamilton–Jacobi equations in Hilbert spaces
V Barbu and G Da Prato

87 Harmonic maps with symmetry, harmonic morphisms and deformations of metric
P Baird

88 Similarity solutions of nonlinear partial differential equations
L Dresner

89 Contributions to nonlinear partial differential equations
C Bardos, A Damlamian, J I Díaz and J Hernández

90 Banach and Hilbert spaces of vector-valued functions
J Burbea and P Masani

91 Control and observation of neutral systems
D Salamon

92 Banach bundles, Banach modules and automorphisms of C^*-algebras
M J Dupré and R M Gillette

93 Nonlinear partial differential equations and their applications: Collège de France Seminar. Volume V
H Brezis and J L Lions

94 Computer algebra in applied mathematics: an introduction to MACSYMA
R H Rand

95 Advances in nonlinear waves. Volume I
L Debnath

96 FC-groups
M J Tomkinson

97 Topics in relaxation and ellipsoidal methods
M Akgül

98 Analogue of the group algebra for topological semigroups
H Dzinotyiweyi

99 Stochastic functional differential equations
S E A Mohammed

100 Optimal control of variational inequalities
V Barbu
101 Partial differential equations and dynamical systems
W E Fitzgibbon III
102 Approximation of Hilbert space operators Volume II
C Apostol, L A Fialkow, D A Herrero and D Voiculescu
103 Nondiscrete induction and iterative processes
V Ptak and F-A Potra
104 Analytic functions – growth aspects
O P Juneja and G P Kapoor
105 Theory of Tikhonov regularization for Fredholm equations of the first kind
C W Groetsch
106 Nonlinear partial differential equations and free boundaries. Volume I
J I Díaz
107 Tight and taut immersions of manifolds
T E Cecil and P J Ryan
108 A layering method for viscous, incompressible L_p flows occupying R^n
A Douglis and E B Fabes
109 Nonlinear partial differential equations and their applications: Collège de France Seminar. Volume VI
H Brezis and J L Lions
110 Finite generalized quadrangles
S E Payne and J A Thas
111 Advances in nonlinear waves. Volume II
L Debnath
112 Topics in several complex variables
E Ramírez de Arellano and D Sundararaman
113 Differential equations, flow invariance and applications
N H Pavel
114 Geometrical combinatorics
F C Holroyd and R J Wilson
115 Generators of strongly continuous semigroups
J A van Casteren
116 Growth of algebras and Gelfand–Kirillov dimension
G R Krause and T H Lenagan
117 Theory of bases and cones
P K Kamthan and M Gupta
118 Linear groups and permutations
A R Camina and E A Whelan
119 General Wiener–Hopf factorization methods
F-O Speck
120 Free boundary problems: applications and theory. Volume III
A Bossavit, A Damlamian and M Fremond
121 Free boundary problems: applications and theory. Volume IV
A Bossavit, A Damlamian and M Fremond
122 Nonlinear partial differential equations and their applications: Collège de France Seminar. Volume VII
H Brezis and J L Lions
123 Geometric methods in operator algebras
H Araki and E G Effros
124 Infinite dimensional analysis–stochastic processes
S Albeverio
125 Ennio de Giorgi Colloquium
P Krée
126 Almost-periodic functions in abstract spaces
S Zaidman
127 Nonlinear variational problems
A Marino, L Modica, S Spagnolo and M Degliovanni
128 Second-order systems of partial differential equations in the plane
L K Hua, W Lin and C-Q Wu
129 Asymptotics of high-order ordinary differential equations
R B Paris and A D Wood
130 Stochastic differential equations
R Wu
131 Differential geometry
L A Cordero
132 Nonlinear differential equations
J K Hale and P Martinez-Amores
133 Approximation theory and applications
S P Singh
134 Near-rings and their links with groups
J D P Meldrum
135 Estimating eigenvalues with *a posteriori/a priori* inequalities
J R Kuttler and V G Sigillito
136 Regular semigroups as extensions
F J Pastijn and M Petrich
137 Representations of rank one Lie groups
D H Collingwood
138 Fractional calculus
G F Roach and A C McBride
139 Hamilton's principle in continuum mechanics
A Bedford
140 Numerical analysis
D F Griffiths and G A Watson
141 Semigroups, theory and applications. Volume I
H Brezis, M G Crandall and F Kappel
142 Distribution theorems of L-functions
D Joyner
143 Recent developments in structured continua
D De Kee and P Kaloni
144 Functional analysis and two-point differential operators
J Locker
145 Numerical methods for partial differential equations
S I Hariharan and T H Moulden
146 Completely bounded maps and dilations
V I Paulsen
147 Harmonic analysis on the Heisenberg nilpotent Lie group
W Schempp
148 Contributions to modern calculus of variations
L Cesari
149 Nonlinear parabolic equations: qualitative properties of solutions
L Boccardo and A Tesei
150 From local times to global geometry, control and physics
K D Elworthy

151 A stochastic maximum principle for optimal control of diffusions
U G Haussmann
152 Semigroups, theory and applications. Volume II
H Brezis, M G Crandall and F Kappel
153 A general theory of integration in function spaces
P Muldowney
154 Oakland Conference on partial differential equations and applied mathematics
L R Bragg and J W Dettman
155 Contributions to nonlinear partial differential equations. Volume II
J I Díaz and P L Lions
156 Semigroups of linear operators: an introduction
A C McBride
157 Ordinary and partial differential equations
B D Sleeman and R J Jarvis
158 Hyperbolic equations
F Colombini and M K V Murthy
159 Linear topologies on a ring: an overview
J S Golan
160 Dynamical systems and bifurcation theory
M I Camacho, M J Pacifico and F Takens
161 Branched coverings and algebraic functions
M Namba
162 Perturbation bounds for matrix eigenvalues
R Bhatia
163 Defect minimization in operator equations: theory and applications
R Reemtsen
164 Multidimensional Brownian excursions and potential theory
K Burdzy
165 Viscosity solutions and optimal control
R J Elliott
166 Nonlinear partial differential equations and their applications: Collège de France Seminar. Volume VIII
H Brezis and J L Lions
167 Theory and applications of inverse problems
H Haario
168 Energy stability and convection
G P Galdi and B Straughan
169 Additive groups of rings. Volume II
S Feigelstock
170 Numerical analysis 1987
D F Griffiths and G A Watson
171 Surveys of some recent results in operator theory. Volume I
J B Conway and B B Morrel
172 Amenable Banach algebras
J-P Pier
173 Pseudo-orbits of contact forms
A Bahri
174 Poisson algebras and Poisson manifolds
K H Bhaskara and K Viswanath
175 Maximum principles and eigenvalue problems in partial differential equations
P W Schaefer
176 Mathematical analysis of nonlinear, dynamic processes
K U Grusa
177 Cordes' two-parameter spectral representation theory
D F McGhee and R H Picard
178 Equivariant K-theory for proper actions
N C Phillips
179 Elliptic operators, topology and asymptotic methods
J Roe
180 Nonlinear evolution equations
J K Engelbrecht, V E Fridman and E N Pelinovski
181 Nonlinear partial differential equations and their applications: Collège de France Seminar. Volume IX
H Brezis and J L Lions
182 Critical points at infinity in some variational problems
A Bahri
183 Recent developments in hyperbolic equations
L Cattabriga, F Colombini, M K V Murthy and S Spagnolo
184 Optimization and identification of systems governed by evolution equations on Banach space
N U Ahmed
185 Free boundary problems: theory and applications. Volume I
K H Hoffmann and J Sprekels
186 Free boundary problems: theory and applications. Volume II
K H Hoffmann and J Sprekels
187 An introduction to intersection homology theory
F Kirwan
188 Derivatives, nuclei and dimensions on the frame of torsion theories
J S Golan and H Simmons
189 Theory of reproducing kernels and its applications
S Saitoh
190 Volterra integrodifferential equations in Banach spaces and applications
G Da Prato and M Iannelli
191 Nest algebras
K R Davidson
192 Surveys of some recent results in operator theory. Volume II
J B Conway and B B Morrel
193 Nonlinear variational problems. Volume II
A Marino and M K V Murthy
194 Stochastic processes with multidimensional parameter
M E Dozzi
195 Prestressed bodies
D Iesan
196 Hilbert space approach to some classical transforms
R H Picard
197 Stochastic calculus in application
J R Norris
198 Radical theory
B J Gardner
199 The C^*-algebras of a class of solvable Lie groups
X Wang
200 Stochastic analysis, path integration and dynamics
K D Elworthy and J C Zambrini

201 Riemannian geometry and holonomy groups
S Salamon
202 Strong asymptotics for extremal errors and polynomials associated with Erdős type weights
D S Lubinsky
203 Optimal control of diffusion processes
V S Borkar
204 Rings, modules and radicals
B J Gardner
205 Two-parameter eigenvalue problems in ordinary differential equations
M Faierman
206 Distributions and analytic functions
R D Carmichael and D Mitrovic
207 Semicontinuity, relaxation and integral representation in the calculus of variations
G Buttazzo
208 Recent advances in nonlinear elliptic and parabolic problems
P Bénilan, M Chipot, L Evans and M Pierre
209 Model completions, ring representations and the topology of the Pierce sheaf
A Carson
210 Retarded dynamical systems
G Stepan
211 Function spaces, differential operators and nonlinear analysis
L Paivarinta
212 Analytic function theory of one complex variable
C C Yang, Y Komatu and K Niino
213 Elements of stability of visco-elastic fluids
J Dunwoody
214 Jordan decomposition of generalized vector measures
K D Schmidt
215 A mathematical analysis of bending of plates with transverse shear deformation
C Constanda
216 Ordinary and partial differential equations. Volume II
B D Sleeman and R J Jarvis
217 Hilbert modules over function algebras
R G Douglas and V I Paulsen
218 Graph colourings
R Wilson and R Nelson
219 Hardy-type inequalities
A Kufner and B Opic
220 Nonlinear partial differential equations and their applications: Collège de France Seminar. Volume X
H Brezis and J L Lions
221 Workshop on dynamical systems
E Shiels and Z Coelho
222 Geometry and analysis in nonlinear dynamics
H W Broer and F Takens
223 Fluid dynamical aspects of combustion theory
M Onofri and A Tesei
224 Approximation of Hilbert space operators. Volume I. 2nd edition
D Herrero
225 Operator theory: proceedings of the 1988 GPOTS–Wabash conference
J B Conway and B B Morrel
226 Local cohomology and localization
J L Bueso Montero, B Torrecillas Jover and A Verschoren
227 Nonlinear waves and dissipative effects
D Fusco and A Jeffrey
228 Numerical analysis 1989
D F Griffiths and G A Watson
229 Recent developments in structured continua. Volume III
D De Kee and P Kaloni
230 Boolean methods in interpolation and approximation
F J Delvos and W Schempp
231 Further advances in twistor theory. Volume I
L J Mason and L P Hughston
232 Further advances in twistor theory. Volume II
L J Mason and L P Hughston
233 Geometry in the neighborhood of invariant manifolds of maps and flows and linearization
U Kirchgraber and K Palmer
234 Quantales and their applications
K I Rosenthal
235 Integral equations and inverse problems
V Petkov and R Lazarov
236 Pseudo-differential operators
S R Simanca
237 A functional analytic approach to statistical experiments
I M Bomze
238 Quantum mechanics, algebras and distributions
D Dubin and M Hennings
239 Hamilton flows and evolution semigroups
J Gzyl
240 Topics in controlled Markov chains
V S Borkar
241 Invariant manifold theory for hydrodynamic transition
S Sritharan
242 Lectures on the spectrum of $L^2(\Gamma\backslash G)$
F L Williams
243 Progress in variational methods in Hamiltonian systems and elliptic equations
M Girardi, M Matzeu and F Pacella
244 Optimization and nonlinear analysis
A Ioffe, M Marcus and S Reich
245 Inverse problems and imaging
G F Roach
246 Semigroup theory with applications to systems and control
N U Ahmed
247 Periodic-parabolic boundary value problems and positivity
P Hess
248 Distributions and pseudo-differential operators
S Zaidman
249 Progress in partial differential equations: the Metz surveys
M Chipot and J Saint Jean Paulin
250 Differential equations and control theory
V Barbu

251 Stability of stochastic differential equations with respect to semimartingales
X Mao
252 Fixed point theory and applications
J Baillon and M Théra
253 Nonlinear hyperbolic equations and field theory
M K V Murthy and S Spagnolo
254 Ordinary and partial differential equations. Volume III
B D Sleeman and R J Jarvis
255 Harmonic maps into homogeneous spaces
M Black
256 Boundary value and initial value problems in complex analysis: studies in complex analysis and its applications to PDEs 1
R Kühnau and W Tutschke
257 Geometric function theory and applications of complex analysis in mechanics: studies in complex analysis and its applications to PDEs 2
R Kühnau and W Tutschke
258 The development of statistics: recent contributions from China
X R Chen, K T Fang and C C Yang
259 Multiplication of distributions and applications to partial differential equations
M Oberguggenberger
260 Numerical analysis 1991
D F Griffiths and G A Watson
261 Schur's algorithm and several applications
M Bakonyi and T Constantinescu

C Bandle
University of Basel, Switzerland
J Bemelmans
University of Saarlandes, Germany
M Chipot
University of Metz, France
M Grüter
University of Saarlandes, Germany

and

J Saint Jean Paulin
University of Metz, France

(Editors)

Progress in partial differential equations: elliptic and parabolic problems

Longman Scientific & Technical

Copublished in the United States with
John Wiley & Sons, Inc., New York

Longman Scientific & Technical
Longman Group UK Limited
Longman House, Burnt Mill, Harlow
Essex CM20 2JE, England
and Associated companies throughout the world.

Copublished in the United States with
John Wiley & Sons Inc., 605 Third Avenue, New York, NY 10158

© Longman Group UK Limited 1992

All rights reserved; no part of this publication may be reproduced, stored in a retrieval system, or transmitted in any form or by any means, electronic, mechanical, photocopying, recording, or otherwise, without the prior written permission of the Publishers, or a licence permitting restricted copying in the United Kingdom issued by the Copyright Licensing Agency Ltd, 90 Tottenham Court Road, London, W1P 9HE

First published 1992

AMS Subject Classification: (Main) 35XX, 49XX, 65XX
(Subsidiary) 76XX

ISSN 0269-3674

ISBN 0 582 07252 2

British Library Cataloguing in Publication Data

A catalogue record for this book is
available from the British Library

Library of Congress Cataloging-in-Publication Data

Progress in partial differential equations. Elliptic and parabolic
problems / C. Bandle ... [et al.] (editors)
p. cm. -- (Pitman research notes in mathematics series, ISSN
0269-3674 ; 266)
Texts of conferences given at Pont-à-Mousson, June 1991 during the
First European Conference on Elliptic and Parabolic Problems.
1. Differential equations, Elliptic--Congresses. 2. Differential
equations, Parabolic--Congresses. 3. Bifurcation theory-
- Congresses. I. Bandle, Catherine, 1943- . II. Title: Elliptic
and parabolic problems. III. Series.
QA377.P75 1992
515'.353--dc20 92-18952
CIP

Printed and bound in Great Britain
by Biddles Ltd, Guildford and King's Lynn

Cstack
SCIMON
12-3-92

Contents

Preface

This book is a collection of texts from conferences that were given in Pont-à-Mousson in June 1991 during the First European Conference on Elliptic and Parabolic Problems.

The subjects addressed in this volume include Equations and Systems of Elliptic and Parabolic Type, Bifurcation, and various applications in physics, mechanics and engineering.

We would like to thank all the participants to this meeting for their help in making it successful. Special thanks go to the contributors of this volume.

The meeting has been made possible by grants from the "Caisse d'Epargne de Lorraine Nord", the C.N.R.S., the "Région de Lorraine", the Universities of Basel, Metz and Saarbrücken. We express our deep appreciation to them.

Finally, we thank Longman for helping us to publish these proceedings.

C. Bandle, J. Bemelmans, M. Chipot, M. Grüter, J. Saint Jean Paulin

J I DIAZ

Symmetrization of nonlinear elliptic and parabolic problems and applications: a particular overview

Introduction.

The symmetrization process will be illustrated by the consideration of different classes of nonlinear partial differential equations. In Section 1 we consider the Dirichlet problem associated to the elliptic equation

$$-\operatorname{div}(Q(|\nabla u|)\nabla u) + \beta(u) = f.$$

The parabolic formulation

$$\gamma(u)_t - \operatorname{div}(Q(|\nabla u|)\nabla u) + \beta(u) = f$$

with $Q(r) = r^{p-2}$, will be the object of Section 2. Some variational inequalities will be considered in Section 3. Among them we can mention the obstacle problem, the Stefan problem and a suitable formulation related with the study of Bingham fluids. Finally, Section 4 contains an extension of some of the precedent results to the case of systems of nonlinear equations, as, for instance, the system arising in chemical adsorption.

1. The symmetrization process for nonlinear elliptic equations.

Let Ω be a bounded regular open set of $\mathbf{R}^N$. We consider the Dirichlet (or Plateau) problem

$$-\operatorname{div}(Q(|\nabla u|)\nabla u) + \beta(u) = f \quad \text{in } \Omega \tag{1}$$

$$u = 0 \quad \text{on } \partial\Omega. \tag{2}$$

We assume $Q \in C^2((0,\infty))$, $Q(r)r \to 0$ if $r \to 0$, $Q(r)r^2$ convex strictly increasing and β continuous non-decreasing such that $\beta(0) = 0$.

To introduce the notion of weak solution we first define the auxiliary function

$$A(r) = \int_0^r Q(s)s\,ds$$

and the Orlicz and Orlicz-Sobolev spaces $L^A(\Omega)$, $W^{1,A}(\Omega)$ and $W_0^{1,A}(\Omega)$ associated to A as usual (see [50]). Notice that if $Q(r) = r^{p-2}$ then $W^{1,A}(\Omega) = W^{1,p}(\Omega)$ (the usual Sobolev space).

By using variational techniques it is possible to show ([43],[23]) the following existence result

Theorem 0.

Let $f \in \left(W_0^{1,A}(\Omega)\right)'$ *and assume*

$$\max(|\,\Omega\,|^{\frac{1}{N}},\ |f|_{L^N(\Omega)}) \leq N(\omega_N)^{\frac{1}{N}} \lim_{r\to\infty} \mathcal{B}(r) \tag{3}$$

where

$$\omega_N \quad \textit{is the measure of the unit ball in } \mathbf{R}^N \tag{4}$$

and

$$\mathcal{B}(r) = Q(r)r \qquad \textit{for any } r > 0. \tag{5}$$

Then there exists a unique $u \in W_0^{1,A}(\Omega)$ *with* $\beta(u) \in L^1(\Omega)$ *weak solution of* (1),(2) *in the sense that*

$$\int_\Omega Q(|\,\nabla u\,|)\nabla u \cdot \nabla v dx + \int_\Omega \beta(u) v dx =< f, v >$$

for any $v \in W_0^{1,A}(\Omega) \cap L^\infty(\Omega)$.

Remarks.

1. If $Q(r) = r^{p-2}$ then $\lim_{r\to\infty} \mathcal{B}(r) = +\infty$ and assumption (3) is trivially satisfied for any $f \in (W_0^{1,A})'$.
2. If $Q(r) = \frac{1}{\sqrt{1+r^2}}$ assumption (3) is "almost" necessary (see [49],[33] and their references)
3. The "comparison principle" holds in this class of solutions and thus if, for instance, $f \geq 0$ in Ω then $u \geq 0$ *a.e.* in Ω ([26]).

We also consider the "symmetrized problem"

$$-\text{div}(Q(|\,\nabla U\,|)\nabla U) + \beta(U) = F \quad \text{in } \Omega^* \tag{6}$$

$$U = 0 \quad \text{on } \partial\Omega^* \tag{7}$$

where Ω^* is the ball $\text{B}_R(0)$ centered at the origin and with equal measure than Ω. As we want to find solutions of (6),(7) as simple as possible, we assume that $F : \Omega^* \to \mathbf{R}$ is radially symmetric and decreasing along the radii. In this way the solution U of (6) and (7) have a simple structure: U is a function that is radially symmetric and decreasing along the radii. Our main goal is to collect several results comparing (in some sense) the solutions u and U of (1), (2) and (6), (7) respectively, assuming a suitable relation between the data f and F.

A first choice leading to comparison results is $F = f_*$: the symmetric decreasing rearrangement of f.

Definitions. Let $f : \Omega \to \mathbf{R}$ measurable. We define the functions

$$\mu(t) = |\,\{x \in \Omega : f(x) > t\}\,| \qquad \text{(the distribution function)}\ ,$$

$\tilde{f} : [0,\infty) \to \mathbf{R},\ \tilde{f}(s) = \inf\{t \geq 0 : \mu(t) \leq s\}$ (the decreasing rearrangement),

and finally

$f_* : \Omega^* \to \mathbf{R}, f_*(x) = \tilde{f}(\omega_N \mid x \mid^N)$ (the symmetric decreasing rearrangement).

This notion have been extensively treated in the literature in the last years. The reader can find an exhaustive treatment in the books [47], [7], [45] and [41]. Concerning the comparison between u and U, assumed $F = f_*$, the situation is quite different for the cases $\beta \equiv 0$ and $\beta \not\equiv 0$.

Theorem 1 ([50]).

Let $\beta \equiv 0$ and assume $f \in L^1(\Omega) \cap (W_0^{1,A}(\Omega))'$ and $f \geq 0$ on Ω. Then $U_* = U$ and

$$u_*(x) \leq U(x) \quad a.e. \quad x \in \Omega^*. \tag{8}$$

When $\beta \not\equiv 0$ the pointwise comparison (8) fails ([42], [45]) and the comparison between u and U must be established in a more complicated way (see [17], [42], [51], [52],[45], [19]). The following result shows that as a consequence of the "stability" of the symmetrization process:

Theorem 2 ([21]).

Let f and F be nonnegative functions in the spaces $(W_0^{1,A})' \cap L^1$ associated to the domains Ω and Ω^* respectively. We assume F symmetric and decreasing along the radii. Let u and U the solutions of (1),(2) and (6),(7). For $s \in (0, \mid \Omega \mid]$ define the auxiliary "mass" functions

$$l(s) = \int_0^s \tilde{f}(\sigma)d\sigma, \qquad L(s) = \int_0^s \tilde{F}(\sigma)d\sigma,$$

$$k(s) = \int_0^s \beta(\tilde{u}(\sigma))d\sigma, \quad K(s) = \int_0^s \beta(\tilde{U}(\sigma))d\sigma.$$

Then the following estimate holds

$$\| (k-K)_+ \|_{L^\infty(0,|\Omega|)} \leq \| (l-L)_+ \|_{L^\infty(0,|\Omega|)}, \tag{9}$$

where $(.)_+$ denotes the positive part of the corresponding functions. In particular, if we assume

$$\int_{B_r(0)} f_*(x)dx \leq \int_{B_r(0)} F(x)dx \quad for\ any\ r \in [0,R] \tag{10}$$

then

$$\int_{B_r(0)} \beta(u_*(x))dx \leq \int_{B_r(0)} \beta(U(x))dx \quad for\ any\ r \in [0,R]. \tag{11}$$

In spite of its "sophisticated" statement, the above result has many relevant applications:

Corollary 1.

Assume f and F as in Theorem 2 and satisfying (10). *Then*

$$\int_{\Omega} \Phi(\beta(u(x)))dx \leq \int_{\Omega^*} \Phi(\beta(U(x)))dx$$

for any convex nondecreasing real function Φ. In particular

$$\| \beta(u) \|_{L^q(\Omega)} \leq \| \beta(U) \|_{L^q(\Omega^*)} \qquad \textit{for any} \quad 1 \leq q \leq \infty.$$

Corollary 2 ([21]).

Let $F = f_*$ and assume that

$$\textit{dist (support } U, \ \partial\Omega^*) > 0$$

Then

$$| \{x \in \Omega : \ u(x) = 0\} | \geq | \{x \in \Omega^* : \ U(x) = 0\} | .$$

Remarks.

1. The idea of the proof of Theorem 2 is the following: By using the Fleming-Rishel formula and the De Giorgi's isoperimetric theorem in a similar way to [50] we obtain that

$$a(s)\mathcal{B}\left(-a(s)\frac{\mathrm{d}}{\mathrm{d}s}\beta^{-1}\left(\frac{\mathrm{d}k}{\mathrm{d}s}(s)\right)\right) + k(s) \leq l(s) \quad \text{in } (0, |\,\Omega\,|)$$

$$a(s)\mathcal{B}\left(-a(s)\frac{\mathrm{d}}{\mathrm{d}s}\beta^{-1}\left(\frac{\mathrm{d}K}{\mathrm{d}s}(s)\right)\right) + K(s) = L(s) \quad \text{in } (0, |\,\Omega\,|)$$

$$k(0) = K(0) = 0$$

$$k'(|\,\Omega\,|) = K'(|\,\Omega\,|) = 0$$

   where $\mathcal{B}$ is given by (5) and $a(s) = N(\omega_N)^{\frac{1}{N}} s^{\frac{N-1}{N}}$. The conclusion comes by L^∞-techniques for fully nonlinear elliptic problems (see details in [21]).
2. The proof of Corollary 1 uses a classical result due to Hardy-Littlewood-Polya [40].
3. Corollary 2 is of special interest for the study of free boundary problems arising if, for instance, we assume $Q(r) = r^{p-2}$, $\beta(r) = r^q$ and $0 < q < p - 1$ (see the monograph [19]).

We end this section by the consideration of the case of nonhomogeneous Dirichlet boundary conditions. The symmetrization process can be successfully applied at least for two special cases of interest in the applications:

(i) the case of u constant on $\partial\Omega$,

(ii) capacity type problems.

The treatment of the first class of those problems can be carried out by a direct approach ([9]) or by a homogenization argument and the application of Theorem 2 ([19]). The following conclusions hold

Corollary 3 ([19]).
Let $h \in \mathbf{R}_+$ and let v, V solutions of the problems

$$-\text{div}(Q(\mid \nabla v \mid)\nabla v) + \beta(v) = 0 \quad \text{in } \Omega$$

$$v = h \quad \text{on } \partial\Omega$$

and

$$-\text{div}(Q(\mid \nabla V \mid)\nabla V) + \beta(V) = 0 \quad \text{in } \Omega^*$$

$$V = h \quad \text{on } \partial\Omega^*$$

Then

$$\int_{B_r(0)} \beta(v_*(x))dx \geq \int_{B_r(0)} \beta(V(x))dx \quad \text{for any } r \in [0, R].$$

Moreover, $V > 0$ in Ω^* implies that $v > 0$ in Ω and, otherwise, the following estimate holds

$$\mid \{x \in \Omega : v(x) = 0\} \mid \leq \mid \{x \in \Omega^* : \; V(x) = 0\} \mid . \tag{16}$$

The case of "capacity" type problems leads to another comparison results. Let ω be an open bounded regular set of $\mathbf{R}^N$ such that $\overline{\omega} \subset \Omega$. We consider the problem

$$-\text{div}(Q(\mid \nabla u \mid)\nabla u) + \beta(u) = f \quad \text{in } \Omega - \overline{\omega} \tag{17}$$

$$u = 1 \quad \text{on } \partial\omega \tag{18}$$

$$u = 0 \quad \text{on } \partial\Omega \tag{19}$$

and its symmetrized version

$$-\text{div}(Q(\mid \nabla U \mid)\nabla U)) + \beta(U) = F \quad \text{in } \Omega^* - \overline{\omega}^* \tag{17}$$

$$U = 1 \quad \text{on } \partial\omega^* \tag{18}$$

$$U = 0 \quad \text{on } \partial\Omega^* \tag{19}$$

where ω^* is the ball centered at the origin and with equal measure than ω. We have

Theorem 3 ([20]).

Assume $f \in L^\infty(\Omega - \overline{\omega})$ and $F \in L^\infty(\Omega^* - \overline{\omega}^*)$ be nonnegative functions. Let $\underline{f}$ and $\underline{F}$ the extension of these functions to Ω and Ω^* by means of $\sup f$ *and* $\sup F$ *on ω and ω^* respectively. Assume also that F is symmetric and decreasing along the radii and*

$$\int_{\mathrm{B}_r(0)} \underline{f}_*(x)dx \leq \int_{\mathrm{B}_r(0)} \underline{F}(x)dx \quad \text{for any } r \in [0, R]. \tag{23}$$

Then

$$\int_{\mathrm{B}_r(0)} \beta(\underline{u} * (x))dx - C_u \leq \int_{B_r(0)} \beta(\underline{U}(x))dx - C_U, \tag{24}$$

where

$$C_u = \int_{\partial\omega} Q(|\nabla u|)\nabla u \cdot n dx, \quad C_U = \int_{\partial\omega^*} Q(|\nabla U|)\nabla U \cdot n dx. \tag{25}$$

Remark.

If $\beta \equiv 0$ the conclusion (24) is replaced by the pointwise comparison

$$\frac{\underline{u}_*(x)}{C_u} \leq \frac{\underline{U}(x)}{C_U} \quad a.e. \quad x \in \Omega^*. \tag{26}$$

([47],[45]). In the linear case, $Q(r) = 1$, the solution u represents the capacity potential in the condenser $\overline{\Omega - \omega}$, and C_u is the electrostatic capacity of ω relative to Ω (see [47],[48],[30] and [34]).

2. The symmetrization process for nonlinear parabolic equations.

The formulation of the class of parabolic problems we shall consider is the following

$$\gamma(u)_t - \Delta_p u + \beta(u) = f(t,x) \quad \text{in } (0,T) \times \Omega \tag{27}$$

$$u = 0 \quad \text{on } (0,T) \times \partial\Omega, \tag{28}$$

$$\gamma(u(0,x)) = \gamma(u_0(x)) \quad \text{on } \Omega, \tag{29}$$

where Ω is an open bounded regular set of $\mathbf{R}^N$, the diffusion operator is

$$\Delta_p u = \operatorname{div}(|\nabla u|^{p-2} \nabla u)$$

and β and γ are continuous functions with γ strictly increasing. The existence and uniqueness of weak solutions have been largely studied by different authors and methods. Many references can be found in the articles [5], [12], [15], [27].

The symmetrized problem is formulated as

$$\gamma(U)_t - \Delta_p U + \beta(U) = F(t,x) \quad \text{in } (0,T)\times\Omega^*, \tag{30}$$

$$U = 0 \quad \text{on } (0,T)\times\partial\Omega^*, \tag{31}$$

$$\gamma\big(U(0,x)\big) = \gamma\big(U_0(x)\big) \quad \text{on } \Omega^*, \tag{32}$$

where, again, Ω^* is the ball $\mathrm{B}_R(0)$ centered at the origin and with equal measure than Ω and $F(t,.)$ (for t fixed) and U_0 are real symmetric functions defined on Ω^* and decreasing along the radii.

The following result contain a comparison result as a consequence of a "stability" (or continuous dependence) estimate.

Theorem 4 ([22]).
Let f, u_0 and U_0 integrable nonnegative and bounded functions. Assume

$$\beta \circ \gamma^{-1} = \phi_1 + \phi_2 \tag{33}$$

with ϕ_1 convex and ϕ_2 concave
Let u and U the solutions of (27) (28) (29) and (30) (31) (32) respectively and assume

$$u \text{ and } U \text{ are bounded functions} \tag{34}$$

For $t \in [0,T]$ and $s \in \big[0, |\,\Omega\,|\big]$ we define the auxiliary functions

$$k(t,s) = \int_0^s \gamma\big(\tilde{u}(t,\sigma)\big)d\sigma, \quad K(t,s) = \int_0^s \gamma\big(\tilde{U}(t,\sigma)\big)d\sigma$$

$$l(t,s) = \int_0^s \tilde{f}(t,\sigma)d\sigma, \quad L(t,s) = \int_0^s \tilde{F}(t,\sigma)d\sigma.$$

Then, there exists a constant C such that,

$$\|\,[k(t,.) - K(t,.)]_+\,\|_{L^\infty(0,|\Omega|)} \le e^{Ct}\,\|\,[k(0,.) - K(0,.)]_+\,\|_{L^\infty(0,|\Omega|)}$$

$$+ \int_0^t e^{C(t-\tau)}\,\|\,[l(\tau,.) - L(\tau,.)]_+\,\|_{L^\infty(0,|\Omega|)}\,d\tau$$

for any $t \in [0,T]$. In particular, if we assume

$$\int_{\mathrm{B}_r(0)} \gamma\big(u_{0*}(x)\big)dx \le \int_{\mathrm{B}_r(0)} \gamma\big(U_0(x)\big)dx \tag{35}$$

and $\big($ *for a.e.* $t \in (0,T)\big)$

$$\int_{\mathrm{B}_r(0)} f_*(t,x)dx \le \int_{\mathrm{B}_r(0)} F(t,x)dx \tag{36}$$

for any $r \in [0, R]$, *then*

$$\int_{B_r(0)} \gamma(u_*(t,x))dx \leq \int_{B_r(0)} \gamma(U(t,x))dx \qquad (37)$$

for any $t \in [0, T]$ *and any* $r \in [0, R]$.

Remark.

It is possible to show that the pointwise comparison $u_* \leq U$ is not true for parabolic equations (even for the linear heat equation). The comparison given in (37) have been established by many different authors by using different methods. See [5],[6],[51],[1],[10],[11],[46] ... It seems that the stability estimate was first obtained in [22]. Such kind of estimate appears in the treatment of fully nonlinear parabolic equations.

Theorem 4 has many applications, some of them of special interest when solutions exhibit some peculiar behaviors.

Corollary 4.

Assume u_0, f, U_0 *and* F *as in Theorem 4 and satisfying* (35) *and* (36). *Then*

$$\| \gamma(u(t,.)) \|_{L^q(\Omega)} \leq \| \gamma(U(t,.)) \|_{L^q(\Omega^*)}$$

for any $t \in [0, T]$ *and any* $q \in [1, +\infty]$.

Corollary 5 ([6],[22]. Blow-up problems).

Let u_0, f, U, F *as in Corollary 4. Define the blow-up time* T_Ω *(resp.* T_{Ω^*}*) in* $L^r(\Omega)$ *by means of*

$$\lim_{t \nearrow T_\Omega} \| u(t,.) \|_{L^r(\Omega)} = +\infty$$

and

$$\| u(t,.) \|_{L^\infty(\Omega)} < +\infty \quad if \quad t \in [0, T_\Omega),$$

(analogously for T_{Ω^*} *by replacing* Ω *and* u *by* Ω^* *and* U*). Then* $T_\Omega \geq T_{\Omega*}$.

Corollary 6 ([22]. Finite time extinction problems).

Let u_0, f, U, F *as in Corollary 4. Define the extinction time* $T^\#_\Omega$*(resp.* $T^\#_{\Omega*}$*) by means of*

$$\| u(t,.) \|_{L^1(\Omega)} = 0 \quad for\ any \quad t \geq T^\#_\Omega$$

and

$$\| u(t,.) \|_{L^1(\Omega)} > 0 \quad if \quad t \in [0, T^\#_\Omega),$$

(analogously for $T^\#_\Omega$ *by replacing* Ω *and* u *by* Ω^* *and* U *). Then* $T^\#_\Omega \leq T^\#_{\Omega^*}$.

Corollary 7 ([10],[22]. Free boundary problems).
Assume u_0, f, U and F as in Corollary 4. Then

$$|\{x \in \Omega : u(t,x) = 0\}| \leq |\{x \in \Omega^* : \; U(t,x) = 0\}|$$

for any $t \in [0,T]$.

Remark.
The phenomena mentioned in Corollaries 5-7 arises under suitable conditions on the nonlinear terms of the equation (27). So, for instance, the blow-up property holds if, for instance, $F = f = 0,\ p = 2,\ \gamma(u) = u$ and $\beta(u) = -u^q$ with $q > 1$. The existence of a finite extinction time can be proved, for instance, if $p = 2,\ \gamma(u) = u^{\frac{1}{m}}$ with $0 < q < 1$ and $m > 0$. The existence of a free boundary is typical of slow diffusion problems ($\gamma(u) = u^{\frac{1}{m}},\ \beta \equiv 0$ and $(p-1)m > 1$) and also when the absorption is strong enough (e.g. $p = 2,\ \gamma(u) = u^{\frac{1}{m}},\ \beta(u) = u^q$ and $0 < q < m$).

The case of nonhomogeneous boundary conditions has been also considered in the literature. So, if the solution is constant on the boundary $(0,T)\times\partial\Omega$ conclusions similar to Corollary 3 have been obtained in [10] and [11]. The parabolic capacity problems was considered in [39] (for $p = 2$ and $\beta \equiv 0$) where it was proved that the comparison (24) must be now stated in terms of

$$\int_{\mathrm{B}_r(0)} \gamma(u^*(t,x))dx - \int_0^t C_u(\tau)d\tau \leq \int_{\mathrm{B}_r(0)} \gamma(U(t,x))dx - \int_0^t C_U(\tau)d\tau \qquad (38)$$

where $t \in [0,T],\ r \in [0,R]$ and

$$C_u(t) = \int_{\partial\omega} \nabla u(t,x)\cdot n dx \quad and \quad C_U(t) = \int_{\partial\omega^*} \nabla U(t,x)\cdot n dx$$

3. The symmetrization process for variational inequalities.
Many different problems of natural sciences are formulated in terms of variational inequalities instead of equations ([28]). Here we shall comment some result illustrating the symmetrization process in that context.

We start by the consideration of the Obstacle Problem. As we shall see the treatment is different according inferior or superior obstacles, $u \geq \psi$ or $u \leq \psi$, respectively. To simplify the exposition le us consider the obstacle problem associated to $u \geq 0$ on Ω. Let $f \in L^{p'}(\Omega)$ and $u \in K(\Omega)$ be the solution of

$$\inf_{w\in K(\Omega)} \left\{ \frac{1}{p}\int_\Omega |\nabla w|^p dx - \int_\Omega fwdx \right\} \qquad (39)$$

where

$$K(\Omega) = \{w \in W_0^{1,p}(\Omega) : w \geq 0 \quad a.e. \text{ on } \Omega\}. \qquad (40)$$

The symmetrized problem is

$$\inf_{w \in K(\Omega^*)} \left\{ \frac{1}{p} \int_\Omega |\nabla w|^p \, dx - \int_\Omega F w dx \right\} \tag{41}$$

where $K(\Omega^*)$ is given by (40) replacing Ω by Ω^* and F is a radially symmetric and decreasing along the radii function. We have

Theorem 5 ([8]).
Assume $f \in L^{p'}(\Omega)$ and $F = f_*$. Then

$$u*(x) \leq U(x) \quad \text{a.e. } x \in \Omega^* \tag{42}$$

Moreover

$$|\{x \in \Omega : \ u(x) = 0\}| \geq |\{x \in \Omega^* : \ U(x) = 0\}|, \tag{43}$$

and

$$|\{x \in \Omega^* : \ U(x) = 0\}| > 0$$

if and only if there exists $s_0 \in (0, |\Omega|]$ such that

$$\int_0^{s_0} \tilde{f}(\sigma) d\sigma = 0.$$

An "equivalent" formulation of problem (39)-(40) is the given by

$$-\Delta_p u + \beta(u) \ni f \quad \text{in } \Omega \tag{44}$$

$$u = 0 \quad \text{on } \partial\Omega \tag{45}$$

which has a great similarity with the formulation (1)-(2) but where now β is the maximal monotone graph of $\mathbf{R}^2$ given by

$$\beta(r) = \begin{cases} 0, & \text{if } r > 0, \\ (-\infty, 0], & \text{if } r = 0, \\ \emptyset, & \text{if } r < 0. \end{cases}$$

(see [16],[19]).

As an example of the superior obstacle problem we can take the associated to the condition $u \leq 1$. We introduce the closed and convex sets

$$K(\Omega) = \{w \in W_0^{1,p}(\Omega) : \ w \leq 1 \quad a.e. \text{ on } \Omega\}. \tag{46}$$

The variational formulation is given by (39), (46) and the symmetrized version by (41), (46). When, $f \in L^{p'}(\Omega)$ and $F \in L^{p'}(\Omega^*)$ it is possible to show ([19]) that the corresponding solutions $u \in W_0^{1,p}(\Omega)$ and $U \in W_0^{1,p}(\Omega^*)$ are "characterized" by the existence of two functions $b \in L^1(\Omega)$ and $B \in L^1(\Omega^*)$such that

$$-\Delta_p u + b = f \quad \text{in } \Omega \qquad \text{and} \qquad b(x) \in \beta\big(u(x)\big) \quad a.e. \quad x \in \Omega \tag{47}$$

$$-\Delta_p U + B = F \quad \text{in } \Omega^* \qquad \text{and} \qquad B(x) \in \beta\big(U(x)\big) \quad a.e. \quad x \in \Omega^* \tag{48}$$

where β is now the maximal monotone graph of $\mathbf{R}^2$ given by

$$\beta(r) = \begin{cases} 0, & \text{if } r < 1, \\ [1, +\infty), & \text{if } r = 1, \\ \emptyset, & \text{if } r > 1. \end{cases} \tag{49}$$

The comparison result is now stated in terms similar to Theorem 2:

Theorem 6 ([21]).
Assume f and F nonnegative functions in $L^{p'}(\Omega)$ and $L^{p'}(\Omega^*)$ satisfying (10) *and F symmetric and decreasing along the radii. Then*

$$\int_{B_r(0)} b_*(x)dx \leq \int_{B_r(0)} B(x)dx \tag{50}$$

Moreover

$$|\{x \in \Omega : \; u(x) = 1\}| \leq |\{x \in \Omega^* : \; U(x) = 1\}| \,. \tag{51}$$

Remarks.

1. The study of the measure of the coincidence set $\{x \in \Omega^* : \; U(x) = 1\}$ (for $p = 2$) was carried out in [44] where the superior obstacle problem was connected with the inferior obstacle problem $u \geq 0$ but under the nonhomogeneous boundary condition $u = 1$ on $\partial\Omega$. An extension of Theorem 5 to more general maximal monotone graphs was given in [19] (see Theorem 2.22).
2. The parabolic (inferior) obstacle problem was considered in [25]. By discretization in the time variable similar conclusion to the stability inequality of Theorem 4 was proved.

Another example of variational inequality for which the symmetrization process leads to interesting conclusions is the one-phase Stefan problem

$$\gamma(u)_t - \Delta u \ni 0 \qquad \text{in } (0,T) \times \Omega$$

$$u = h(t) \qquad \text{on } (0,T) \times \partial\Omega$$

$$\gamma\big(u(0,.)\big) = \gamma\big(u_0(.)\big) \qquad \text{on } \Omega$$

where γ is any maximal monotone graph such that $\gamma(0) = [-L, 0]$ (see [39] and [24]). A related formulation corresponds to the Hele-Shaw flows ([38]).

Our last example of variational inequality arises in the study of special formulations of Bingham fluids. We define $V = H_0^1(\Omega)$,

$$a(u,v) = \int_\Omega \nabla u \cdot \nabla v dx, \quad j(v) = \int_\Omega \mid \nabla v \mid dx. \tag{52}$$

Given two real positive constants μ and g we consider the problem of finding $u \in V$ such that

$$\mu a(u, v-u) + g(j(v) - j(u)) \geq < f, v-u > \quad \forall v \in V, \tag{53}$$

where $f \in V'$ is given. The existence and uniqueness of a solution u of (53) was shown in [28]. Let U be the solution of the symmetrized problem, *i.e.* (53) but replacing Ω and f as in above problems. We have

Theorem 7 ([18]).
Assume $f \in L^2(\Omega), f \geq 0$ *and* $F = f_*$. *Then*

$$u_*(x) \leq U(x) \quad a.e. \quad x \in \Omega^*. \tag{54}$$

Remark.
The idea of the proof is to start by showing

$$u = \lim_{p \to 1} u_p,$$

where $u_p \in H_0^1(\Omega)$ satisfies

$$-\mu \Delta u_p - g \Delta_p u_p = f \quad \text{in } \Omega.$$

Then, the conclusion holds by applying Theorem 1 to u_p. We point out that (54) is one of the main ingredients in order to get estimates on the location and measure of the "rigid region" $\{x \in \Omega : \nabla u(x) = 0\}$ (see [28] and [37]).

4. On the application of the symmetrization process to systems of equations.

A very simple nonlinear system to which the symmetrization process can be applied is the following

$$u_t - \Delta u + f_1(u) + g_1(v) = 0 \quad \text{in } (0,T) \times \Omega$$

$$v_t - d\Delta v + f_2(u) + g_2(v) = 0 \quad \text{in } (0,T) \times \Omega,$$

$$u = v = 0 \quad \text{on } (0,T) \times \partial\Omega$$

$$u(0,x) = u_0(x) \quad \text{on } \Omega$$

$$v(0,x) = v_0(x) \quad \text{on } \Omega.$$

Theorem 8 ([22]).

Assume f_1 *and* g_2 *Lipschitz (or nondecreasing) functions satisfying the property* (33), f_2 *and* g_1 *nonincreasing and concave functions.*
Let $u_0, v_0 \in L^\infty(\Omega)$ *be nonnegative functions and let* (U,V) *be the solution of the symmetrized system replacing* Ω *by* Ω^* *and* u_0, v_0 *by* u_{0*} *and* v_{0*}. *Then*

$$\int_{B_r(0)} u_*(t,x)dx \leq \int_{B_r(0)} U(t,x)dx$$

and

$$\int_{B_r(0)} v_*(t,x)dx \leq \int_{B_r(0)} V(t,x)dx$$

for any $t \in [0,T]$ *and any* $r \in [0,R]$.

Remarks.

1. The idea of the proof is to show that $u = \lim u_n$, $v = \lim v_n$ with (u_n, v_n) given by the iterative algorithm

$$u_t^n - \Delta u^n + f_1(u^n) = -g_1(v^{n-1})$$

$$v_t^n - d\Delta v^n + g_2(v^n) = -f_2(u^{n-1}).$$

After that the conclusion comes from the application of Theorem 4 to u_n and v_n.

2. By making $d \to 0$ and showing that $u_1 \to u$ as $d \to 0$ the conclusion of Theorem 8 remains true for system of PDE-ODE equations. Such is the case of the system arising in chemical adsorption for which $d = 0$, $f_1(u) = -f_2(u)$ and $g_1(v) = -g_2(v) = -v$ (see [29]).

References.

[1] Abourjaily, C. and Benilan, Ph.: Article in preparation.

[2] Alvino, A. and Trombetti, G.: Su una classe di equazioni elliptiche non lineari degeneri. Rend. Accad. Naz. Lincei,*29*. 1980, 193-212.

[3] Alvino, A., Lions, P.L. and Trombetti, G.: Comparison results for elliptic and parabolic equations via Schwarz symmetrization. Ann. Inst. Henri Poincaré, *7*. 1990, 37-65.

[4] Alvino, A., Lions, P.L. and Trombetti, G.: Comparison results for Elliptic and Parabolic Equations via symmetrization: a new approach, Differential and Integral Equations, *4*, 1991, 25-50.

[5] Alt, H.W. and Luckhaus, S.: Quasilinear Elliptic-Parabolic Differ.Equations, Math. Z. *183*, 1983, 311-341.

[6] Bandle, C.: Isoperimetric Inequalities for a Class of Nonlinear Parabolic Equations, ZAMP *27*, 1976, 377-384.

[7] Bandle, C.: *Isoperimetric Inequalities and Applications*, Pitman, London, 1980.

[8] Bandle, C. and Mossino, J.: Rearrangement in Variational Inequalities. Ann di Mat. Pura et Applicata *88*, 1984, 1-14.

[9] Bandle, C., Sperb, R.P. and Stakgold, I.: Diffusion and reaction with monotone kinetics, Nonlinear Analysis Th. Meth. and Appl. *8*, 1984, 321-333.

[10] Bandle, C. and Stakgold, I.: The formation of the dead core in parabolic reaction-diffusion equations. Trans. Amer. Math. Soc. *286*, 1984, 275-293.

[11] Bandle, C. and Stakgold, I.: Isoperimetric Inequality for the Effectiveness in Semilinear Parabolic Problems, International Series of Numerical Mathematics, Vol. 71, 1984, 289-295.

[12] Benilan, Ph.: *Evolution equations and accretive operators.* Lecture Notes, Univ. of Kentucky, 1981.

[13] Betta, M.F. and Mercaldo, A.: Comparison and regularity results for a nonlinear elliptic equations, To appear in Nonlinear Analysis T.M.A.

[14] Betta, M.F. and Mercaldo, A.: Existence and regularity results for a nonlinear elliptic equations, To appear in Rend. di Matematica.

[15] Blanchard, D. and Francfort, G.: Study of a doubly nonlinear heat equation with no growth assumptions on the parabolic term, SIAM J. Math. Anal. *19*, 1988, 1032-1056.

[16] Brezis, H.: Problémes Unilateraux, J. Math. Pures Appl. *51*, 1972,1-164.

[17] Chiti, G.: Norme di Orlicz delle soluzioni di una classe di equazioni ellittiche. Boll. Unione Mat. It. *16-A*, 1979, 178-185.

[18] Cirmi, R. and Díaz, J.I.: Article in preparation.

[19] Díaz, J.I.: *Nonlinear Partial Differential Equations and Free Boundaries. Vol. 1. Elliptic Equations.* Pitman, London, 1985.

[20] Díaz, J.I.: Applications of symmetric rearrangement to certain nonlinear elliptic equations with a free boundary. In *Nonlinear Differential Equations*, J. Hale and P. Martinez (eds), 1985, 155-181.

[21] Díaz, J.I.: Desigualdades de tipo isoperimétrico para problemas de Plateau y Capilaridad. Revista Academia Canaria de Ciencias, 1991.

[22] Díaz, J.I.: Simetrización de problemas parabólicos no lineales: Aplicación a ecuaciones de reacción-difusión. Revista Real Academia de Ciencias, Madrid, 1992.

[23] Díaz, J.I.: Article in preparation.

[24] Díaz, J.I., Fasano, A. and Meirmanov, A.: On the disappearance of the mushy region in multidimensional Stefan Problems. To appear in *Free Boundary Problems: Theory and Applications.* Pitman, 1992.

[25] Díaz, J.I. and Mossino, J.: Inegalité isoperimetrique dans un probléme d'obsta cle parabolique. C.R. Acad. Sc. Paris, *305*, 1987, 737-740.

[26] Díaz, J.I., Saa, J.E. and Thiel, U.: Sobre la ecuación de curvatura media prescrita y otras ecuaciones elípticas con soluciones anulándose localmente. Rev. Unión Mat. Argentina, 1991.

[27] Díaz, J.I. and de Thelin, F.: On doubly nonlinear parabolic equations arising in some models related to turbulent flows. To appear.

[28] Duvaut, G. and Lions, J.L.: *Les Inéquations en Mécanique et en Physique.* Dunod, Paris, 1972.

[29] Van Duijn, C.J. and Knaber, P.: Solute Transport Through Porous Media with Slow Adsorption. In *Free Boundary Problems: Theory and Applications* (K.H. Hoffmann and J. Sprekels eds). Pitman, London, 1990, 375-388.

[30] Ferone, V.: Symmetrization results in electrostatic problems, Richerche di Matematica *37*, 1988, 359-370.

[31] Ferone, V. and Posteraro, M.R.: Maximization on classes of functions with fixed rearrangement, Differential and Integral Equations *4*, 1991.

[32] Ferone, V. and Posteraro, M.R.: On a class of quasilinear elliptic equations with quadratic growth in the gradient, To appear in Nonlinear Analysis T.M.A.

[33] Finn, R. *Equilibrium Capillary Surfaces.* Springer, 1986.

[34] Fraenkel, L.E.: A lower bound for electrostatic capacity in the plane. Proc. Royal Soc. Edinburgh, *88A*, 1981, 267 – 273.

[35] Giarruso, E. and Nunziante, D.: Regularity theorems in limit cases for solutions of linear and nonlinear equations, Rend. Ist. Mat. Univ. Trieste, *20*, 1988.

[36] Giarruso, E. and Trombetti, G.: Estimates for solutions of elliptic equations in a limit case, Bull. Australian Math. Soc. *36*, 1987.

[37] Glowinski, R., Lions, J.L. and Trémolières, R.: *Analyse Numérique des Inequations Variationelles.* Vol 2. Dunod, Paris, 1976.

[38] Gustafsson, B. and Mossino, J.: Some Isoperimetric Inequalities in Electrochemistry and Hele-Shaw Flows. IMA J. Appl. Math. 1987.

[39] Gustafsson, B. and Mossino, J.: Isoperimetric Inequalities for the Stefan Problem. SIAM J. Math. Anal. *20*, 1989, 1095-1108.

[40] Hardy, G.H., Littlewood, J.E. and Polya, G.: Some simple inequalities satisfied by convex functions, Messenger Math. *58*,1929, 145-152.

[41] Kawohl, B. *On Rearrangements, Symmetrization and Maximum Principles*, Lectures Notes in Math. Springer-Verlag, 1985.

[42] Lions, P.L.: Quelques remarques sur la symmetrization de Schwarz. In *Nonlinear Partial Differential Equations and Their Applications.* Vol.1, (H. Brezis and J.L. Lions eds.), Pitman, London, 1980, 308-319.

[43] Maderna, C.: Optimal problems for a certain class of nonlinear Dirichlet Problems. Suppl. Bull. UMI, *I*, 1980, 31-34.

[44] Maderna, C. and Salsa, S.: Some special properties of solutions to obstacle problems. Rend. Sem. Mat. Univ. Padova, *71*, 1984, 121- 129.

[45 Mossino, J.: *Inegalités Isoperimetriques et Applications en Physique.* Herman, Paris, 1984.

[46] Mossino, J. and Rakotoson, J.M.: Isoperimetric inequalities in parabolic equations. Ann. Scuola Norm. Sup. Pisa *Cl*, Sci, (4), *13*, 1986, 51-73.

[47] Polya, G. and Sezgo, G.: *Isoperimetric Inequalities in Mathematical Physics.* Ann. Math. Stud. *27*, Princeton Univ. Press.1952.

[48] Sarvas, J.: Symmetrization of condensers in n-space. Ann. Acad. Sci. Fenn. Ser. A1 *522*, 1972.

[49] Serrin, J.: The problem of Dirichlet for quasilinear elliptic differential equations with many independent variables. Philos. Trans. Roy. Soc. London, Ser A *264*, 1969, 413-496.

[50] Talenti, G.: Nonlinear elliptic equations, rearrangements of functions and Orlicz spaces. Ann. Mat. Pura Appl. IV, 120, 1977, 159-184.

[51] Vazquez, J.L.: Symmetrization pour $u_t = \Delta\varphi(u)$ et applications, C.R. Acad. Sci. Paris, *295*, 1982, 71-74.

[52] Vazquez, J.L.: Symmetrization in nonlinear parabolic equation, Port. Math. *41*, 1982, 171-200.

J.I. Díaz.
Departamento de Matemática Aplicada.
Universidad Complutense de Madrid. 28040 Madrid.
Spain.

A FRIEDMAN†

Elliptic and parabolic systems associated with semiconductor modeling*

§1. Introduction.

Semiconductor modeling is a rich source of problems in nonlinear elliptic and parabolic equations. Consider for example the standard drift-diffusion model consisting of an elliptic system

$$\sigma\nabla^2\psi = q(n - p - N(x)) \;, \tag{1.1}$$

$$\nabla(\mu_n n\nabla\psi - D_n\nabla n) = 0 \;, \tag{1.2}$$

$$\nabla(\mu_p p\nabla\psi + D_p\nabla p) = 0 \;, \tag{1.3}$$

with appropriate mixed Dirichlet and Neumann boundary conditions, where σ is the conductivity, q is the unit charge of electrons, μ_n and μ_p are the mobility coefficients of electrons and holes, and D_n, D_p are the diffusion coefficients for electrons and holes; $N(x)$ is the doping function (For details see, for instance, [11] [12]). It has been shown numerically, for x 1-dimensional, that if $N(x)$ changes sign three times then the solution is not unique. The system (1.1) – (1.3) with such $N(x)$ is a simple model of the thyristor, and the non-uniqueness phenomenon corresponds to a non-monotone $I-V$ curve [5; Chap. 6]. However, in spite of the fundamental importance of this semiconductor device, there is no rigorous mathematical proof of non-uniqueness! (cf. [5; Chap. 6] and the references given there).

Here we shall describe several problems associated with semiconductor, that were presented within the last year in the "Seminar on Industrial Problems" held at the Institute for Mathematics and its Applications. We shall indicate some partial solutions, and emphasize open problems.

§2. Semiconductor lasers.

This is a device consisting of two conductor materials, p and n, with photoactive layer sandwiched between them; See Figure 1.

---

*Partially supported by National Science Foundation Grant DMS–86–12880

†University of Minnesota, Institute for Mathematics and its Applications, Minneapolis, Minnesota 55455

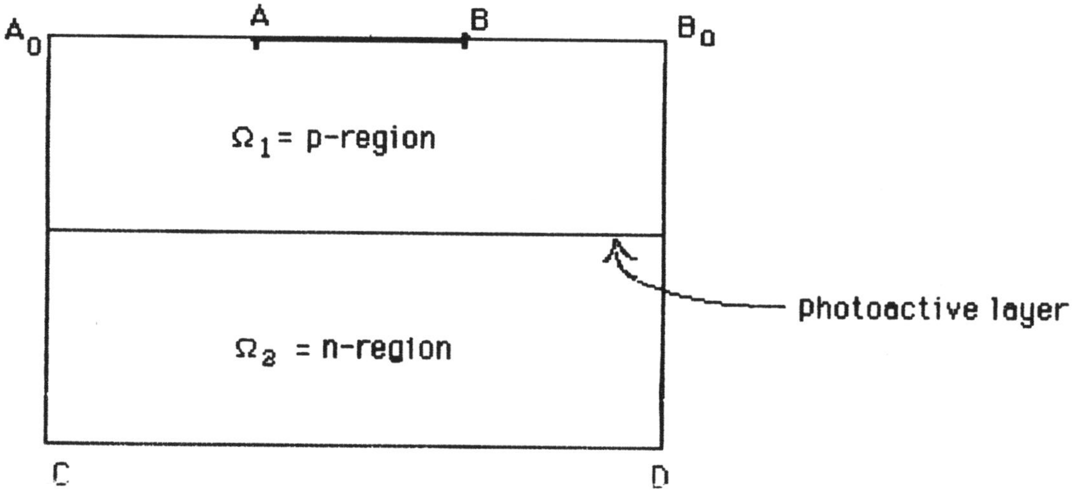

FIGURE 1

The electric potential $\varphi = \varphi_i$ in Ω_i satisfies:

$$\nabla \cdot (\sigma_i \nabla \varphi_i) = 0 \quad \text{in} \quad \Omega_i \ , \tag{2.1}$$

$$\begin{cases} \varphi_1 = V & \text{on} \quad AB \\ \varphi_2 = 0 & \text{on} \quad CD \ , \end{cases} \tag{2.2}$$

$$\varphi_1 = 0 \quad \text{or} \quad \frac{\partial \varphi_1}{\partial y} = 0 \quad \text{on} \quad \overline{A_0A} \cup \overline{B_0B} \tag{2.3}$$

$$\frac{\partial \varphi}{\partial n} = 0 \qquad \text{on the remaining boundary of } \ \Omega \ , \tag{2.4}$$

where $\Omega = \text{int } \overline{\Omega_1 \cup \Omega_2}$. Finally, on the photoactive layer

$$\Gamma = \{-a < x < a \ , \ y = 0\} \ ,$$

φ_1 and φ_2 satisfy rather complicated jump conditions [6] [14]. After some simplifications and non-dimensionalization, these conditions can be reduced to $(n = n(x,0))$

$$
\begin{aligned}
&-\frac{1}{k}\, n'' + f(n) = \frac{\partial \varphi_1(x,0)}{\partial y} \ , \\
&\varphi_1 - \varphi_2 = V_1 + \varepsilon \log n(n+1) \ , \\
&\frac{\partial \varphi_1}{\partial y} - \alpha \frac{\partial \varphi_2}{\partial y} = (1-\delta) f(n) \ , \\
&n_x(\pm a, 0) = 0 \ ,
\end{aligned}
\tag{2.5}
$$

where typically $\alpha \sim 2$, $k \sim 10, \delta \sim \dfrac{1}{10}$, $0 < V_1 < V$, and ε is positive and small. The dimensions of Ω_i in the new scale are $O(1)$.

X. Chen, A. Friedman and L.S. Jiang [2] proved existence and uniqueness, in case $\varphi_1 = 0$ on $\overline{A_0 A} \cup \overline{B_0 B}$, provided σ_i are constants and $s \to f(s)$ is monotone increasing with $f(0) < 0$ (the conditions on f are fulfilled in the physical case). For the boundary condition $\dfrac{\partial \varphi_1}{\partial y} = 0$ on $\overline{A_0 A} \cup \overline{B_0 B}$, additional assumptions are needed.

The proof of existence is based on a fixed point argument which depends upon the following estimate:

LEMMA 2.1. *Suppose*

$$
\begin{aligned}
&\Delta \psi = 0 \quad in \quad \Omega_1 \cup \Omega_2 \ , \\
&\psi(x, 0+) = \psi(x, 0-) \ , \\
&\psi_y(x, 0+) - \psi(x, 0-) = h(x) \ ,
\end{aligned}
$$

and ψ satisfies the boundary conditions

$$
\begin{aligned}
&\psi = 0 \qquad on \quad \overline{A_0 B_0} \cup \overline{CD} \ , \\
&\frac{\partial \psi}{\partial n} = 0 \quad on\ the\ remaining\ part\ of \quad \partial\Omega \ .
\end{aligned}
$$

Suppose $\overline{A_0 B_0}$ lies on $y = 1$ and $\overline{CD}$ lies on $y = -d$. Then

$$
\frac{d}{d+1} \quad \min h \le \psi_y(x, 0+) \le \frac{d}{d+1} \quad \max h \ . \tag{2.6}
$$

The case where σ_1 is non-constant is of special interest (see [6] [14]), but the proof does not extend to this case (except if σ_1 is "near" a constant).

§3. Augmented-drift diffusion model.

The drift-diffusion model is not a good approximation for submicron devices. For such devices one faces the dilemma of either going to the full model, which involves the Boltzmann equation (cf. [11]) or somehow revising the drift-diffusion model.

The drift-diffusion model is based on the assumption that the local mean velocity $\vec{v}$ of the electrons is given by

$$\vec{v} = \vec{v}_d(E) - \frac{D(E)}{n} \nabla n \qquad (\vec{E} = -\nabla\psi) \tag{3.1}$$

where $v_d(E)$ is the drift part (often taken linear in $\vec{E}$, the electric field) and $D(E)$ is the diffusion coefficients (often taken to be constant). A similar assumption holds for the velocity of holes, but here, for simplicity, we shall assume that the device has zero hole distribution. One supplements (3.1) with the Poisson equation

$$\sigma\Delta\psi = q(n - N) \tag{3.2}$$

and the conservation of mass (we assume non-stationary flow)

$$\frac{\partial n}{\partial t} = -\nabla(n\vec{v}) . \tag{3.3}$$

For small devices one can still derive from the Boltzmann equations (under some assumptions) the drift-diffusion equation with μ_n, D_n which depend nonlinearly on E; see [10] [13].

Another direction was taken by Blakey, Maziar and Wang [1]; they have added a term $\gamma dE/dt$ to the right-hand side of (3.1) where $\gamma = \gamma(E, dE/dt)$ is a function which is computed experimentally. In the case of one space dimension this leads to the following system (taking for simplicity $N = 0$):

$$\sigma\psi_{xx} = qL^2 n , \quad 0 < x < 1 , \quad t > 0 , \tag{3.4}$$

$$n_t - Dn_{xx} = \left(-\varepsilon D n_x + \frac{1}{L^2}\psi_x + \frac{\gamma}{L^2}\psi_{xt}\right)\frac{n_x}{1+\varepsilon n} - qn^2 \tag{3.5}$$

where $\varepsilon = \gamma q$ is small, and

$$\begin{aligned} &n(x,0) = n_0(x) , \quad n_0 \geq 0 , \\ &-\alpha n_x + \beta n = 0 \quad \text{at } x = 0, \quad \alpha n_x + \beta n = 0 \text{ at } x = 1, \\ &\psi(0,t) = V , \quad \psi(1,t) = 0 , \end{aligned} \tag{3.6}$$

where $\alpha > 0$, $\beta \geq 0$, $V > 0$.

A. Friedman and W. Liu [9] have proved:

THEOREM 3.1. *If γ is constant then there exists a unique smooth solution of (3.4)–(3.6).*

The proof is valid for a range of parameters which includes "most" of the physical range. The proof is based on deriving L^∞ a priori estimates on n_x and then reducing the problem, in several steps, to a parabolic equation of the form

$$n_t - n_{xx} = F(n, n_x)$$

where F is a nonlinear nonlocal functional of n, n_x.

PROBLEM: Extend Theorem 3.1 to the cases where $\gamma = \gamma(\psi_x)$, $\quad \gamma = (\psi_x, \psi_{xt})$.

§4. Semiconductor processing.

In semiconductor processing one encounters several free boundary problems, such as the formation of crystals and the formation of the silicon-oxide domain; cf. [4; Chap. 17]. Here we shall consider one such problem which models titanium silicide film growth; for details we refer to [7; Chap. 8]. Denote by u the concentration of $TiSi$ in Ω_1 and by v the concentration of $TiSi_2$ in Ω_2; see Figure 2.

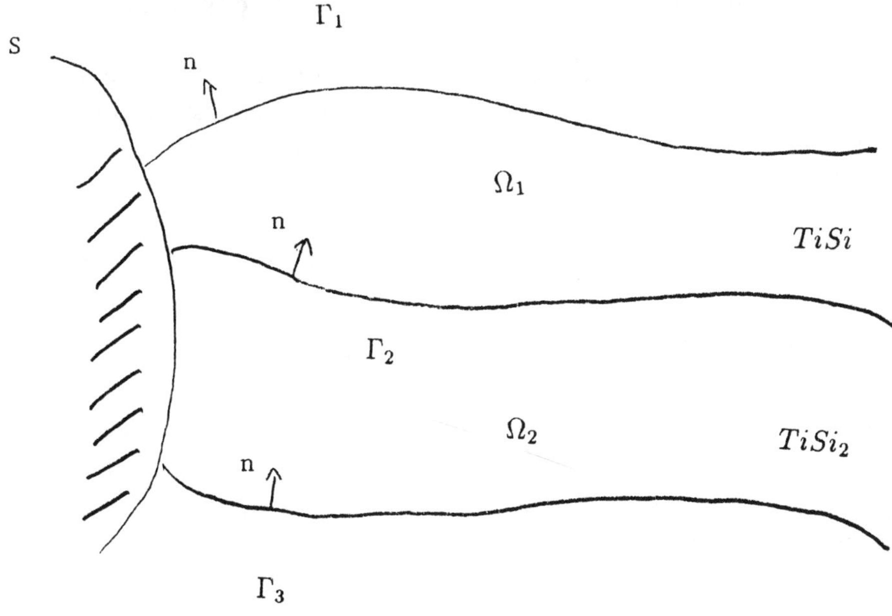

FIGURE 2

Then

$$\nabla^2 u = 0 \quad \text{in} \quad \Omega_1 \ , \tag{4.1}$$

$$\nabla^2 v = 0 \quad \text{in} \quad \Omega_2 \ , \tag{4.2}$$

$$\frac{\partial u}{\partial n} + \alpha_1 u = 0 \quad \text{on} \quad \Gamma_1 \ , \tag{4.3}$$

$$V_{\Gamma_1} = \beta_1 u \quad \text{on} \quad \Gamma_1 \ , \tag{4.4}$$

$$-D_1 \frac{\partial u}{\partial n} + D_2 \frac{\partial v}{\partial n} = k_1(u - v) \quad \text{on} \quad \Gamma_2 \ , \tag{4.5}$$

$$\alpha_2 v = u \quad \text{on} \quad \Gamma_2 \ , \tag{4.6}$$

$$V_{\Gamma_2} = \beta_2(u - v) \quad \text{on} \quad \Gamma_2 \ , \tag{4.7}$$

$$v = k_2 \quad \text{on} \quad \Gamma_3 \ , \tag{4.8}$$

$$-V_{\Gamma_3} = k_3 \frac{\partial v}{\partial n} \quad \text{on} \quad \Gamma_3 \ , \tag{4.9}$$

where all the constants are positive, and V_{Γ_i} is the velocity of the free boundary Γ_i.

The free boundary conditions (4.8), (4.9) are of the same type as for the Stefan problem [3]. Thus, if we consider Γ_2 as fixed, the problem (4.2), (4.8), (4.9) will have a unique solution. On the other hand the free boundary conditions on Γ_1 can be written as

$$V_{\Gamma_1} = -\frac{\beta_1}{\alpha_1} \frac{\partial u}{\partial n} \qquad \text{(Stefan condition)} \tag{4.10}$$

and, instead of $u = 0$ as in the Stefan problem,

$$\frac{\partial u}{\partial n} + \alpha_1 u = 0 \ . \tag{4.11}$$

Suppose Γ_2 is fixed and consider only (4.1) with (4.10) (4.11). On the free boundary $y = g(x,t)$, where

$$\Gamma_1 = \{0 < y < g(x,t), \ \ x \in \mathbf{R}^1\} \ .$$

We assume that

$$u(x,0,t) = h(x,t) \qquad (h \quad \text{in} \quad L^\infty) \tag{4.13}$$

is given and that $g(x,0)$ is given and is in $C^{2,\alpha}$. A. Friedman and Bei Hu have recently proved [8] that this problem has a unique local solution (u,g). Furthermore, if

$$(1+t^{\gamma})|h(\cdot,t)-h_0(t)|_{L^\infty}<\varepsilon\ , \quad \text{for some} \quad \gamma>1\ ,$$

$$\|g(\cdot)-g(0)\|_{C^{2,\alpha}}<\varepsilon$$

where ε is sufficiently small, then there exists a unique global solution.

REFERENCES

1. P.A. Blakey, L.M. Maziar and X.-L. Wang, *A generalized formulation of augmented drift-diffusion for use in semiconductor device modeling*, submitted to IEEE Trans. Electron Devices.
2. X. Chen, A. Friedman and L.S. Jiang, *Mathematical modeling of semiconductor lasers*, to appear.
3. A. Friedman, *Variational Principles and Free Boundary Problems*, Wiley–Interscience, New York, 1982.
4. A. Friedman, *Mathematics in Industrial Problems*, (IMA Volume 16), Springer–Verlag, Heidelberg, 1988.
5. A. Friedman, *Mathematics in Industrial Problems, Part 2*, (IMA Volume 24), Springer–Verlag, Heidelberg, 1989.
6. A. Friedman, *Mathematics in Industrial Problems, Part 3*, (IMA Volume 31), Springer–Verlag, Heidelberg, 1990.
7. A. Friedman, *Mathematics in Industrial Problems, Part 4*, (IMA Volume 38), Springer–Verlag, Heidelberg, 1991.
8. A. Friedman and B. Hu, *The Stefan problem with kinetic condition at the free boundary*, to appear.
9. A. Friedman and W. Liu, *An augmented drift-diffusion model in semiconductor device*, J. Math. Anal. Appl., to appear.
10. P.S. Hagan, R. Cox and B. Wagner, *High field semiconductor equations*, in preparation.
11. P.A. Markowich, C.A. Ringhofer and C. Schmeiser, *Semiconductor Equations*, Springer–Verlag, Vien New York, 1990.
12. M.S. Mock, *Analysis of Models of Semiconductor Devices*, Boole Press, Dublin (1983).
13. R. Stratton, *Diffusion of hot and cold electrons in semiconductor barriers*, Physics Reviews, 126 (1962), 2002–2014.
14. D.P. Wilt and A. Yariv, *A self-consistent static model of double heterostructure laer*, J. Quantum Electronics, vol. QE–17 (1981), 1941–1949.

P HESS

On the asymptotically periodic Fisher equation

0. Introduction

We investigate the behavior as $t \to +\infty$ of the solutions of the Fisher equation with asymptotically periodic fitnesses, subject to no-flux boundary conditions

$$(*) \qquad \begin{cases} \partial_t u - \Delta u = m(x,t)h(u) & \text{in } \Omega \times (0,\infty) \\ \dfrac{\partial u}{\partial n} = 0 & \text{on } \partial\Omega \times (0,\infty) \ . \end{cases}$$

We assume that $\Omega \subset \mathbb{R}^N$ $(N \geq 1)$ is a bounded domain with smooth boundary $\partial\Omega$, and that the function $h : [0,1] \to \mathbb{R}$ of class C^2 satisfies

$$(0.1) \qquad \begin{cases} h \text{ is strictly concave} \\ h(0) = h(1) = 0 \ . \end{cases}$$

The function m is assumed to be Hölder continuous and asymptotically T-periodic in t (with given period $T > 0$), i.e. m admits the decomposition

$$m(x,t) = m_0(x,t) + m_1(x,t)$$

where m_0 is T-periodic and

$$(0.2) \qquad \|m_1(\cdot,t)\|_{C(\overline{\Omega})} \to 0 \quad \text{as } t \to +\infty \ .$$

We remark that $(*)$ has the trivial solutions $u \equiv 0$ and $u \equiv 1$. We shall only be interested in solutions having initial values in $[0,1]$. If

$$(0.3) \qquad 0 \lneqq u(\cdot,0) \lneqq 1 \ ,$$

such solutions exist for all $t \geq 0$, and $0 < u(x,t) < 1$ for $(x,t) \in \overline{\Omega} \times (0,\infty)$ by the maximum principle. [To be more precise, we take $X = L^p(\Omega), p > N$, and choose $u(\cdot,0)$ in the fractional power space $E = X_\alpha$ with $\frac{1}{2} + \frac{N}{2p} < \alpha < 1$; then the solutions are classical.] $(*)$ is a well-known model of population genetics, in which u represents the fraction of one of two alleles of a migrating diploid species, which changes by diffusion and by selective advantage or disadvantage. Therefore m may change sign.

Our main result is

Theorem A. (i) *If $m_0 = m_0(t)$ is independent of $x \in \Omega$ and $\int_0^T m_0(t)dt = 0$, the limiting periodic Fisher equation (with m replaced by m_0) has a one- parameter family $\mathcal{V} = \{v_\tau : 0 \le \tau \le 1\}$ of spatially constant periodic solutions, and each solution of $(*)$ converges in $C^1(\overline{\Omega})$ to a subset of $\mathcal{V}$ as $t \to +\infty$. If m_1 decays such that*

$$\int_0^\infty \|m_1(\cdot,t)\|_{C(\overline{\Omega})}dt < \infty \, , \tag{0.4}$$

each solution of $(*)$ converges in $C^1(\overline{\Omega})$ to an element of $\mathcal{V}$.

(ii) *In all the other possible cases for m_0, there is a stable periodic solution u^* of the limiting periodic Fisher equation which globally attracts the solutions of $(*)$ satisfying* (0.3). *More precisely, if the trivial solution 0 is linearly stable or linearly neutrally stable (for the limiting periodic equation), $u^* = 0$. A symmetric result holds of course for the trivial periodic solution $\mathbb{1}$. If both 0 and $\mathbb{1}$ are linearly unstable, $0 < u^* < \mathbb{1}$.*

Remark. It is possible in case (i) that a solution of $(*)$ approaches more than one periodic solution of the limiting equation. Indeed, assume also $m_1 = m_1(t)$ is independent of x, and that $m_1(t) \to 0$ $(t \to +\infty)$ in such a way that

$$\limsup_{n\to\infty} \int_0^{nT} m_1(t)dt = 1 \ , \quad \liminf_{n\to\infty} \int_0^{nT} m_1(t)dt = -1 \ .$$

Taking initial conditions independent of x, we look at solutions of the

$$\dot{u} = (m_0(t) + m_1(t))h(u) \ , \quad u(0) = u_0 \ . \tag{ODE}$$

Let $H(s)$ be a primitive of the function $\frac{1}{h(s)}$. Then there are subsequences $n_k \to \infty, n_{k'} \to \infty$ such that

$$H(u(n_kT)) \to H(u_0) + 1 \ , \quad H(u(u_{k'}T)) \to H(u_0) - 1 \ .$$

Both initial values $H^{-1}(H(u_0) \pm 1)$ give rise to periodic solutions of the limiting equation $\dot{v} = m_0(t)h(v)$.

The proof of Theorem A is based on abstract results on discrete dynamical processes which are asymptotic to strongly order-preserving discrete semigroups, and on results for the periodic Fisher equation. The abstract results are contained in section 1 and slightly improve those of [2]. The periodic Fisher equation has been treated in much greater generality in [7], rather by ad-hoc methods. For the present special case we state the results in section 2 and refer to [7,

5, 6] for proofs. Section 3 contains the proof of Theorem A. An important observation for the applicability of the results of section 1 is the fact that unstable periodic solutions of the periodic Fisher equation occur only at the boundary of the considered order interval $[0, \mathbb{1}]$ and can be eliminated by the construction of suitable sub- and supersolutions.

For notations, the general set up and the underlying results on linear and semilinear periodic-parabolic boundary value problems we refer to the author's notes [5].

In the setting of almost periodic functions, Vuillermot [8] obtained partial results which motivated the present studies. Our time-dependence of the nonlinearity is of course considerably more restrictive; on the other hand we can allow selection functions m which depend on $x \in \Omega$, and obtain a complete description of all the possible cases.

1. Discrete order-preserving semigroups and asymptotically autonomous discrete dynamical processes

Let E be a strongly ordered Banach space, i.e. a real Banach space provided with an order cone P having nonempty interior. We use the standard notations $x \leq y \Longleftrightarrow y - x \in P$, $x < y \Longleftrightarrow y - x \in P\backslash\{0\}$, $x \ll y \Longleftrightarrow y - x \in int(P)$. Let $U \subset E$ be a subset and $S : U \to U$ a continuous map which is strongly order-preserving: $x < y \Longrightarrow S(x) \ll S(y)$. An element $a \in U$ is called a subequilibrium if $a \leq S(a)$; similarly $b \in U$ is a superequilibrium if $S(b) \leq b$. Assume $a < b$ are ordered sub- and superequilibria, and let $D := [a, b]_E$ denote the generated order-interval in E. Assume that $D \subset U$, and that

$$(1.1) \qquad S(D) \text{ is relatively compact in } E\ .$$

Note that D is mapped into itself by S.

We consider the discrete strongly order-preserving semigroup $(S^n)_{n \in \mathbb{N}}$ on D. For $x \in D$ we let $\gamma^+(x) := \{S^n(x) : n \in \mathbb{N}\}$ denote its semiorbit and $\omega(x) := \{y \in D : \exists n_k \to \infty$ such that $S^{n_k}(x) \to y\}$ its ω-limit set, which is nonempty, compact and invariant [i.e. $S(\omega(x)) = \omega(x)$]. Let $\mathcal{F}$ denote the set of fixed points of S in D. A fixed point x is stable if $\forall \varepsilon > 0\ \exists \delta > 0$ such that $\gamma^+(y) \subset \mathbf{B}_\varepsilon(x)$, $\forall y \in \mathbf{B}_\delta(x) \cap D$ [here $\mathbf{B}_\varepsilon(x)$ denotes the open ball in E with center x, radius ε]. We need the following result of [4], which is also collected in [5, Thm. 3.3] and reproved in a different way in [1].

Proposition 1.1. *Assume all fixed points of S in D are stable. Then all semiorbits converge, i.e. $\omega(x)$ is a singleton $\forall x \in D$. Moreover $\mathcal{F}$ is totally ordered and connected.*

$\gamma(x) := \{x_n : n \in \mathbf{Z}, x_0 = x\}$ is an entire orbit of S through x if $x_{n+1} = S(x_n)$ $\forall n \in \mathbf{Z}$. Set $\alpha(x) := \{z \in D : \exists n_k \to -\infty$ such that $S^{n_k}(x) \to z\}$. Again $\alpha(x)$ is nonempty, compact and invariant.

Lemma 1.2. *Assume all fixed points of S in D are stable. Let $\gamma(x) = \{x_n : n \in \mathbf{Z}, x_0 = x\}$ be an entire orbit, and let $A \subset \overline{\gamma(x)}$ be a nonempty, closed and positively invariant set (i.e. $S(A) \subset A$). Then $A = \overline{\gamma(x)}$.*

Proof. We first show that $\alpha(x) = \overline{\gamma(x)}$. Clearly $\alpha(x) \subset \overline{\gamma(x)}$. To prove that $\alpha(x) \supset \gamma(x)$, we assume by contradiction that there exists $x_{\overline{n}} \in \gamma(x)$ with $x_{\overline{n}} \notin \alpha(x)$. We observe that there exists $q \in \alpha(x) \cap \mathcal{F}$. [Indeed, take any $z \in \alpha(x)$. By Proposition 1.1, its semiorbit $\gamma^+(x)$ converges to some $q \in \mathcal{F}$, and $q \in \alpha(x)$ by (positive) invariance of $\alpha(x)$.] Let $\varepsilon := dist(x_{\overline{n}}, \alpha(x)) > 0$. Since q is stable, there is a $\delta > 0$ such that $S^n(y) \in \mathbf{B}_{\varepsilon/2}(q)$ $\forall y \in \mathbf{B}_\delta(q) \cap D$, $\forall n \in \mathbf{N}$. There exists a subsequence $x_{n_k} \to q$ $(n_k \to -\infty)$. Let $n_{k_0} < \overline{n}$ such that $\|x_{n_{k_0}} - q\| < \delta$. Then

$$x_{n+n_{k_0}} = S^n(x_{n_{k_0}}) \in \mathbf{B}_{\varepsilon/2}(q) \subset \mathbf{B}_{\varepsilon/2}(\alpha(x)) \ , \quad \forall n \in \mathbf{N} \ .$$

For $n := \overline{n} - n_{k_0}$, it follows that $x_{\overline{n}} \in \mathbf{B}_{\varepsilon/2}(\alpha(x))$, contradicting the definition of ε. Thus $\alpha(x) = \overline{\gamma(x)}$.

Next we show that $\gamma(x) \subset A$ for A as in the statement of the lemma. Suppose $\exists x_{\overline{n}} \notin A$. Since $A \subset \overline{\gamma(x)} = \alpha(x)$, we can proceed as above, with $\alpha(x)$ replaced by A, to reach a contradiction. This proves the lemma. □

Setting in particular $A = \omega(x)$, we get

Corollary 1.3. *Let S have the property that all fixed points in D are stable, and let $\gamma(x)$ be an entire orbit in D. Then $\alpha(x) = \omega(x) = \overline{\gamma(x)}$.*

This is essentially Lemma 3.1 of [2]. Indeed, it is a consequence of Proposition 1.1 that the stability of all fixed points implies the stability of all semiorbits, cf. [1, Remark p. 22] and [5, p. 26].

We now consider discrete dynamical processes, defined by a sequence of continuous mappings $S_n : D_n \to D_{n+1}$ $(n \in \mathbf{N})$ by

$$\begin{cases} T_n := S_{n-1} \circ \ldots \circ S_0 : D_0 \to D_n \ (n \geq 1) \ , \\ T_0 := id_{D_0} \ . \end{cases}$$

We assume that $(T_n)_{n \in \mathbf{N}}$ is asymptotic to a strongly order-preserving discrete semigroup $(S^n)_{n \in \mathbf{N}}$ as investigated above: we suppose there exists an order interval $D = [a, b]_E$ with $D \supset D_n$ $\forall n \in \mathbf{N}$, and a strongly order-preserving continuous map $S : D \to D$ satisfying (1.1), such that $S_n \to S$ along trajectories: $\forall x \in D_0$,

$$\|S_n \circ T_n(x) - S \circ T_n(x)\| \to 0 \ (n \to \infty) . \tag{1.2}$$

[Observe that a and b are necessarily sub- and superequilibria for S.] The semiorbit with respect to the dynamical process is denoted by $\tilde{\gamma}^+(x)$, the ω-limit set by $\tilde{\omega}(x)$. We remark that $\tilde{\gamma}^+(x)$ is relatively compact by (1.1) and (1.2).

Lemma 1.4. $S(\tilde{\omega}(x)) = \tilde{\omega}(x)$.

Proof. If $y \in \tilde{\omega}(x)$, there exists a subsequence $n_k \to \infty$ with $T_{n_k}(x) \to y$. For a further subsequence, $T_{n_k+1}(x) \to z \in \tilde{\omega}(x)$. Since

$$\|S \circ T_{n_k}(x) - S_{n_k} \circ T_{n_k}(x)\| \to 0 ,$$

it follows that $S(y) = z \in \tilde{\omega}(x)$.

Conversely, to $z \in \tilde{\omega}(x)$ there exists $y \in \tilde{\omega}(x)$ with $z = S(y)$. Indeed, for some subsequence $S_{n_k-1} \circ T_{n_k-1}(x) = T_{n_k}(x) \to z$, and we may further assume that $T_{n_k-1}(x) \to y \in \tilde{\omega}(x)$. Since

$$\|S_{n_k-1} \circ T_{n_k-1}(x) - S \circ T_{n_k-1}(x)\| \to 0 ,$$

we conclude that $z = S(y)$. □

Our main abstract result is

Theorem B. *Let $(T_n)_{n \in \mathbb{N}}$ be a discrete dynamical process asymptotic to the strongly order-preserving discrete semigroup $(S^n)_{n \in \mathbb{N}}$. Assume all fixed points of S in D are stable. Then $\tilde{\omega}(x) \subset \mathcal{F}$, the fixed point set of S, for all $x \in D_0$.*

Each semiorbit $\tilde{\gamma}^+(x)$ is thus quasiconvergent. We know that we cannot expect convergence in this generality.

Proof. Let $y \in \tilde{\omega}(x)$. Since $\tilde{\omega}(x)$ is invariant under S by Lemma 1.4, there exists an entire orbit $\gamma(y)$ (relative to the limiting semigroup) through y in $\tilde{\omega}(x)$. In particular, $y \in \overline{\gamma(y)} = \omega(y)$ by Corollary 1.3. Further $\omega(y) = \{q\}, q \in \mathcal{F}$, by Proposition 1.1. Hence $y = q \in \mathcal{F}$. □

2. The periodic Fisher equation

We first study the behavior as $t \to +\infty$ of solutions of the *periodic* Fisher equation

$$(2.1)\qquad \begin{cases} \partial_t u - \Delta u = m_0(x,t)h(u) & \text{in } \Omega \times (0,\infty) \\ \dfrac{\partial u}{\partial n} = 0 & \text{on } \partial\Omega \times (0,\infty) \,. \end{cases}$$

We let $X := L^p(\Omega)$ with $p > N$ and $E := X_\alpha$, with $\frac{1}{2} + \frac{N}{2p} < \alpha < 1$, the fractional power space associated to the sectorial operator A in X which is the realization of $(\Delta, \frac{\partial}{\partial n})$. Then $E \hookrightarrow C^1(\overline{\Omega})$ is a strongly ordered Banach space, and by the maximum principle and regularity theory the Poincaré map S, which associates to the initial condition $u_0 \in [0, \mathbb{1}]_E$ the value of the solution of (2.1) at $t = T$, is well-defined on $[0, \mathbf{1}]_E$ and strongly order-preserving. Moreover $S([0, \mathbf{1}]_E)$ is relatively compact in E.

Let u be a T-periodic solution of (2.1). In order to characterize its stability, we look at the linearization

$$(2.2)\qquad \begin{cases} \partial_t w - \Delta w = m_0(x,t)h'((u(x,t))w & \text{in } \Omega \times (0,\infty) \\ \dfrac{\partial w}{\partial n} = 0 & \text{on } \partial\Omega \times (0,\infty) \end{cases}$$

and take the period map Q_u of (2.2). Then Q_u is the Fréchet derivative in E of the Poincaré map S, at $u(\cdot, 0)$. Moreover Q_u is a compact, strongly positive operator in E. By the Krein-Rutman theorem, the spectral radius $r = spr(Q_u)$ is positive and is the principal eigenvalue of Q_u, with positive eigenfunction w_0. The solution w of (2.2) with initial value $w(\cdot, 0) = w_0$ then satisfies $w(\cdot, t+T) = rw(\cdot, t)$ for $t \geq 0$. The periodic solution u of (2.1) is said to be *linearly stable* if $r < 1$, *linearly neutrally stable* if $r = 1$, and *linearly unstable* if $r > 1$. Linearized stability (resp. linearized instability) implies local asymptotic stability (resp. instability) of u for the nonlinear equation. Using the theory of principal eigenvalue for linear periodic-parabolic problems ([3] and [5, sect. 14-16]), the linearized stability property of u can be formulated also in a different way. There exists a unique number $\mu \in \mathbf{R}$ (principal eigenvalue) for which the linear problem

$$(2.3)\qquad \begin{cases} \partial_t \varphi - \Delta\varphi - m_0(x,t)h'(u(x,t))\varphi = \mu\varphi & \text{in } \Omega \times \mathbf{R} \\ \dfrac{\partial \varphi}{\partial n} = 0 & \text{on } \partial\Omega \times \mathbf{R} \\ \varphi \; T-\text{periodic in } t & \end{cases}$$

admits a positive solution φ (principal eigenfunction). The relation between the spectral radius r of Q_u and μ is given by $\mu = -\frac{1}{T} \log r$.

Proposition 2.1. *Each solution of* (2.1) *with values in* $[0, \mathbf{1}]$ *converges to a* T*-periodic solution as* $t \to +\infty$, *in the norm of* X_α. *More precisely,*

(i) *If* $m_0 = m_0(t)$ *with* $\int_0^T m_0(t)dt = 0$, (2.1) *admits a one-parameter family* $\mathcal{V} = \{v_\tau : 0 \leq \tau \leq 1\}$ *of spatially constant stable periodic solutions, and each solution of* (2.1) *converges to a member of* $\mathcal{V}$.

(ii) *In all the other cases there is a globally asymptotically stable periodic solution* u^* [*with respect to initial conditions in* E *satisfying* (0.3)]. *If* 0 (*resp.* $\mathbb{1}$) *is linearly stable or linearly neutrally stable,* $u^* = 0$ (*resp.* $u^* = \mathbb{1}$). *If both* 0 *and* $\mathbb{1}$ *are linearly unstable,* $0 < u^* < \mathbb{1}$.

Proposition 2.1 is a special case of Theorem 1 of [7] and is also proved in [5, sect. 25 and 29] by arguments adapted from [7], which are more of an ad-hoc nature. A different proof, which rests upon results on abstract strongly order-preserving discrete semigroups and on linear periodic-parabolic eigenvalue problems, can be given along the lines of [6].

The different cases in Proposition 2.1 can be characterized by integral conditions; cf. [5, sect. 29].

3. Proof of Theorem A

The discrete dynamical process $(T_n)_{n\in\mathbb{N}}$ associated with $(*)$ is initially defined on $[0,\mathbb{1}]_E$ by

$$u_0 \in [0,\mathbb{1}]_E \longrightarrow T_n(u_0) = \text{ value of the solution } \tilde{u} \text{ of } (*) \text{ at}$$
$$t = nT \text{ having initial value } \tilde{u}(\cdot,0) = u_0 \ .$$

Thus $S_n(v_0)$ = value of the solution $\tilde{u}$ of $(*)$ at $t = (n+1)T$ satisfying $\tilde{u}(nT) = v_0$. By S we still denote the Poincaré map of the limiting periodic Fisher equation (2.1).

Lemma 3.1. *Condition* (0.2) *implies that* $\|S_n(v_0) - S(v_0)\|_E \to 0$ $(n \to \infty)$, *uniformly in* $v_0 \in [0,\mathbb{1}]_E$.

Proof. Let $\tilde{u}$ denote the solution of $(*), u$ the solution of the periodic equation (2.1), both having initial value v_0 at $t = nT$. By the variation of constant's formula,

$$\tilde{u}(t) = e^{(t-nT)A}v_0 + \int_{nT}^t e^{(t-s)A}m(\cdot,s)h(\tilde{u}(s))ds \ ,$$
$$u(t) = e^{(t-nT)A}v_0 + \int_{nT}^t e^{(t-s)A}m_0(\cdot,s)h(u(s))ds \ ,$$

$nT < t \leq (n+1)T$. Thus

$$(\tilde{u} - u)(t) = \int_{nT}^t e^{(t-s)A}[m_1(\cdot,s)h(\tilde{u}(s)) + m_0(\cdot,s)(h(\tilde{u}(s)) - h(u(s)))]ds \ ,$$

which implies that

$$\|(\tilde{u}-u)(t)\|_\alpha \le c\int_{nT}^{t} \|e^{(t-s)A}\|_{0,\alpha}\|m_1\|_{C(\overline{\Omega}\times[nT,(n+1)T])}ds$$
$$+\,c\int_{nT}^{t} \|e^{(t-s)A}\|_{0,\alpha}\|(\tilde{u}-u)(s)\|_\alpha ds\ .$$

Using the estimates

$$\|e^{(t-s)A}\|_{0,\alpha} \le c(t-s)^{-\alpha} \qquad (nT \le s < t \le (n+1)T)\ ,$$
$$\int_{nT}^{t}(t-s)^{-\alpha}ds \le \frac{1}{1-\alpha}T^{1-\alpha} \qquad (nT < t \le (n+1)T)\ ,$$

we conclude by a version of Gronwall's inequality that

$$\|(\tilde{u}-u)(t)\|_\alpha \le c_1\|m_1\|_{C(\overline{\Omega}\times[nT,(n+1)T])}\ .$$

For $t=(n+1)T$ the claim follows. □

Lemma 3.1 implies that the convergence condition (1.2) is satisfied. But we cannot apply directly Theorem B since there are always unstable periodic solutions of (2.1) present. We distinguish between three cases:

(a) both 0 and $\mathbb{1}$ are linearly unstable for (2.1);

(b) 0 is linearly stable or linearly neutrally stable for (2.1), but the exceptional case $m_0 = m_0(t)$, $\int_0^T m_0(t)dt = 0$, is excluded;

(c) $m_0 = m_0(t)$ with $\int_0^T m_0(t)dt = 0$.

Proof of Theorem A in case (a). Let $\varepsilon > 0$ be so small that both

0 is linearly unstable for the periodic problem

$$(3.1)_\varepsilon \qquad \begin{cases} \partial_t u - \Delta u = (m_0(x,t)-\varepsilon)h(u) & \text{in } \Omega\times\mathbf{R} \\ \dfrac{\partial u}{\partial n} = 0 & \text{on } \partial\Omega\times\mathbf{R} \end{cases}$$

and

$\mathbb{1}$ is linearly unstable for the periodic problem

$$(3.2)_\varepsilon \qquad \begin{cases} \partial_t v - \Delta v = (m_0(x,t)+\varepsilon)h(v) & \text{in } \Omega\times\mathbf{R} \\ \dfrac{\partial v}{\partial n} = 0 & \text{on } \partial\Omega\times\mathbf{R}\ ; \end{cases}$$

this is possible since 0 and $\mathbb{1}$ are linearly unstable for (2.1) and the principal eigenvalue is a continuous function: $C(\overline{Q}_T)\to\mathbf{R}$ of the periodic Hölder continuous weight m, $Q_T := \Omega\times[0,T]$; cf.

[5, Lemma 15.7]. Let the initial condition $u_0 \in [0,\mathbb{1}]_E$ be given; we can assume without loss of generality that $0 \ll u_0 \ll \mathbb{1}$. Let $\tilde{u}$ be the solution of $(*)$ with $\tilde{u}(\cdot,0) = u_0$. Moreover let $n_\varepsilon \in \mathbb{N}$ be so large that $\|m_1\|_{C(\bar{\Omega}\times[nT,(n+1)T])} \le \varepsilon$, $\forall n \ge n_\varepsilon$. If $\underline{v}$ and $\overline{v}$ denote the solution of equations $(3.1)_\varepsilon$ and $(3.2)_\varepsilon$, respectively, with initial values $\underline{v}(\cdot,n_\varepsilon T) = \tilde{u}(\cdot,n_\varepsilon T) = \overline{v}(\cdot,n_\varepsilon T)$, it follows that $\underline{v}$ and $\overline{v}$ are order-related sub- and supersolutions for $(*)$. Hence $\underline{v}(\cdot,t) \le \tilde{u}(\cdot,t) \le \overline{v}(\cdot,t)$ for $t \ge n_\varepsilon T$. By Proposition 2.1 there exist (unique) periodic solutions $u^*_{1,\varepsilon}$ and $u^*_{2,\varepsilon}$ of $(3.1)_\varepsilon$ and $(3.2)_\varepsilon$ respectively, with $0 \ll u^*_{1,\varepsilon} \ll u^*_{2,\varepsilon} \ll \mathbb{1}$, and $\|\underline{v}(\cdot,t) - u^*_{1,\varepsilon}(\cdot,t)\|_E \to 0, \|\overline{v}(\cdot,t) - u^*_{2,\varepsilon}(\cdot,t)\|_E \to 0$ as $t \to +\infty$. For $n \ge n_\varepsilon$ we take $D_n := [\underline{v}(\cdot,nT),\overline{v}(\cdot,nT)]_E$ for the discrete dynamical process in E associated with $(*)$. Since both 0 and $\mathbb{1}$ are linearly unstable periodic solutions for (2.1), there exist periodic sub- and supersolutions $\underline{\underline{v}}$ and $\overline{\overline{v}}$ for (2.1) with values in $(0,1)$ as close to 0 and $\mathbb{1}$ as we please, respectively. Thus we find $D := [\underline{\underline{v}}(\cdot,0),\overline{\overline{v}}(\cdot,0)]_E$ which contains D_n $\forall n \ge n_\varepsilon$. Now the hypotheses of Theorem B are satisfied, and convergence in E of $\tilde{u}$ to the global attractor u^* of (2.1) follows, first for $t = nT \to \infty$, and then for $t \to \infty$ by a standard argument using a Gronwall type lemma.

Proof of Theorem A in case (b). As a consequence of assertion (ii) of Proposition 2.1 the trivial solution 0 is stable, while $\mathbb{1}$ is linearly unstable for (2.1). We proceed as in case (a), taking only equation $(3.2)_\varepsilon$ to eliminate the unstable periodic solution $\mathbb{1}$ of (2.1). Convergence to $u^* = 0$ follows as above.

Proof of Theorem A in case (c). Since all periodic solutions of (2.1) are stable, we can take $D_n = D = [0,\mathbb{1}]_E$, $\forall n \in \mathbb{N}$. By Theorem B, for $u_0 \in [0,\mathbb{1}]_E$ it follows that $\tilde{\omega}(u_0) \subset \{v_\tau(0) : 0 \le \tau \le 1\}$. This proves that the solution $\tilde{u}$ of $(*)$ converges to a set of spatially constant periodic solutions of (2.1), i.e. a subset of $\mathcal{V}$.

Assume now the stronger decay (0.4), and let τ be such that $v_\tau(0) \in \tilde{\omega}(u_0)$. We claim that $T_n(u_0) \to v_\tau(0)$ as $n \to \infty$. There is a subsequence $n_k \to \infty$ such that $T_{n_k}(u_0) \to v_\tau(0)$ in E. Given $\varepsilon > 0$, there exists thus $k_\varepsilon \in \mathbb{N}$ such that $\|T_{n_k}(u_0) - v_\tau(0)\|_{L^\infty(\Omega)} < \frac{\varepsilon}{2}$, $\forall k \ge k_\varepsilon$. For $t \ge n_{k_\varepsilon}T$ we compare the solution $\tilde{u}$ of $(*)$ with the solutions $\underline{w}$, $\overline{w}$ of the *ODE's*

$$(3.3)_\varepsilon \qquad \begin{cases} \dot{w} = (m_0(t) \mp \tilde{m}_1(t))h(w) & t \ge n_{k_\varepsilon}T \\ w(n_{k_\varepsilon}T) = v_\tau(0) \mp \dfrac{\varepsilon}{2}\,, \end{cases}$$

respectively, where $\tilde{m}_1(t) := \|m_1(\cdot,t)\|_{C(\bar{\Omega})}$. We infer that the spatially constant functions $\underline{w}$, $\overline{w}$ are ordered sub- and supersolutions for $(*)$, and that

$$(3.4) \qquad \underline{w}(t) \le \tilde{u}(\cdot,t) \le \overline{w}(t) \quad \text{for} \quad t \ge n_{k_\varepsilon}T\,.$$

Let $h_{\max} := \max\{h(\sigma) : 0 \le \sigma \le 1\}$. Looking at $(3.3)_\varepsilon$ and increasing k_ε, if necessary,

$$\left|\int_{w(n_{k_\varepsilon}T)}^{w(nT)} \frac{d\sigma}{h(\sigma)}\right| \le \int_{n_{k_\varepsilon}T}^{\infty} \tilde{m}_1(s)ds < \frac{\varepsilon}{2h_{\max}}$$

for $n \geq n_{k_\varepsilon}$, for either $w = \underline{w}$ and $w = \overline{w}$. By the mean value theorem, $|w(nT) - w(n_{k_\varepsilon}T)| < \frac{\varepsilon}{2}$ and hence $|w(nT) - v_\tau(0)| < \varepsilon$ for $n \geq n_{k_\varepsilon}$. By (3.4) we have proved that $\tilde{u}(\cdot, nT) \to v_\tau(0)$ in $L^\infty(\Omega)$ as $n \to \infty$, and by compactness the convergence in E follows. □

References

[1] N.D. Alikakos and P. Hess: *Ljapunov operators and stabilization in strongly order-preserving dynamical systems.* Differential and Integral Equations 4 (1991), 15 - 24.

[2] N.D. Alikakos, P. Hess and H. Matano: *Discrete order preserving semigroups and stability for periodic parabolic differential equations.* J. Differential Equations 82 (1989), 322 - 341.

[3] A. Beltramo and P. Hess: *On the principal eigenvalue of a periodic-parabolic operator.* Comm. Partial Differential Equations 9 (1984), 919 - 941.

[4] E.N. Dancer and P. Hess: *Stability of fixed points for order-preserving discrete-time dynamical systems.* J. reine angew. Mathematik (to appear).

[5] P. Hess: *Periodic-parabolic boundary value problems and positivity.* Pitman Research Notes in Mathematics Series, Vol. 247, 1991, Longman Scientific & Technical.

[6] P. Hess: *Asymptotics in semilinear periodic diffusion equations with Dirichlet or Robin boundary conditions.* Arch. Rat. Mech. Anal. (to appear).

[7] P. Hess and H. Weinberger: *Convergence to spatial-temporal clines in the Fisher equation with time-periodic fitnesses.* J. Math. Biol. 28 (1990), 83 - 98.

[8] P.A. Vuillermot: *Almost-periodic attractors for a class of nonautonomous reaction-diffusion equations on* $\mathbf{R}^N$, I. *Global stabilization processes.* J. Differential Equations (to appear).

Peter Hess
Mathematics Institute
University of Zurich
Rämistrasse 74

CH 8001 Zurich, Switzerland

G H KNIGHTLY AND D SATHER

Time-periodic states in problems containing a structure parameter

1 Introduction:

In most nonlinear problems of physical interest one or more real parameters occur, corresponding to the structure of the physical setting and the applied forces. Here we investigate a class of problems of the form

$$\frac{dw}{dt}+L(\lambda,\gamma)w+B(w)=0, \quad w(t)\in\mathcal{H} \tag{$*$}$$

in which λ measures the forces driving the system, γ distinguishes different assumptions about the structure of the system, $L(\lambda,\gamma)$ is a linear operator, B a nonlinear operator with $B(0)=0$ and $\mathcal{H}$ is a Hilbert space. We regard solutions, w, of $(*)$ as disturbances of some primary state, $w=0$, of the system and look for secondary states branching from the primary state at some critical value $\lambda=\lambda_c(\gamma)$ of the load.

We shall consider problems $(*)$ for which zero is a double eigenvalue of $L(\lambda,\gamma)$ at $\lambda=\lambda_0=\lambda_c(0)$, $\gamma=0$. Under suitable hypotheses one can prove for each *fixed* γ near zero that $L(\lambda,\gamma)$ has a conjugate pair of imaginary eigenvalues at $\lambda_c(\gamma)$ near λ_0 and that Hopf bifurcation of a time-periodic nontrivial state of $(*)$ occurs as λ crosses $\lambda_c(\gamma)$. In this development, the direction of bifurcation (subcritical or supercritical) and the stability are determined by the sign of a coefficient $c_{0r}(\gamma)$ (see equation (3.2) below); knowledge of this sign for one structure (e.g., $\gamma = 0$) then gives the sign for all nearby structures.

In §2 we formulate our hypotheses on the operators in $(*)$ and present the main lemma concerning the spectral properties of $L(\lambda, \gamma)$. Many authors have studied Hopf bifurcation (e.g., see [1;2;3;4;6;8]); our setting enables us in §3 to use the extensive results of Iooss [1;2] to obtain our theorem on bifurcating periodic states, their stability and form.

Problems of the form $(*)$ arise in the treatment of spiral flows governed by the Navier-Stokes equations (e.g., see [5]). In particular, problems of rotating plane Couette flow and rotating plane Couette-Poiseuille flow lead to this form and satisfy our hypotheses; in these problems the positivity of the coefficient c_{0r} can be verified at $\gamma = 0$ and stable supercritical periodic states obtained. These applications of the theory will appear elsewhere.

2 Preliminaries and Hypotheses

We consider problem $(*)$ in a Hilbert space $\mathcal{H}$ with inner product $(\cdot,\cdot)$ and norm $\|\cdot\|$. We shall also require Hilbert spaces $\mathcal{D}$ with inner product $(\cdot,\cdot)_{\mathcal{D}}$ and $\mathcal{K}$ with inner product $(\cdot,\cdot)_{\mathcal{K}}$ such that the embeddings $\mathcal{D} \hookrightarrow \mathcal{K} \hookrightarrow \mathcal{H}$ are compact.

The linear operator $L(\lambda, \gamma)$ is defined in $\mathcal{H}$ with fixed domain $\mathcal{D}$ for all $(\lambda, \gamma) \in \mathcal{I} = \mathcal{I}_1 \times \mathcal{I}_2$ where $\mathcal{I}_1 = \{\lambda : -\gamma_1 < \lambda - \lambda_0 < \gamma_1\}$ and $\mathcal{I}_2 = \{\gamma : -\gamma_1 < \gamma < \gamma_1\}$ are real intervals. It is convenient to assume $L(\lambda, \gamma)$ is holomorphic for λ and γ in these intervals and therefore in complex neighborhoods $O_j \supset \mathcal{I}_j, j = 1, 2$. We further suppose that the operators in $(*)$ satisfy the following hypotheses, where $\mathcal{L}(X, Y)$ denotes the bounded linear operators from X to Y. (For additional terminology, see [1;7].)

(H.1) For $(\lambda, \gamma) \in \mathcal{I}$, $L(\lambda, \gamma)$ is real, holomorphic of type (A), has compact resolvent and is m-sectorial with vertex b_0 and seminangle θ_0 independent of (λ, γ).

(H.2) The holomorphic semigroup $e^{-L(\lambda,\gamma)t}$ generated by $-L(\lambda,\gamma)$ satisfies

$$\|e^{-L(\lambda,\gamma)t}u\|_{\mathcal{D}} \leq C_1 e^{b_0 t}(1+t^{-\alpha})\|u\|_{\mathcal{K}} \tag{2.1}$$

for all $u \in \mathcal{K}$ and some constants $C_1 > 0$, b_0, and $\alpha \in [0,1)$ independent of (λ,γ).

(H.3) For $(\lambda,\gamma) \in \mathcal{I}$ and some interior point λ_0 of $\mathcal{I}_1$,

$$L(\lambda,\gamma) = L_0 + (\lambda-\lambda_0)L_1 + \gamma\lambda L_2 + \gamma^2 L_3(\lambda,\gamma), \tag{2.2}$$

where $L_0, L_1, L_2, L_3(\lambda,\gamma)$ are real, linear operators in $\mathcal{H}$ with domain $\mathcal{D}$ and satisfying:

(i) $L_0 = L(\lambda_0, 0)$ belongs to $\mathcal{L}(\mathcal{D},\mathcal{H})$, is nonnegative, selfadjoint and has a two dimensional nullspace $\mathcal{N} = \mathcal{N}(L_0)$ spanned by an orthogonal, complex conjugate pair of eigenfunctions $\psi_0, \overline{\psi}_0$ with $\|\psi_0\|^2 = d_0$. The norm $\|u\|_{\mathcal{D}}$ on $\mathcal{D}$ is equivalent to $[\|u\|^2 + \|L_0 u\|^2]^{\frac{1}{2}}$.

(ii) L_1 belongs to $\mathcal{L}(\mathcal{D},\mathcal{K})$, is selfadjoint and satisfies $(L_1\psi_0, \overline{\psi}_0) = (L_1\overline{\psi}_0, \psi_0) = 0$, and

$$(L_1\psi_0, \psi_0) = d_1 \neq 0. \tag{2.3}$$

(iii) $L_2 = M_1 + M_2$, where M_1 and M_2 are real operators belonging to $\mathcal{L}(\mathcal{D},\mathcal{K})$ such that $M_1 : \mathcal{N} \to \mathcal{N}$, $M_1 : \mathcal{D} \cap \mathcal{N}^\perp \to \mathcal{N}^\perp$, $M_2 : \mathcal{N} \to \mathcal{N}^\perp$ and $M_1\psi_0 = -i\eta_0\psi_0$ where $\eta_0 > 0$ and $\mathcal{N}^\perp$ is the orthogonal complement of $\mathcal{N}$ in $\mathcal{H}$.

(iv) $L_3(\lambda,\gamma)$ belongs to $\mathcal{L}(\mathcal{D},\mathcal{K})$ with $\|L_3(\lambda,\gamma)\|_{\mathcal{L}(\mathcal{D},\mathcal{K})} \leq C_2$, for some $C_2 > 0$ independent of (λ,γ).

(H.4) The nonlinear operator $B(u) = \Phi(u,u)$ is quadratic, defined in terms of a real bilinear operator $\Phi : \mathcal{D} \times \mathcal{D} \to \mathcal{K}$ satisfying

$$\|\Phi(u,v)\|_K \le C_3 \|u\|_{\mathcal{D}} \|v\|_{\mathcal{D}} \tag{2.4}$$

for all $u, v \in \mathcal{D}$ and some $C_3 > 0$.

Let $S : \mathcal{H} \to \mathcal{N}$ denote the orthogonal projection of $\mathcal{H}$ onto $\mathcal{N}$. Then $Q = I - S$ is the orthogonal projection of $\mathcal{H}$ onto $\mathcal{N}^\perp$. Note that for $w \in \mathcal{H}$, $Sw = \beta_1\psi_0 + \beta_2\overline{\psi}_0$ where $\beta_1 = (w, \psi_0)d_0^{-1}, \beta_2 = (w, \overline{\psi}_0)d_0^{-1}$. We shall investigate the spectrum, $\sum(L(\lambda,\gamma))$, of $L(\lambda,\gamma)$ for (λ,γ) near $(\lambda_0, 0)$. We seek solutions of

$$L(\lambda,\gamma)w = \zeta w \tag{2.5}$$

in the form

$$w = \gamma(u + \gamma v), \quad u \in \mathcal{N}, \quad v \in \mathcal{N}^\perp, \tag{2.6a}$$

$$\zeta = \gamma\mu = \gamma(\xi + i\eta), \tag{2.6b}$$

$$\lambda = \lambda_0 + \gamma\tau. \tag{2.6c}$$

The following lemma shows that for small γ the operator $L(\lambda,\gamma)$ has a unique pair of complex conjugate eigenvalues with real part passing through zero as λ increases through a critical value $\lambda_c(\gamma)$. The lemma is proved by making the substitution (2.6) in (2.5) and using splitting methods; details will appear elsewhere.

Lemma 2.1. Suppose $L(\lambda,\gamma)$ satisfies hypotheses (H.1),(H.2),(H.3). Given $r_0 > 0$, there exist positive numbers γ_0 and ρ_0 such that for each fixed $\gamma \in (-\gamma_0, \gamma_0), \gamma \ne 0$, the following hold for λ real, $|\lambda - \lambda_0| < \gamma_0$.

(i) $\sum(L(\lambda,\gamma)) = \{\zeta(\lambda,\gamma), \overline{\zeta}(\lambda,\gamma)\} \cup \sum'(L(\lambda,\gamma))$, where the eigenvalues $\zeta(\lambda,\gamma), \overline{\zeta}(\lambda,\gamma)$ are simple and lie in the disc $|\zeta| < \rho_0$, and $\sum' \subset \{\zeta : Re\zeta \ge 2\rho_0\}$. The eigenvalue ζ is analytic in (λ,γ)

for (λ, γ) in $\mathcal{S} = \{(\lambda, \gamma) : \lambda_0 - \gamma r_0 \leq \lambda \leq \lambda_0 + \gamma r_0, |\gamma| < \gamma_0\}$. Moreover, ζ has the form

$$\zeta(\lambda, \gamma) = \gamma[\xi(\lambda, \gamma) + i\eta(\lambda, \gamma)]$$

where $\gamma\xi(\lambda, \gamma)$ is monotone in λ for $\lambda_0 - \gamma r_0 < \lambda < \lambda_0 + \gamma r_0$, increasing if $d_1 > 0$ and decreasing if $d_1 < 0$.

(ii) There is a unique $\lambda_c(\gamma) \in (\lambda_0 - \gamma r_0, \lambda_0 + \gamma r_0)$ such that $\zeta(\lambda_c(\gamma), \gamma) = i\gamma\tilde{\eta}(\gamma)$ is imaginary. In addition,

$$\lambda_c(\gamma) = \lambda_0 - b_1\gamma^2 + 0(\gamma^3), \tag{2.7}$$

with known constant b_1, and
$\xi(\lambda, \gamma) = (\lambda - \lambda_c(\gamma))\tilde{\xi}_1(\gamma) + O(\lambda - \lambda_c(\gamma))^2$,
$\eta(\lambda, \gamma) = \tilde{\eta}(\gamma) + (\lambda - \lambda_c(\gamma))\tilde{\eta}_1(\gamma) + O(\lambda - \lambda_c(\gamma))^2$,
where $\tilde{\eta}(0) = \lambda_0\eta_0$.

(iii) The eigenfunction, ψ_γ, of $L(\lambda_c(\gamma), \gamma)$ corresponding to $\zeta(\lambda_c(\gamma), \gamma)$ and the eigenfunction, ψ_γ^*, of the adjoint $L^*(\lambda_c(\gamma), \gamma)$ corresponding to $\overline{\zeta}(\lambda_c(\gamma), \gamma)$ can be taken in the form

$$\psi_\gamma = \psi_0 + \tilde{\psi}(\gamma) \quad , \quad \psi_\gamma^* = \psi_0 + \tilde{\psi}^*(\gamma) \tag{2.8}$$

with $\tilde{\psi}(\gamma), \tilde{\psi}^*(\gamma) \in \mathcal{D}$, analytic in γ, $\|\tilde{\psi}(\gamma)\|^2 = \|\tilde{\psi}^*(\gamma)\|^2 = d_o$ and $\lim_{\gamma \to 0} \|\tilde{\psi}(\gamma)\|_{\mathcal{D}} = \lim_{\gamma \to 0} \|\tilde{\psi}^*(\gamma)\|_{\mathcal{D}} = 0$.

3 Periodic States

In this section we collect some results for the nonlinear problem $(*)$ that now follow directly from theorems of Iooss [1] because of the properties obtained in §2 for the linear operators. In fact, for fixed $\gamma \neq 0$ the hypotheses of [1,§2] are just (H.1), (H.2), (H.3) in our §2, while the supplementary hypotheses

(H.1), (H.2), (H.3) in [1, §4(2)] are direct consequences of our (H.3)(iii) and Lemma 2.1. Theorems 1-4 of [1] on the bifurcation and stability of periodic states of the nonlinear problem therefore apply directly to equation (*) for each $\gamma \neq 0$ in $\gamma_0 < \gamma < \gamma_0$, γ_0 sufficiently small.

Briefly, the idea is to look for a solution of (*) in the form $w = u+v$ with $u = E_{\lambda\gamma}w$ and $v = P_{\lambda\gamma}w$, where $E_{\lambda\gamma}$ is the projection of $\mathcal{H}$ onto the subspace spanned by $\{\psi(\lambda,\gamma), \overline{\psi}(\lambda,\gamma)\}$ and $P_{\lambda\gamma}$ is the projection corresponding to the rest of the spectrum of $L(\lambda,\gamma)$; here $\psi(\lambda,\gamma)$ is the eigenfunction corresponding to $\zeta(\lambda,\gamma)$. One obtains $v = v(u,\lambda,\gamma)$, e.g., by a contraction argument for the projection of (*) on $P_{\lambda\gamma}\mathcal{H}$. This leads to a problem of the form

$$\frac{du}{dt} + L(\lambda,\gamma)u = f(u), \quad u \in E_{\lambda\gamma}\mathcal{H}. \tag{3.1}$$

One now splits $u = \psi + U$, with ψ in the nullspace $\mathcal{M}$ of the operator on the left in (3.1), and solves (3.1) for U, again by contraction, for small ψ provided $f(u)$ has zero projection onto $\mathcal{M}$. The latter condition, the bifurcation equation, is then solved for ψ and the period T. One obtains expansions for the unknowns in powers of $\varepsilon = \sqrt{|\lambda - \lambda_c(\gamma)|}$ provided that $c_{0r}(\gamma) \neq 0$, where $c_{0r}(\gamma)$ is the real part of $c_0(\gamma) = c_{0r}(\gamma) + ic_{0i}(\gamma)$ given by (see(2.8))

$$\begin{aligned} c_o(\gamma) &= -(\{\Phi(\psi_\gamma, L^{-1}(\lambda_c(\gamma),\gamma)[\Phi(\psi_\gamma,\overline{\psi}_\gamma) + \Phi(\overline{\psi}_\gamma,\psi_\gamma)]) \\ &+\Phi(L^{-1}(\lambda_c(\gamma),\gamma)[\Phi(\psi_\gamma,\overline{\psi}_\gamma) + \Phi(\overline{\psi}_\gamma,\psi_\gamma)],\psi_\gamma) \\ &+\Phi(\overline{\psi}_\gamma,(L(\lambda_c(\gamma),\gamma) + 2i\gamma\tilde{\eta}(\gamma))^{-1}\Phi(\psi_\gamma,\psi_\gamma)) \\ &+\Phi((L(\lambda_c(\gamma),\gamma) + 2i\gamma\tilde{\eta}(\gamma))^{-1}\Phi(\psi_\gamma,\psi_\gamma),\overline{\psi}_\gamma)\},\psi_\gamma^*). \end{aligned} \tag{3.2}$$

We refer to [1] for the extensive details and state the result.

Theorem 3.1. Suppose $L(\lambda,\gamma)$ and B satisfy hypotheses (H.1)-(H.4) and assume $d_1 < 0$, $c_{0r}(\gamma) \neq 0$. For each *fixed* $\gamma \neq 0$, $|\gamma| < \gamma_0$, γ_0 sufficiently small, there is a branch $w(\lambda,\gamma)$ of T-periodic nontrivial solutions of (*)

bifurcating from $w = 0$ at $\lambda = \lambda_c(\gamma)$. The solution has the following properties.

(i) For fixed λ, $w(\lambda,\gamma) \in C^0(-\infty,\infty;\mathcal{D})$ and is unique among small nontrivial solutions of $(*)$, up to arbitrary translations in time.

(ii) If $c_{0r}(\gamma) > 0$ the bifurcation is "to the right" $(\lambda > \lambda_c(\gamma))$ and the solution orbits are asymptotically stable. If $c_{0r}(\gamma) < 0$ the bifurcation is "to the left" $(\lambda < \lambda_c(\gamma))$ and the solutions are unstable.

(iii) $w(\lambda,\gamma)$ is analytic in $\varepsilon = \sqrt{|\lambda - \lambda_c(\gamma)|}$ and the period $T(\lambda,\gamma)$ is analytic in ε^2. In fact, the following expansions are valid

$$w(\lambda,\gamma) = \sum_{n=1}^{\infty} w_n(\gamma)\varepsilon^n, \quad T(\lambda,\gamma) = \sum_{n=0}^{\infty} T_n(\gamma)\varepsilon^{2n},$$

where the $w_n(\gamma)$ are $T(\lambda,\gamma)$ - periodic in time. In particular,

$$\begin{aligned} w_1(\gamma)(t) &= \beta(t)\psi_\gamma + \overline{\beta}(t)\overline{\psi}_\gamma, \\ w_2(\gamma)(t) &= |\beta(t)|^2 L^{-1}(\lambda_c(\gamma),\gamma)[\Phi(\psi_\gamma,\overline{\psi}_\gamma) + \Phi(\overline{\psi}_\gamma,\psi_\gamma)] \\ &\quad +\beta^2(t)[L(\lambda_c(\gamma),\gamma) + 2i\gamma\eta_0 I]^{-1}\Phi(\psi_\gamma,\psi_\gamma) \\ &\quad +\overline{\beta}^2(t)[L(\lambda_c(\gamma),\gamma) - 2i\gamma\eta_0 I]^{-1}\Phi(\overline{\psi}_\gamma,\overline{\psi}_\gamma), \end{aligned}$$
$$\beta(t) = \beta_0 e^{\frac{i2\pi t}{T}}, \quad |\beta_0| = [\gamma\tilde{\eta}(\gamma)\tilde{\xi}_1(\gamma)/|c_{0r}(\gamma)|]^{\frac{1}{2}}$$

and

$$T_0 = 2\pi/\gamma\tilde{\eta}(\gamma) \ , \ T_1 = \gamma\left(\frac{c_{0i}(\gamma)}{c_{0r}(\gamma)}\tilde{\xi}_1(\gamma) - \tilde{\eta}_1(\gamma)\right)T_0,$$

with $\tilde{\xi}_1(\gamma)$ and $\tilde{\eta}_1(\gamma)$ as in (ii) of Lemma 2.1.

Remark 3.2. The assumption $0 > d_1 = (L_1\psi_0, \psi_0)$ corresponds to $Re\zeta$ (in (i) of Lemma 2.1) decreasing through zero as λ increases through $\lambda_c(\gamma)$. This is the expected situation for $w = 0$ to lose stability at $\lambda = \lambda_c(\gamma)$.

Remark 3.3. From part (ii) of Lemma 2.1 we see for small γ that $\lambda_c(\gamma) > \lambda_0$ if $b_1 < 0$ and $\lambda_c(\gamma) < \lambda_0$ if $b_1 > 0$. If we regard $\gamma[\lambda L_2 + \gamma L_3(\lambda, \gamma)]$ as a structural modification of a basic system $L(\lambda, 0)$, the above observation suggests this modification is destabilizing when $b_1 > 0$ and stabilizing when $b_1 < 0$.

Acknowledgement. The research of Knightly was supported in part by ONR Grant N00014-90-J-1031; that of Sather was supported in part by ONR Grant N00014-90-J-1336.

References

[1] G. Iooss, Existence et stabilité de la solution périodique secondaire intervenant dans les problèmes d'évolution du type Navier-Stokes, Arch. Rational Mech. Anal. 47, 301-329 (1972).

[2] G. Iooss, Bifurcation and transition to turbulence in hydrodynamics, Bifurcation Theory and Applications, L. Salvadori, ed., Lecture Notes in Mathematics #1057, Springer Verlag, New York, 152-201 (1984).

[3] G. Iooss & A. Mielke, Bifurcating time-periodic solutions of Navier-Stokes equations in infinite cylinders, J. Nonlinear Sci., 1 107-146 (1991).

[4] V. I. Iudovich, The onset of auto-oscillations in a fluid, Prikl. Mat. Mek. 35 638-655 (1971).

[5] D. D. Joseph, Stability of Fluid Motions I, Springer Tracts in natural Philosophy 28, Spring Verlag, New York, 1976.

[6] D. D. Joseph & D. H. Sattinger, Bifurcating time periodic solutions and their stability, Arch. Rational Mech. Anal. 45, 79-109 (1972).

[7] T. Kato, Perturbation Theory for Linear Operators, Springer Verlag, New York 1966.

[8] D. H. Sattinger, Bifurcation of periodic solutions of the Navier Stokes equations, Arch. Rational Mech. Anal. 41, 66-80 (1971).

GEORGE H. KNIGHTLY
DEPARTMENT OF MATHEMATICS
UNIVERSITY OF MASSACHUSETTS
AMHERST, MA 01003

D. SATHER
DEPARTMENT OF MATHEMATICS
UNIVERSITY OF COLORADO
BOULDER, CO 80309-0426

H A LEVINE

Fujita type theorems for weakly coupled parabolic systems

0. Introduction.

In [Fu1,Fu2], Fujita considered nonnegative solutions of

$$\text{(F)}\qquad \begin{aligned} &u_t = \Delta u + u^p && (x,t) \in R^N \times (0,T) \\ &u(x,0) = u_0(x) \geq 0 && x \in R^N, \quad u_0 \in L^\infty(R^N) \end{aligned}$$

where $p > 0$. He proved that if $1 < p < 1 + 2/N$ then either $u_0 \equiv 0$ or else $T < +\infty$. That is, global, nontrivial solutions are not possible for $p \in (1, 1+2/N)$. (This is known as the blow up or subcritical case, $p_c = 1 + 2/N$ being the so-called critical blow up exponent.) Later, several authors proved, by various arguments, that p_c is in the blow up case. See [We] for a particularly elegant proof. He also showed that if $p > 1 + 2/N$, both global, nontrivial (small data) solutions and nonglobal solutions exist. This is the super critical or global existence case. He also showed that for $0 < p \leq 1$, u is necessarily global. (Uniqueness fails if $0 < p < 1$ [FuW]. We shall say more about uniqueness later.)

The purpose of this talk is to discuss some extensions to a weakly coupled system of parabolic equations obtained by several writers. A survey of results for single equations is given in [L1].

Let us consider the following rather trivial reformulation of (F):

$$(\text{F}_s)\qquad \begin{aligned} u_t &= \Delta u + u^p \\ v_t &= \Delta v + v^q \end{aligned} \qquad (x,t) \in R^N \times (0,T)$$

$$\begin{aligned} u(x,0) &= u_0(x) \\ v(x,0) &= v_0(x) \end{aligned} \qquad x \in R^N$$

where now $p, q > 0$, $u_0, v_0 \geq 0$, $u_0, v_0 \in L^\infty(R^N)$.

Theorem I. *Let*

$$A = \begin{bmatrix} p & 0 \\ 0 & q \end{bmatrix},$$
$$\delta = \det(A - I)$$

and, whenever $\delta \neq 0$, $X = (\alpha, \beta)^t$ *be the unique solution of*

$$(A - I)X = \begin{pmatrix} 1 \\ 1 \end{pmatrix}.$$

We have

1. *Suppose* $\delta \neq 0$.
   (a) *If* $\min(\alpha, \beta) \geq \frac{1}{2}N$, $(\mathbf{F}_s)$ *has no nontrivial global solutions.*
   (b) *If* $\min(\alpha, \beta) < \frac{1}{2}N$ *and* $\max(\alpha, \beta) > 0$, $(\mathbf{F}_s)$ *has both nontrivial global and nonglobal solutions.*
   (c) *If* $\max(\alpha, \beta) < 0$ *all solutions of* $(\mathbf{F}_s)$ *are global.*
2. *Assume* $\delta = 0$. *Let* λ_1, λ_2 *denote the eigenvalues of* A.
   (a) *If* $\lambda_1 > \lambda_2 = 1$, *then* $(\mathbf{F}_s)$ *has both nontrivial global and nonglobal solutions.*
   (b) *If* $1 = \lambda_1 \geq \lambda_2$, *then all solutions of* $(\mathbf{F}_s)$ *are global.*

Here

$$\max(\alpha, \beta) = \frac{1}{\min(p, q) - 1}$$
$$\min(\alpha, \beta) = \frac{1}{\max(p, q) - 1}.$$

In [EH], Escobedo and Herrero considered the system

$$\text{(EH)} \qquad \begin{aligned} u_t &= \Delta u + v^p \\ u_t &= \Delta v + u^p \end{aligned} \qquad (x, t) \in R^N \times (0, T)$$

$$\begin{aligned} u(x, 0) &= u_0(x) \\ v(x, 0) &= v_0(x) \end{aligned} \qquad x \in R^N$$

with u_0, v_0 as above.

Taking

$$A = \begin{pmatrix} 0 & p \\ q & 0 \end{pmatrix}$$

with the same definitions of δ, α, β as above, they proved:

Theorem II. *Consider nonnegative solutions of* (EH).

1. *Suppose* $\delta \neq 0$.
   (a) *If* $\max(\alpha, \beta) \geq \frac{1}{2}N$, (EH) *has no nontrivial global solutions.*
   (b) *If* $0 \leq \max(\alpha, \beta) < \frac{1}{2}N$, (EH) *has both nontrivial global and nonglobal solutions.*
   (c) *If* $\max(\alpha, \beta) < 0$, *all solutions of* (EH) *are global.*
2. *If* $\delta = 0$, *then* $\lambda_1 = -\lambda_2 = 1$ *and all solutions are global.*

For (EH)

$$\max(\alpha, \beta) = \frac{\max(p, q) + 1}{pq - 1}$$
$$\min(\alpha, \beta) = \frac{\min(p, q) + 1}{pq - 1}.$$

Remark 1. If $p, q > 1$

$$\min\left(\frac{1}{p-1}, \frac{1}{q-1}\right) = \frac{1}{\max(p, q) - 1} \leq \frac{\max(p, q) + 1}{pq - 1}$$

with equality if and only if $p = q$. Thus (F_s) can have global, nontrivial solutions while (EH) does not, for the same pair (p, q). This would suggest that coupling has a destabilizing effect on true systems (EH) over single equations (F_s). However, when $\delta = 0$, blow up can occur for (F_s) but never for (EH). This suggests that coupling can also have a destabilizing effect over single equations. (However, $\delta = 0$ for both (EH) and (F_s) $\Longleftrightarrow p = q = 1$.)

1. Existence and Nonexistence.

We may extend Theorems I and II in a number of directions. For example, we could consider the more general system

$$(\mathrm{S}_1) \qquad \begin{aligned} u_t &= \Delta u + u^{p_1} v^{q_1} \\ v_t &= \Delta v + u^{p_2} v^{q_2} \end{aligned}$$

in a cylinder

$$D \times (0,T)$$

where now $D \subset R^N$ is some unbounded domain with a piecewise smooth boundary, ∂D, and where $p_i, q_j \geq 0$, but $p_1 + q_1 > 0$, $p_2 + q_2 > 0$.

We could also consider more general semilinear systems such as

$$(\mathrm{S}_2) \qquad \begin{aligned} u_t &= \Delta(u^m) + u^{p_1} v^{q_1} \\ v_t &= \Delta(u^n) + u^{p_2} v^{q_2} \end{aligned}$$

with $m, n > 0$, or quasilinear systems such as

$$(\mathrm{S}_3) \qquad \begin{aligned} u_t &= \nabla \cdot \left(|\nabla u|^{p-1} \nabla u\right) + u^{p_1} v^{q_1} \\ v_t &= \nabla \cdot \left(|\nabla v|^{q-1} \nabla v\right) + u^{p_2} v^{q_2}. \end{aligned}$$

There is some recent work on the Cauchy problem for the "porous medium system" (S_2) when $p_1 = q_2 = 0$ ([LQi]) but nothing has been established for quasilinear systems.

For most of the remainder of this talk we will consider only (S_1).

The only domain, other than R^N, which we consider here is that of a cone with vertex at the origin: That is if $\Omega \subset S^{N-1}$ is a given open (in the topology of S^{N-1}) subset of S^{N-1}, we say D is a cone with cross section Ω and vertex at the origin if

$$D = \{x \in R^N \mid x \neq 0, \; x/|x| \equiv \hat{x} \in \Omega\}.$$

Associated with Ω is the following Dirichlet eigenvalue problem:

$$\begin{aligned} \Delta_{\underset{\sim}{\theta}} \psi + \omega \psi &= 0 \qquad \text{on } \Omega \\ \psi &= 0 \qquad \text{on } \partial\Omega \end{aligned}$$

where $\Delta_{\underset{\sim}{\theta}}$ is the Laplace-Beltrami operator. Let ω_1, ψ_1 denote the first eigenvalue and corresponding (necessarily positive) eigenfunction of this problem. We shall assume

$$\int_\Omega \psi_1 dS_{\underset{\sim}{\theta}} = 1.$$

Let $\gamma(D)$ denote the positive root of

$$x(x + N - 2) = \omega_1.$$

In a recent series of papers, [BL,LM1,LM2] the authors showed that Theorem I.1a,b holds in a cone with $\frac{1}{2}N$ replaced by $\frac{1}{2}(N+\gamma(D))$. It is not too hard to see, by using their results well as comparison arguments, that Theorem I.1c and Theorem I.2 also hold in cones.

The number $\frac{1}{2}(N+\gamma(D))$ arises in a natural way if one makes the standard "hot spot" change of variables in (F). Then this number is the first eigenvalue of the linear operator $L \equiv -\Delta y - \frac{1}{2}\vec{y}\cdot\vec{\nabla}_y$ in a suitable weighted Hilbert space. See [EK, K] for details.

Levine [L2] considered (EH) in a cone. He was forced to restrict his attention to the case $p \geq 1$, $q \geq 1$. He showed that Theorem II.1a,b holds with $\frac{1}{2}N$ replaced by $\frac{1}{2}(N+\gamma(D))$. By comparison theorems using the corresponding result for R^N, Theorem II.1c and II.2 hold for cones also.

In order to obtain results for the full problem

$$\text{(EL)} \qquad \begin{aligned} u_t &= \Delta u + u^{p_1}v^{q_1} \\ v_t &= \Delta v + u^{p_2}v^{q_2} \end{aligned} \qquad (x,t)\in D\times(0,T)$$

$$u = v = 0 \qquad \text{on } \partial D\times(0,T)$$

$$\begin{aligned} u(x,0) &= u_0(x) \geq 0 \\ v(x,0) &= v_0(x) \geq 0 \end{aligned} \qquad x\in D$$

we have to overcome several difficulties. These are

1. Solutions of (EL) need not be unique. This difficulty is resolved by extending the notion of maximal and minimal solutions for single equations introduced in [Fu2,FuW] to systems.

2. If, for example, $q_1 > 0$, $q_2 > 0$, then $v \equiv 0$ solves the second equation while, for any u_0, $S(t)u_0 = w$ solves the first equation where $S(t)u_0$ is the solution of the heat equation with $u_0(\cdot)$ as initial values. We will say, therefore, that $(u,v)^t$ is

a trivial solution of (EL) if one component of $(u,v)^t$ vanishes while the other solves the (linear) heat equation.

Thus, to prove that all nontrivial solutions are nonglobal, we need only prove this for nontrivial minimal solutions. To prove that there exist global solutions, we need only prove that a maximal solution is global.

3. We have to distinguish somehow between Theorem I and II in the case of (EL). That is, our theorem must take into account the potential conflict between the two statements. To do this, we shall say that a system of the form

$$u_t = \Delta u + f(u,v)$$
$$v_t = \Delta v + g(u,v)$$

where f, g are C^1 in (u,v) for $u,v > 0$, is completely coupled if $f_v(u,v) \neq 0$ and $g_u(u,v) \neq 0$ for all $u,v > 0$. Otherwise we say the system is incompletely coupled. For (EL), the system is completely coupled if and only if $p_2, q_1 > 0$.

4. We would like the result for the general system (EL) to apply to cones. In particular we would like to remove the restriction that $p, q \geq 1$ required by Levine when $p_1 = q_2 = 0$ and $q_1 = p$, $p_2 = q$ (the system (EH) in cones).

We now let

$$A = \begin{pmatrix} p_1 & q_1 \\ p_2 & q_2 \end{pmatrix}$$

with $\delta = \det(A - I)$ and (α, β) as before.

The precise statement which Escobedo and Levine claim to hold ([EL]) is stated below. The reader is cautioned that the result is not completely established at this writing. Moreover, the statements do not exhaust all cases in the completely coupled case. In particular, the case when $p_! + q_1 < 1$ is not fully understood. The remarks following the statement of the theorem discuss some of the current limitations of the statements.

Theorem III. (Completely coupled case.) *Assume that* $p_2q_1 > 0$, *D is a cone or* $D = R^N$ *and that U is nontrivial.*

A. $p_1 \geq 1$. *Then* $p_1 + q_1 > 1$ *and we have:*

A.1 *If* $(p_1 + q_1 - 1)^{-1} < \frac{1}{2}(N + \gamma)$, *then there are choices of* (u_0, v_0) *for which U is global and for which U is nonglobal.*

A.2 *If* $(p_1 + p_1 - 1)^{-1} \geq \frac{1}{2}(N + \gamma)$, *then U is nonglobal.*

B. $0 \leq p_1 < 1$.

B.1 *Assume* $\delta < 0$.

B.1.a. *Suppose* $p_1 + q_1 \geq 1$ *and* $\max(\alpha, \beta) \geq \frac{1}{2}(N + \gamma)$. *Then U cannot be global.*

B.1.b. *If* $\max(\alpha, \beta) < \frac{1}{2}(N + \gamma)$ *then there are choices of* (u_0, v_0) *for which U is global and choices for which U is not global.*

B.2. *Suppose* $\delta = 0$. *Then since* $p_1 < 1$ *we may also assume* $q_2 < 1$. *In this case we must have* $\lambda_1 = 1 \geq \lambda_2$ *and U must be global.*

B.3. *Suppose* $\delta > 0$. *Since* $p_1 < 1$, *we may assume* $q_2 < 1$ *also. Consequently* $\max(\alpha, \beta) = \alpha$. *We have* $\alpha \leq 0$ *in this case and U must be global.*

Theorem IV. (Incompletely coupled case) *Assume* $p_2q_1 = 0$, $p_1 + q_2 > 0$. *(This last we may assume in view of Theorem 1(s).)*

A. *Assume* $\delta \neq 0$

A.1. *If* $\min(\alpha, \beta) \geq \frac{1}{2}(N + \gamma)$, *$U$ is nonglobal.*

A.2. *Assume* $\max(\alpha, \beta) \geq 0$ *and* $\min(\alpha, \beta) < \frac{1}{2}(N + \gamma)$

A.1.a. *If* $0 \leq \max(\alpha, \beta) < \frac{1}{2}(N + \gamma)$, *there are choices of initial values for which U will be global and for which U will be nonglobal.*

A.1.b. *Assume* $\max(\alpha, \beta) \geq \frac{1}{2}(N + \gamma)$ *and* $\delta < 0$. *If* $q_1 = 0$ *and* $p_2 > 0$, *then U will be logbal for some initial values and nonglobal for others. If* $q_1 > 0$ *and* $p_2 = 0$, *U will be nonglobal.*

A.1.c. *Assume* $\max(\alpha, \beta) \geq \frac{1}{2}(N + \gamma)$ *and* $\delta > 0$. *If* $p_2 = 0$ *and* $q_1 > 0$, *then U will be global for some initial values and nonglobal for* *others. If $p_2 > 0$ and $q_1 = 0$, the U will be nonglobal.*

A.3. If $\max(\alpha, \beta) < 0$, U must be global.

B. *Assume $\delta = 0$.*

B.1. Assume $\lambda_1 > \lambda_2 = 1$.

B.1.a. $p_2 = 0$, $q_1 = 1$. *Then $q_2 > 1$ and U is nonglobal if $1 < q_2 < 1 + 2/(N + \gamma)$. If $q_2 > 1 + 2/(N + \gamma)$, U will be global for some choices of $(u_0, v_0)^t$ and nonglobal for others. If $q_2 = 1$, U is global.*

B.1.b. $q_1 = 0$, $q_2 = 1$. *Then $p_1 \geq 1$ and we have the same conclusions as in* B.1.a *with q_2 replaced by p_1.*

B.1.c. $q_1 = 0$, $p_1 = 1$. *If $q_2 > 1$, U is never global. If $q_2 = 1$, U is always global.*

B.2. If $1 = \lambda_1 \geq \lambda_2$, then U must be global.

Lower bounds for $S(t)u_0$ are easily found. In general

$$S(t)u_0(0) \geq ct^{-N/2}$$

when $D = R^N$ while

$$(S(t)u_0, \psi_1)_\Omega \left(\sqrt{t}, t\right) \geq ct^{-\frac{1}{2}(N+\gamma)}$$

when D is a cone.

In order to prove blow up for large data, when $\delta \neq 0$, one derives a lower bound (when $u_0 = v_0$)

$$u(t) \geq \sum_{j=1}^{n} I_j(t)$$

$$v(t) \geq \sum_{j=1}^{n} J_j(t)$$

valid for $n = 1, 2, 3, \ldots$, where

$$I_j(t) = c_1^{\lambda_1^j} t^{\tilde{\alpha}_j} \left(S(t)u_0^\varepsilon\right)^{\pi_j}$$

$$J_j(t) = c_2^{\lambda_1^j} t^{\tilde{\beta}_j} \left(S(t)u_0^\varepsilon\right)^{\theta_j}$$

where, for $j = 1, 2, 3, \ldots$

$$\begin{pmatrix}\pi_j\\ \theta_j\end{pmatrix} = A\begin{pmatrix}\pi_{j-1}\\ \theta_{j-1}\end{pmatrix}, \qquad \begin{pmatrix}\pi_0\\ \theta_0\end{pmatrix} = \begin{pmatrix}1/\varepsilon\\ 1/\varepsilon\end{pmatrix}$$

$$\begin{pmatrix}\tilde{\alpha}_j\\ \tilde{\beta}_j\end{pmatrix} = A\begin{pmatrix}\tilde{\alpha}_{j-1}\\ \tilde{\beta}_{j-1}\end{pmatrix} + \begin{pmatrix}1\\ 1\end{pmatrix}, \qquad \begin{pmatrix}\alpha_0\\ \beta_0\end{pmatrix} = \begin{pmatrix}0\\ 0\end{pmatrix}$$

and where $\varepsilon = p_1 + q_1$. The idea for this iteration scheme was introduced in [AW] for single equations and extended in [EH] for system (EH). From these inequalities and the relevant matrix asymptotics, we can show that $(u, v)^t$ will not be global if $\lambda_1 > 1$ and the initial values are sufficiently large.

We have used a variant of this scheme to prove Theorem III parts A.1. and B.1.a.

The claims of global existence in Theorem III are established by demonstrating the existence of global super solution of the form

$$\begin{pmatrix}\bar{u}\\ \bar{v}\end{pmatrix} = \begin{pmatrix}\beta_1(t)\\ \beta_2(t)\end{pmatrix} W(x,t)$$

where

$$W(x,t) = c(t+t_0)^{-(\gamma+\frac{1}{2}N)}|x|^\gamma e^{-|x|^2/4(t+t_0)}\psi_1(\underset{\sim}{\theta})$$

when $\max(\alpha, \beta) \geq 0$ and of the form

$$\begin{pmatrix}\bar{u}\\ \bar{v}\end{pmatrix} = \begin{pmatrix}A(t+t_0)^a\\ B(t+t_0)^b\end{pmatrix}$$

when $\max(\alpha, \beta) < 0$.

Remark 2. The function $W(x,t)$ is a lower bound for the first term in the orthogonal expansion of the Green's function for the heat equation in cones given by

$$G(x,y,t) = (2t)^{-1}(r\rho)^{-\frac{1}{2}(N-2)}e^{-\frac{(r^2+\rho^2)}{4t}}\sum_{n=1}^{\infty} c_n I_{\nu_n}\left(\frac{r\rho}{2t}\right)\psi_n(\underset{\sim}{\theta})\psi_n(\underset{\sim}{\theta})$$

where $x = (r, \underset{\sim}{\theta})$, $y = (\rho, \underset{\sim}{\varphi})$, $\{\psi_n\}_{n=1}^{\infty}$ is the sequence of orthogonal eigenfunctions for $\Delta_{\underset{\sim}{\theta}}$, and where the $\{c_n\}_1^\infty$ is such that for all $f \in L^2(\Omega)$

$$f(\underset{\sim}{\theta}) = \sum_{n=1}^{\infty} c_n(f, \psi_n)_\Omega \psi_n(\underset{\sim}{\theta}).$$

evaluated at $y = (1, \underset{\sim}{\theta}_0)$ for fixed $\underset{\sim}{\theta}_0 \in \Omega$ and $t = t + t_0$. (The $c_n \neq 1$ in [LM1,LM2,EL] because the authors took $\int_\Omega \psi_1 dS_\theta = 1$. If we abandon this normalization we may take the $c_n = 1$.)

Similar arguments are invoked for the proofs of the claims in Theorem IV.

2. Nonuniqueness. If $p_i + q_i \leq 1$ for $i = 1,2$ and $\max(\alpha, \beta) < 0$ it is fairly easy to see that for fixed $t_0 > 0$,

$$\tilde{u}(t,t_0) = \begin{cases} 0 & 0 \leq t \leq t_0 \\ c_1(t-t_0)^{-\alpha} & t \geq t_0 \end{cases}$$

$$\tilde{v}(t,t_0) = \begin{cases} 0 & 0 \leq t \leq t_0 \\ c_2(t-t_0)^{-\beta} & t > t_0 \end{cases}$$

where

$$c_1 = \left[(-\alpha)^{q_2-1}(-\beta)^{-q_1}\right]^{1/\delta}$$
$$c_2 = \left[(-\alpha)^{-p_2}(-\beta)^{p_1-1}\right]^{1/\delta}$$

is a solution of (EH) when $D = \mathbb{R}^N$.

In order to construct a nontrivial global solution of (EH) which vanishes at $t = 0$ and on ∂D when D is a cone, we try to construct a nontrivial global subsolution with these properties of the form

$$\begin{pmatrix} \underline{u} \\ \underline{v} \end{pmatrix} = \begin{pmatrix} y(t) \\ z(t) \end{pmatrix} \varphi(x)$$

where

$$\varphi(x) = \left(\frac{ke}{m}\right)^m r^m e^{-kr} \psi_1(\underset{\sim}{\theta})$$

where now $\max \psi_1 = 1$, $k > 0$, $m > 0$, $m^2 + (N-2)m - \omega_1 > 0$,

$$\lambda = k^2 \left(m + \omega_1 + \frac{1}{4}(N-1)^2\right) / \left(m^2 + (N-2)m - \omega_1\right)$$

and, in either case $y(0) = z(0) = 0$ while

$$(*) \qquad \begin{aligned} y' &= -\lambda y + y^{p_1} z^{q_1} \\ z' &= -\lambda z + y^{p_2} z^{q_2}. \end{aligned}$$

One can show that a nontrivial solution of this system exists with

$$\lim_{t \to +\infty} y(t) = \lambda^\alpha$$
$$\lim_{t \to +\infty} z(t) = \lambda^\beta.$$

(See the Figure for motivation.)

Consequently, since $\tilde{u} \geq \underline{u}$, $\tilde{v} \geq \underline{v}$, we see that

$$\underline{u} \leq u_k \leq u_{k+1} \leq \tilde{u}, \quad \underline{v} \leq v_k \leq v_{k+1} \leq \tilde{v}$$

where $\{u_k\}_{k=0}^{\infty}$, $\{v_k\}_{k=0}^{\infty}$ are given by

$$u_{k+1} = S(t)u_k + \int_0^t S(t-s)u_k^{p_1}v_k^{q_1}\,ds$$

$$v_{k+1} = S(t)v_k + \int_0^t S(t-s)u_k^{p_2}v_k^{q_2}\,ds$$

and where

$$u_0(t) = \begin{cases} 0 & 0 \leq t \leq t_0 \\ \underline{u}(t-t_0) & t > t_0 \end{cases}$$

$$v_0(t) = \begin{cases} 0 & 0 \leq t \leq t_0 \\ \underline{v}(t-t_0) & t > 0. \end{cases}$$

The limit

$$\begin{pmatrix} u \\ v \end{pmatrix} = \lim_{k\to\infty} \begin{pmatrix} u_k \\ v_k \end{pmatrix}$$

exists, satisfies the boundary condition, the system (EH), vanishes at $t = 0$ and is nontrivial as required.

Phase plane diagram for system (*).

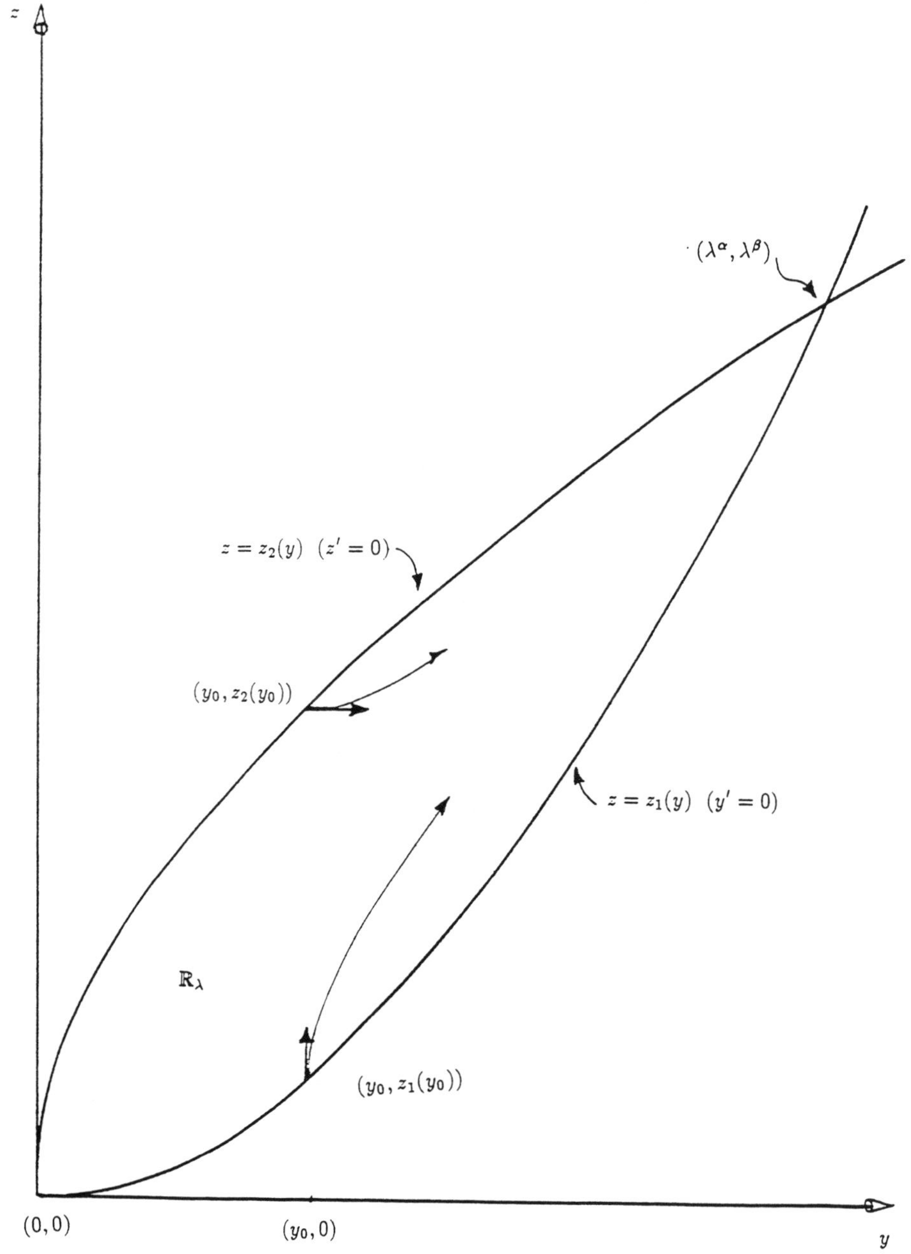

References

[AW] D. G. Aronson and H. F. Weinberger, *Multidimensional nonlinear diffusion arising in population genetics*, Advances in Math **30** (1978), 33–76.

[BL] C. Bandle and H. A. Levine, *On the existence and nonexistence of global solutions of reaction-diffusion equations in sectorial domain*, TAMS **655** (1989), 595–624.

[EH] M. Escobedo and M. A. Herrero, *Boundedness and blow up for a semilinear reaction-diffusion system*, J.D.E. **89** (1991), 176–202.

[EK] M. Escobedo and O. Kavian, *Asymptotic behavior of positive solutions of a nonlinear heat equation*, (in press).

[EL] M. Escobedo and H. A. Levine, *Critical blow up and global existence numbers for a weakly coupled system of reaction- diffusion equations*, (in preparation).

[Fu1] H. Fujita, *On the blowing up of solutions of the Cauchy porblem for* $u_t = \Delta u + u^{1+\alpha}$, J. Fac. Sci. Univ. Tokyo Sect. IA Math **16** (1966), 105–113.

[Fu2] ———, *On some nonexistence and nonuniqueness theorems for nonlinear parabolic equations*, Proc. Symp. Pure Math., 18, Part I, AMS, Providence, RI, 1970.

[FuW] H. Fujita and S. Watanabe, *On the uniqueness and nonuniqueness of solutions of initial value problems for some quasilinear parabolic equations*, Comm. Pure Appl. Math. **21** (1968), 631–652.

[K] O. Kavian, *Remarks on the large time behavior of a nonlinear diffusion equation*, Ann. de l'Inst. Henri Poicaré, Analyse nonlineaire **4** (1987), 423–452.

[KST] K. Kobayoshi, T. Sirao and H. Tanaka, *On the growing up problems for semilinear heat equations*, J. Math. Soc. Japan **29** (1977), 407–429.

[L1] H. A. Levine, *The role of critical exponents in blow up theorems*, SIAM Review **32** (1990), 262–288.

[L2] ———, *A Fujita type global existence – global nonexistence theorem for a weakly coupled system of reaction-diffusion equations*, Zeit Ang. Mat. Phys. **42** (1991), 408–430.

[LM1] H. A. Levine and P. Meier, *The value of the critical exponent for reaction-diffusion equations in cones*, Arch. Rat. Mech. Anal. **109** (1990), 73–90.

[LM2] ———, *a blow up result for the critical exponent in cones*, Israel J. Math. **67** (1989), 1–7.

[LQi] H. A. Levine and Y.-W. Qi, *The critical exponent for degenerate parabolic systems*, ZAMP (submitted).

[We] F. B. Weissler, *Existence and nonexistence of global solutions for a semilinear heat equation*, Israel J. Math. **38** (1981), 29–40.

Department of Mathematics
Iowa State University
Ames, Iowa 50011

S POHOZAEV

On entire solutions of semilinear elliptic equations

Asymptotic estimates and existence are obtained for solutions of some classes semilinear elliptic problems in $\mathbb{R}^N$ ($N\geq 3$), including the supercritical Emden-Fowler type in nonautonomous cases. There is an important and well-known literature about entire solutions for semilinear elliptic equations. We refer for instance [1-3].

Our approach differ from well-known ones by using methods. We apply two methods for an investigation of these problems. The first method based on the variational identities. The second method based on the existence of positive classes of functions for semilinear elliptic of second order operators in the supercritical case.

PART I. THE RADIAL SOLUTIONS IN THE SUPERCRITICAL CASE.

We consider the following equations in $\mathbb{R}^N$ with $N\geq 3$.

(1.1) $$\Delta u + q(|x|)\ |u|^{p-2}\ u = h(|x|)$$

were $h \in C_{loc}(\mathbb{R}_+)$, $\mathbb{R}_+ = [0,+\infty)$.

The function q is required to satisfy the following conditions

(q_1) $q \in C_{loc}(\mathbb{R}_+)$ and there exists $q'(r)$ such that $r\, q'(r) \in C_{loc}(\mathbb{R}_+)$.

(q_2) $q(r)>0$ for $r>0$.

(q_3) There exists $\alpha < \frac{N-2}{2}$ such that

$q_1(2) = (\alpha - \frac{N}{p}) q(r) - \frac{1}{p} r q'(r) > 0$ for $r > 0$.

Remark. Evidently (q_2) and $(q_3) \Rightarrow p > p_c = \frac{2N}{N-2}$.

We study (1.1) by using the well-known variational identity [4], see also [5].

The integral identity.

Let (q_1) holds. Then we have for radial $C^2_{loc}(\mathbb{R}_+)$ - solutions of (1.1) the following identity

$$r^N [\frac{1}{2} u'^2(r) + \frac{1}{p} q(r)|u(r)|^p] +$$

$$+ (\frac{N}{2} - 1 - \alpha) \int_0^r u'^2(s) s^{N-1} ds + \int_0^r [(\alpha - \frac{N}{p}) q(s) -$$

$$- \frac{1}{p} s q'(s)] |u(s)|^p s^{N-1} ds = - \alpha r^{N-1} u(r) u'(r) + \tag{1.2}$$

$$+ \int_0^r h(s) [s u'(s) + \alpha u(s)] s^{n-1} ds.$$

The apriory estimates. The asymptotic behavior.

From (1.2) we obtain the following result.

Theorem 1.1 Let (q_1)-(q_3) hold and assume that $h \in C_{loc}(\mathbb{R}_+)$. Let u be a $C^2_{loc}(\mathbb{R}_+)$ - radial solution of (1.1). Then we have

$$u'^2(r) \le \text{Const } r^{-N} [Q(r)+H_0(r)+H_1(r)],$$

$$|u(r)|^p \le \text{Const } r^{-N} q^{-1}(r) [Q(r)+H_0(r)+H_1(r)],$$

$$\int_0^r u'^2(s) s^{N-1} ds \le \text{Const } [Q(r) +H_0(r)+H_1(r)],$$

$$\int_0^r |u(s)|^p q_1(s) s^{N-1} ds \le \text{Const } [Q(r) +H_0(r)+H_1(r)]$$

for $r>0$.

Here $Q(r) = r^{\frac{(N-2)p-2N}{p-2}} [q(s)]^{-\frac{2}{p-2}}$,

$H_0(r) = \int_0^r h^2(s)\, s^2\, s^{N-1}\, ds$,

$H_1(r) = \int_0^r |h(s)|^{p'} [q_1(s)]^{-\frac{1}{p-1}} s^{N-1}\, ds$.

These estimates are a sharp form with respect to both

· the "nonlinear" coefficient q and

· the function h.

Example 1. (to coefficient q)

Consider the equation

(1.3) $$\Delta u + q(r)\, |u|^{p-2} u = 0 \quad \text{in } \mathbb{R}^N,\ N\geq 3,$$

were $p > 2\frac{N+2}{N-2} > p_c$ and

$$q(r) = \frac{4\lambda^2 N}{p-2}\left[1 + \frac{(N-2)p-2N}{(p-2)N}\lambda^2 r^2\right], \quad \lambda\neq 0.$$

Then we have

$$q_1(r) = \text{Const}\ r^{\frac{(N-2)p-2(N+2)}{p-2}} (1+o(1))$$

as $r\to\infty$ and $H_0(r) \equiv H_1(r) \equiv 0$.

From the theorem 1.1 we obtain

$$|u'(r)| \leq \text{Const}\ r^{-\frac{p+2}{p-2}} (1+o(1)),$$

$$|u(r)| \leq \text{Const}\ r^{-\frac{4}{p-2}} (1+o(1)),$$

(1.4)

$$\int_0^r u'^2(s)\, s^{N-1} ds \le \text{Const } r^{\frac{(N-2)p-2(N+2)}{p-2}} (1+o(1)),$$

$$\int_0^r |u(s)|^p\, q_1(s)\, s^{N-1} ds \le \text{Const } r^{\frac{(N-2)p-2(N+2)}{p-2}} (1+o(1))$$

as $r \to \infty$.

On the other hand we have the exact solution of (1.3)

$$u(r) = (\, 1 + \lambda^2 r^2\,)^{-\frac{2}{p-2}}$$

satisfying estimates (1.4).

Example 2. (to function h)

Consider the equation

$$(1.5) \qquad \Delta u + |u|^{p-2} u = h(|x|) \quad \text{in } \mathbb{R}^N,\ N\ge 3,$$

were $h(r) = h_o(1+r^2)^{-\frac{2p-3}{p-2}}$, $p > p_c$, $h_o \ne 0$.

From theorem 1.1. we obtain

$$(1.6) \qquad \begin{aligned} & u'^2(r) \le \text{Const } r^{-\frac{2p}{p-2}} (1+o(1)), \\ & |u(r)|^p \le \text{Const } r^{-\frac{2p}{p-2}} (1+o(1)), \\ & \int_0^r u'^2(s)\, s^{N-1} ds \le \text{Const } r^{\sigma} (1+o(1)), \\ & \int_0^r |u(s)|^p\, s^{N-1} ds \le \text{Const } r^{\sigma} (1+o(1)) \end{aligned}$$

as $r \to \infty$,

were $\sigma = \frac{N-2}{2N} \frac{1}{p-2} (p-p_c)$.

On the other hand we have the exact solution of (1.5)

$$u = u_o(1+r^2)^{-\frac{1}{p-2}}$$

satisfying estimates (1.6).

These examples show that general "supercritical" equation (1.1) has not solution in Sobolev space $H^1(\mathbb{R}^N)$.

It is the essential difference from the subcritical case $1<p<p_c$.

<u>Existence of solutions.</u>

Consider the following problem

$$(1.7)\qquad \begin{cases} \Delta u + q(|x|)\ |u|^{p-2}\ u = h(|x|) & \text{in } \mathbb{R}^N,\ N\geq 3, \\ \lim\limits_{|x|\to\infty} u(x) = 0 \end{cases}$$

<u>Theorem 1.2</u> Let (q_1)-(q_3) hold and assume that

$$\lim_{|x|\to\infty} |x|^2\ q(|x|) = +\infty.$$

Let h be a $C_{loc}(\mathbb{R}_+)$ - function and assume that

$$\lim_{|x|\to\infty} \frac{H_o(|x|)+H_1(|x|)}{|x|^N q(|x|)} = 0.$$

Then there exists at least a continuum of entire radial $C^2_{loc}(\mathbb{R}_+)$ - solutions of (1.7).

<u>Proof</u>. The proof is based on an apriori estimates of theorem 1.1. and on the theory of ODE.

<u>Example</u>. We consider the problem

$$\begin{cases} \Delta u + |u|^{p-2}\ u = h(|x|) & \text{in } \mathbb{R}^N,\ N\geq 3, \\ \lim\limits_{|x|\to\infty} u(x) = 0 \end{cases}$$

were $p > p_c$ and $h \in C_{loc}(\mathbb{R}_+)$.
This problem has a continuum solutions if

$$\lim_{r\to\infty} \frac{1}{r^N} \left[\int_0^r h^2(s)\, s^{N+1}\, ds + \int_0^r |h(s)|^{p'} s^{N-1}\, ds \right] = 0.$$

The Sign of Radial Solutions of (1.1).

Let (q_1) holds and $h \in C^1_{loc}(\mathbb{R}_+)$. Then we rewrite the identity (1.2) in the following form

$$\begin{aligned} & r^N \left[\frac{1}{2} u'^2(r) + \frac{1}{p} q(r)|u(r)|^p \right] + \\ & + \left(\frac{N}{2} - 1 - \alpha \right) \int_0^r u'^2(s)\, s^{N-1}\, ds + \int_0^r \Big[\left(\alpha - \frac{N}{p} \right) q(s) - \\ & - \frac{1}{p} s\, q'(s)\Big]\, |u(s)|^p\, s^{N-1}\, ds + \alpha\, r^{N-1}\, u(r)\, u'(r) = \\ & = r^N u(r)\, h(r) + \int_0^r u(s)\, [(\alpha-N)\, h(s) - s\, h'(s)]\, s^{N-1}\, ds \end{aligned} \tag{1.8}$$

for $r>0$.

By this identity we obtain the following result.

Theorem 1.3 Suppose $N\geq 3$ and that q satisfies (q_1)-(q_2). Assume furthermore that $h \in C^1_{loc}(\mathbb{R}_+)$ and there exists $\alpha<\frac{N-2}{2}$ such that

$$(\alpha - \frac{N}{p})\, q(r) - \frac{1}{p}\, r\, q'(r) > 0$$

and

$$(N-\alpha)\, h(r) + r\, h'(r) \geq 0$$

for $r>0$.

Then if u is a radial $C^2_{loc}(\mathbb{R}_+)$ - solution of (1.1) satisfy $u(0)>0$, it follows that $u(r)>0$ and

$\alpha\, u'(r) < r\, h(r)$ for $r>0$.

Existence of Positive Solutions.

Theorem 1.4 Let the conditions of theorem 1.3 hold.

Then there exists a continuum of positive radial $C^2_{loc}(\mathbb{R}_+)$ - solutions of (1.1).

Theorem 1.5 Let the conditions of theorem 1.3 hold.

Furthermore assume that

$r^2 q(r) \to +\infty,$

$r^{-N} q^{-1}(r) [H_0(r)+H_1(r)] \to 0 \quad \text{as} \quad r \to +\infty.$

Then there exists a continuum of positive radial $C^2_{loc}(\mathbb{R}_+)$ - solutions of (1.7).

Remark. The analogous results are obtained for the equation

$$\Delta u + f(|x|,u) = h(|x|) \qquad \text{in } \mathbb{R}^N,\ N\geq 3. \tag{1.9}$$

These results are deducted from the identity

$$r^N \left[\frac{u'^2(r)}{2} + F(r,u) \right] + \left(\frac{N}{2} - 1 - \alpha \right) \int_0^r u'^2(s)\, s^{N-1}\, ds +$$

$$+ \int_0^r [\, \lambda\, u\, f(s,u) - N\, F(s,u) - s\, F_s(s,u)\,]\, s^{N-1}\, ds +$$

$$+ \alpha\, r^{N-1}\, u'(r)\, u(r) = \int_0^r h(s)\, [\, s\, u'(s) + \alpha\, u(s)\,]\, s^{N-1}\, ds,$$

were $F(r,u) = \int_0^u f(r,v)\, dv.$

PART II. THE SUB- AND SUPERCRITICAL CASES.

Here we again consider the equation (1.1) and the problem (1.7) but without an assumption $p>p_c$.

Under necessary conditions of smoothness we use the new identity for (1.1)

$$r^2 \left[\frac{1}{2} u'^2(r) + \frac{1}{p} q(r)|u(r)|^p \right] +$$

$$+ (N - 2 - \alpha) \int_0^r u'^2(s)\, s\, ds + \int_0^r [(\alpha - \frac{2}{p}) q(s) -$$

(2.1)

$$- \frac{1}{p} s\, q'(s)]\ |u(s)|^p\, s\, ds + \alpha\, r\, u'(r)\, u(r) +$$

$$+ \frac{\alpha}{2} (N-2)\, u^2(r) - \frac{\alpha}{2} (N-2)\, u^2(0) =$$

$$= \int_0^r h(s) [s\, u'(s) + \alpha\, u(s)]\, s\, ds.$$

This identity is different from the previous identity (1.2) and "works" for $p \geq \frac{2}{N-2}$, instead $p>p_c= \frac{2N}{N-2}$ if $q(x) = \text{Const} > 0$. On the base of this identity we obtain:

- the corresponding apriori and asymptotic estimates for (1.1) in the sharp form;
- the existence theorems for (1.1) and for (1.7);
- the existence of the positive solutions of (1.1) and of (1.7).

We give as an example on application of (2.1) to the following equation

(2.2) $$\Delta u + q(|x|)\ |u|^{p-2}\, u = 0 \quad \text{in } \mathbb{R}^N,\ N\geq 3,$$

were $p>1$.

The function q is required to satisfy (q_1) and the following conditions:

(q_2') $\qquad q(r) \geq 0 \qquad$ for $\quad r>0.$

(q_3') There exists α: $0 \le \alpha \le N-2$ such that

$$(\alpha - \frac{2}{p}) q(r) - \frac{1}{p} r q'(r) \ge 0 \quad \text{for} \quad r>0.$$

Then every radial $C^2_{loc}(\mathbb{R}_+)$ - solution of (2.2) satisfies

$$r^2 q(r) |u(r)|^p \le \frac{\alpha p}{2} (N-2) u^2(0). \tag{2.3}$$

This estimate is exact (unimprovable) estimate in indicated class of equations (2.2) in regard to both

- the behavior of solutions as $r \to \infty$

and

- the Const = the right hand side of (2.3)

Example. Let us consider the equation

$$\Delta u + \omega^2 |x|^{p-2} |u|^{p-2} u = 0 \quad \text{in} \quad \mathbb{R}^3, \tag{2.4}$$

were $\omega \neq 0$.

Then the conditions (q_1), (q_2') and (q_3') hold with $\alpha=1$ and (2.3) has the form

$$\omega^2 r^p |u(r)|^p \le \frac{p}{2} u^2(0)$$

for radial $C^2_{loc}(\mathbb{R}_+)$ - solution of (2.4).

On the other hand we have solutions

$$u(r) = \frac{w(r)}{r}$$

were

$$\begin{cases} w'' + \omega^2 |w|^{p-2} w = 0, \\ w(0) = 0, \qquad w'(0) = A \neq 0 \end{cases}$$

The calculation for these solutions shows the same inequality

$$\omega^2 r^p |u(r)|^p \le \text{Const} = \frac{p}{2} u^2(0)$$

and the existence of points $r_1, r_2, \ldots, r_k, \ldots$ $(r_k \to \infty)$ were

$$\omega^2 r_k^p |u(r_k)|^p \approx \frac{p}{2} u^2(0).$$

It is evidently for p=2. In this case we have

$$\Delta u + \omega^2 u = 0 \qquad \text{in } \mathbb{R}^3, \quad \omega \neq 0.$$

For radial $C^2_{loc}(\mathbb{R}_+)$ - solutions of this equation

$$u(r) = u_o \frac{\sin \omega r}{\omega r}, \qquad r>0.$$

We obtain

$$\omega^2 r^2 |u(r)|^2 = u_o^2 \sin^2 \omega r \le \frac{p}{2} u^2(0)\big|_{p=2} = u^2(0).$$

Remark. The analogous results are obtained for the equation (1.9). These results are deduce from the identity

$$r^2 [\tfrac{1}{2} u'^2(r) + F(r,u)] + \alpha r u'(r) u(r) +$$

$$+ \frac{\alpha}{2} (N-2) u^2(r) - \frac{\alpha}{2} (N-2) u^2(0) +$$

$$+ (N - 2 - \alpha) \int_0^r u'^2(s) s \, ds + \int_0^r [\alpha u f(s,u) -$$

$$- 2 F(s,u) - s F_s(s,u)] ds =$$

$$= \int_0^r h(s) [s u'(s) + \alpha u(s)] s \, ds,$$

were $F(r,u) = \int_0^r f(r,v) \, dv.$

PART III. CLASSES OF POSITIVITY OF OPERATOR A_o

$$A_o(u) \equiv -\Delta u - |u|^{p-2} u \qquad \text{in } \mathbb{R}^N,\ N\geq 3,$$

IN THE SUPERCRITICAL CASE: $p>p_c$.

It is well-known that the operator A_o and A_o^{-1} (when it exists) are not positive operators.

But there are special classes of functions defined on $\mathbb{R}^N$ in which the operator A_o is positive one if $p>p_c$. The certain classes of positivity of A_o exists in the case

$$p > p_* = \frac{2N-2}{N-2}.$$

I-Class of functions $(p>p_c)$.

This class is defined by functions of the form

$$u_1(x) = C_1\ \lambda^{\frac{2}{p-2}} (1+\lambda^2|x|^2)^{\beta_1}, \quad x \in \mathbb{R}^N,\ N\geq 3,$$

were $\beta_1 = -\frac{N-2}{2}$, $\lambda>0$, and

$$0 < C_1 \leq C_1^* = [(N-2)N]^{\frac{1}{p-2}}.$$

The following functions from Im A_o correspond to this class

$$h_1 \equiv A_o(u_1) = C_1\ \lambda^{2\frac{p-1}{p-2}} (1+\lambda^2|x|^2)^{-\frac{N+2}{2}} \times$$

$$\times \left[(N-2)N - C_1^{p-2} (1+\lambda^2|x|^2)^{-\frac{N-2}{2}(p-p_c)} \right].$$

II-Class of functions $(p>p_c)$.

This class is defined by functions of the form

$$u_2(x) = C_2 \lambda^{\frac{2}{p-2}} (1+\lambda^2|x|^2)^{\beta_2}, \quad x \in \mathbb{R}^N,\ N\geq 3,$$

were $\beta_2 = -\frac{2}{p-2}$, $\lambda>0$, and

$$0 < C_2 \leq C_2^* = \left[4 \frac{(N-2)(p-p_c)+2p}{(p-2)^2} \right]^{\frac{1}{p-2}}$$

The functions of the form $h_2 = A_0(u_2)$ from Im A_0 correspond to this class.

III-Class of functions ($p > p_* = \frac{2N-2}{N-2}$).

This class is defined by functions of the form

$$u_3(x) = C_3 \lambda^{\frac{2}{p-2}} (1+\lambda^2|x|^2)^{\beta_3}, \quad x \in \mathbb{R}^N,\ N\geq 3,$$

were $\beta_3 = -\frac{1}{p-2}$, $\lambda>0$, and

$$0 < C_3 \leq C_3^* = \left[2 \frac{(N-2)(p-p_*)}{(p-2)^2} \right]^{\frac{1}{p-2}}$$

The functions of the form $h_3 = A_0(u_3)$ from Im A_0 correspond to this class. Thus we considered three classes of functions

(3.1) $$u(x) = C \lambda^{\frac{2}{p-2}} (1+\lambda^2|x|^2)^{\beta}$$

with $\beta_1 = -\frac{1}{N-2}$, $\beta_2 = -\frac{2}{p-2}$ and $\beta_3 = -\frac{1}{p-2}$.

There exists a continuous scale $\beta : \beta_1 \leq \beta_2 \leq \beta_3$ of classes of functions (3.1) with $0 < C \leq C^*(N,p,\beta)$ on which operator A_0 is positive one.

Now let us consider the following problem

$$\begin{cases} \Delta u + |u|^{p-2} u + f_1(x,u) = h(x), & x \in \mathbb{R}^N,\ N\geq 3, \\ \lim_{|x|\to\infty} u(x) = 0. \end{cases}$$

Solvability of (3.2) in the k-th class of functions with forgiven asymptotic β_k ($k = 1,2,3$).

Theorem 3.1(k) Suppose $N\geq 3$, $p>p_c$ and that

(i) $f_1 \in C^{\nu}(\mathbb{R}^{N+1})$, $\nu \in (0,1)$, and f_1 is loc-Lip-continuous with respect to $u \in \mathbb{R}$ on $\mathbb{R}^N\times\mathbb{R}$

(ii) $h \in C^{\nu}(\mathbb{R}^N)$

Further, let there exists $\lambda>0$ such that

$f_1(x_1 -u_k^*(x)) + h_k^*(x) \geq h(x) \geq f_1(x,u_k^*(x))-h_k^*(x)$, $x \in \mathbb{R}^N$.

Here the functions u_k^* and h_k^* are defined by the corresponding formulas with $C_k = C_k^*$.

Then the problem (3.2) has a $C^2_{loc}(\mathbb{R}_+)$ - solution $u(x)$ and the inequality

$|u(x)| \leq u_k^*(x)$ in $\mathbb{R}^N$ holds.

Positive Solvability of (3.2) in the k-th class of functions ($k=1,2,3$).

Theorem 3.2(k) Suppose $N\geq 3$, $p>p_c$ and the conditions (i)-(ii) from theorem 3.1(k) hold.

Further, let there exists $\lambda>0$ such that

$f_1(x,0) \geq h(x) \geq f_1(x,u_k^*(x)) - h_k^*(x)$, $x \in \mathbb{R}^N$.

Then the problem (3.2) has a nonnegative C^2_{loc} - - solution $u(x)$ and the inequality

$0 \leq u(x) \leq u_k^*(x)$ in $\mathbb{R}^N$, holds.

REFERENCES

[1] H. Berestycki, P.L. Lions, L.A. Peletier, On ODE - approach to existence of positive solutions for semilinear problems in $\mathbb{R}^N$. Indiana Univ. Math. J. 30, N1 (1981), 141-157

[2] H. Berestycki, P.L. Lions, Nonlinear Scalar Field Equations I. II. Arch. Rat. Mech. Anal. 82, N4 (1983), 313-345, 347-375

[3] E.S. Noussair, Ch.A. Swanson, An L^q ($\mathbb{R}^N$) - theory of subcritical semilinear elliptic problems, Journal Dif. Eq. 84, N1 (1990), 52-61

[4] С.И. Похожаев, О собственных функциях уравнения $\Delta u+\lambda f(u)=0$, Доклады АН СССР, т.165, N1 (1965), 36-39

[5] С.И. Похожаев, О собственных функциях квазилинейных эллиптических задач, Математический сборник, т.82, N2 (1970), 192-212

[6] С.И. Похожаев, Об эллиптических задачах в $\mathbb{R}^N$ с суперкритическим показателем нелинейности, Математический сборник, т.182, N4 (1991), 467-489

S. Pohozaev
Steklov Mathematical Institut,
ul. Vavilova, 42,
117966 Moscow, USSR

J-F RODRIGUES

Strong solutions for quasi-linear elliptic-parabolic problems with time-dependent obstacles

1 – Introduction

We consider the existence, uniqueness and comparison properties for strong solutions of certain quasi-linear second order elliptic-parabolic problems with time-dependent unilateral constraints. In order to obtain strong solutions we impose the "natural" condition on the obstacle associated with the elliptic-parabolic differential operator and we prove the extended Lewy–Stampacchia inequalities. We discuss the interior and the boundary obstacle problems from the unifying point of view of the 'a *priori*' dual estimates for unilateral problems.

The main source of physical problems leading to elliptic-parabolic equations of the type

$$Qu \equiv \partial_t b(u) - \nabla \cdot a(b(u), \nabla u) = f \tag{1}$$

arises in the theory of flow through porous media (see [2], [3], [24], [33] or [46]). Here, $\partial_t = \partial/\partial t$, ∇ denotes the gradient in space variables $x \in \Omega \subset \mathbf{R}^n$, $\nabla\cdot = \text{div}$, $u = u(x,t)$ is essentially a renormalized pressure and $b(u)$ the saturation, which is given by a constitutive relation of the type indicated in the figure.

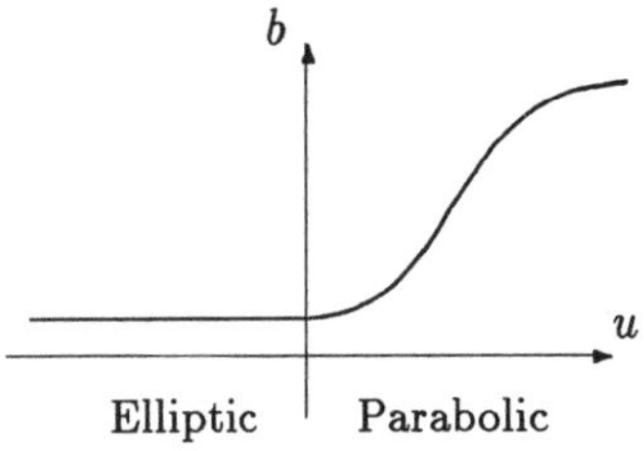

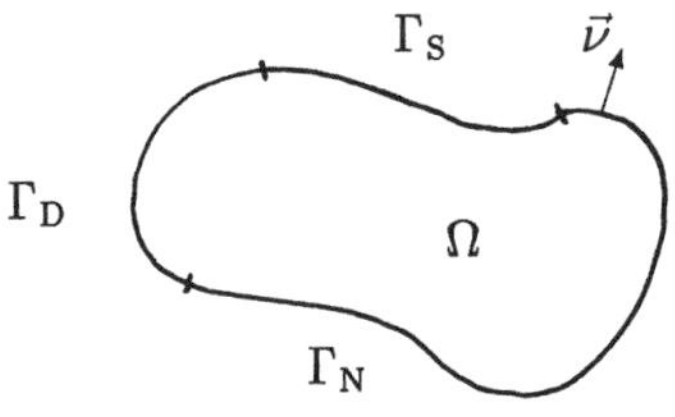

In these problems, the equation (1) is elliptic in the so-called saturated region, where the renormalized pressure is below some critical value, and parabolic in the unsaturated region. The corresponding boundary value problem on the boundary $\partial\Omega = \Gamma_D \cup \Gamma_N \cup \Gamma_S$ takes into account pervious or impervious layers (Neumann condition) and reservoirs (Dirichlet condition)

$$Nu \equiv a(b(u), \nabla u) \cdot \vec{\nu} = g(x,t) \quad \text{on } \Gamma_N\,, \tag{2}$$

$$u = h(x,t) \quad \text{on } \Gamma_D\,, \tag{3}$$

where the Nu denotes the flux through the boundary. Seepages faces are modelled by an overflow condition (Signorini condition)

$$u \geq \varphi(x,t)\,, \quad Nu \geq 0\,, \quad (u-\varphi)Nu = 0 \ \text{ on } \Gamma_S\,. \tag{4}$$

Notice that in our notations the renormalized pressure u has the opposite sign with respect to the usual physical pressure, in which case the (u, b)-graph and the inequalities in (4) are reversed. This time-dependent problem for $t > 0$, requiring the initial condition

$$b(u) = b_0 \quad \text{at } t = 0 , \tag{5}$$

has been considered for the special physical case

$$a(b(u), \nabla u) = \nabla u + k(b(u))\, e , \tag{6}$$

where $-e$ represents the direction of gravity and k the conductivity (see [24], [4], [2] and [31]) and is related to the well-known "dam problem", which may be regarded as a limit case when b is an Heaviside function (see [2], [17]).

In the last decade, a large number of mathematical works have been motivated by this model. Without being exaustive, we refer to [21], [20], [35] or [1] for one-dimensional problems and to the important work of Alt and Lukhaus [3], where a general mathematical theory was develloped for multidimensional systems of elliptic-parabolic type. In [3] the method of backward time differences was combined with two main arguments based on an integration by parts formula ($B(u) = \int_0^u [b(u) - b(v)]\, dv$)

$$\int_\Omega B(u(t)) - \int_\Omega B(u(s)) = \int_s^t \int_\Omega u\, \partial_t b(u) \tag{7}$$

and a compactness property for $\{b(u_n)\}$ in $L^1(\Omega \times (0, T))$ with no other regularity on b than its continuity. This method applies to obtain the existence of weak solutions to certain variational inequalities [3] and to systems in Orlicz–Sobolev spaces [27].

In contrast with parabolic problems, the equation (1) does not provide, in general, any estimate on $\partial_t u$ and, therefore, we cannot expect to obtain very regular solutions to elliptic-parabolic problems. In [3], for Lipschitz continuous b, a regularity of $\partial_t b(u)$ in $L^2(\Omega \times (0, T))$ was obtained, in which case, uniqueness and comparison properties were also proved for equations. Exploiting these properties, the asymptotic stabilization as $t \to \infty$ towards the stationary solution was proved in [36] and the existence and the stability of the periodic solution was studied in [30]. Some results for the corresponding Neumann problem were first given in [25] and generalized in [28] and [29]. On the other hand, the continuity of the solution u is not expected in general, as a counterexample of [3] shows for the simple equation $\partial_t u^+ = \Delta u$. Nevertheless, the conjecture of [3] on the local continuity of $b(u)$ was proved in [18].

We should remark that elliptic-parabolic problems of the type (1) are also free boundary problems, in the sense that the interface separating the elliptic and the parabolic domains is not '*a priori*' known. The only known results on this free boundary for the multidimensional equation $\partial_t b(u) = \Delta u$ are given in [26] for time-monotone flows in two special situations where that interface is Lipschitz continuous.

However, elliptic-parabolic equations are also of mathematical interest in other areas (see [11] and its references, for instance) and can be regarded as special cases of general abstract doubly nonlinear evolution equations of the form

$$\frac{d}{dt}B(u) + A(u) \ni f \tag{8}$$

in a functional space H with initial condition $B(u(0)) = B_0$. In this form, this problem was considered in [42] and [40] for the case $b(u) = |u|^{\alpha-2}u$, $A(u) = -\nabla \cdot (|\nabla u|^{p-2}\nabla u)$, in a Hilbertian framework in [10] and for $B = \partial\Phi$ and $A = \partial\Psi$ as subdifferentials of convex functions Φ and Ψ in [22]. Recently, these results were extended in [8] and [9] for the case without growth condition on b, in [48] and in [16], where comparison and asymptotic stabilization results extending [36] are given. The abstract form (8) also include related degenerate elliptic-parabolic and pseudoparabolic variational inequalities of the type considered for instance in [12], [37], or [45].

By letting $w = b(u)$ be the unknown function, a different approach was used in [32] and [33] to study the elliptic-parabolic Signorini problem (1)–(5) in the case $a(b(u), \nabla u) = \nabla u$. The corresponding problem is proved to be equivalent to an abstract equation of the form

$$\frac{dw}{dt} + \partial\Psi^t(u) \ni f\,, \qquad u \in B^{-1}w\,,$$

which is solved in the Hilbert space $H = L^2(\Omega)$ by nonlinear semigroups methods. This method is well adapted to problems without convection, allowing the study of the asymptotic behaviour (see [34]) and it admits an application to (1)–(5) in the convective case (6) (see [31]).

In this work we are interested in the mathematical problem that consists in associate an interior obstacle to the quasi-linear elliptic-parabolic operator Q defined in (1), with mixed boundary conditions in a domain Ω of $\mathbf{R}^n$. Setting $\Omega_T = \Omega \times (0,T)$, $\Sigma_T = \Gamma_D \times (0,T)$ and $S_T = (\partial\Omega \backslash \overline{\Gamma}_D) \times (0,T)$, for a given function $\psi = \psi(x,t)$ on $\overline{\Omega}_T$, we consider the unilateral problem

$$u \geq \psi, \quad Qu \geq f, \quad (u-\psi)(Qu-f) = 0 \quad \text{in } \Omega_T\,, \tag{9}$$

$$u \geq \psi, \quad Nu \geq g, \quad (u-\psi)(Nu-g) = 0 \quad \text{on } S_T\,, \tag{10}$$

with Dirichlet condition (3) on Σ_T and initial condition (5), where Nu is defined as in (2). Our approach to the existence of strong solutions is based on the interior dual estimate ($f \vee g = \sup\{f, g\}$)

$$Qu \leq f \vee Q\psi \quad \text{in } \Omega_T\,, \tag{11}$$

with the corresponding boundary dual estimate

$$Nu \leq g \vee N\psi \quad \text{on } S_T\,. \tag{12}$$

We notice that for elliptic obstacle problems, the estimate (11) was obtained by Lewy and Stampacchia [39], for $Q = -\Delta$ with Dirichlet condition, and extended in [23] for the Neumann problem, where $Nu = \partial u/\partial n$ in (12) (see [44], for an exposition and more consequences on the regularity of stationary obstacle problems). These inequalities were extended to the parabolic case (i.e., $b(u) = u$) for linear operators in [13] and for nonlinear operators in [19], in order to obtain strong solutions to parabolic variational inequalities. In contrast with the notion of weak solution (see [40] or [6]), where uniqueness may fail as it was observed in [41], the notion of strong solution has been used in parabolic unilateral problems by several authors with the help of the dual estimates in different frameworks (see, for instance, [13], [19], [7], [43] or [47]).

The rigorous meaning of (11) and (12) may be quite delicate under optimal assumptions on the obstacle ψ. In order to avoid technical difficulties, we shall limit ourselves to conditions that are sufficient to satisfy (11) and (12) in the a.e. sense. The main assumptions and results are given in the Section 2 and the proofs are briefly described in the Section 3.

2 – Assumptions and main results

Let Ω be an open bounded subset of $\mathbf{R}^n$, $n \geq 1$, with Lipschitz boundary $\partial\Omega = \Gamma_D \cup \overline{S}$, $\operatorname{meas}\Gamma_D > 0$ (the case $S = \emptyset$ is not excluded), and for $0 < T < \infty$, we set

$$\Omega_T = \Omega \times (0,T), \quad \Sigma_T = \Gamma_D \times (0,T), \quad S_T = S \times (0,T), \quad \Omega_0 = \Omega \times \{0\} ,$$

and for a given obstacle $\psi = \psi(x,t)$, we define

$$\mathbf{K} = \mathbf{K}(\psi) = \left\{ v \in L^2(0,T;V) \colon v \geq \psi \text{ a.e. in } \Omega_T \right\} , \tag{13}$$

where, using the standard notation for Sobolev spaces [38], [40], [6],

$$H_0^1(\Omega) \subseteq V \equiv \left\{ v \in H^1(\Omega) \colon v = 0 \text{ on } \Gamma_D \right\} \subsetneq H^1(\Omega) = W_2^1(\Omega) . \tag{14}$$

We consider the elliptic-parabolic quasilinear obstacle problem (9), (10), (3) with $h = 0$ and (5) for a given function $b_0 = b_0(x)$, in the following strong variational formulation: find $u = u(x,t)$ such that

$$u \in \mathbf{K}, \quad \partial_t b(u) \in L^2(\Omega_T), \quad b(u(0)) = b_0 \text{ in } \Omega_0 , \tag{15}$$

$$\int_{\Omega_T} \partial_t b(u)(v-u) + \int_{\Omega_T} a(b(u),\nabla u)\cdot\nabla(v-u) \geq \int_{\Omega_T} f(v-u) + \int_{S_T} g(v-u), \quad \forall v \in \mathbf{K}. \tag{16}$$

We assume the following conditions on the operator $Qu = \partial_t b(u) - \nabla \cdot a(b(u), \nabla u)$: b is a monotone Lipschitz continuous function, i.e.

$$b \in C^{0,1}(\mathbf{R}), \quad b(0) = 0 \quad \text{and} \quad 0 \leq b' \leq \beta \ \text{ a.e. in } \mathbf{R} \ ; \tag{17}$$

the vector field $a = a(x, b, \xi) : \Omega \times \mathbf{R} \times \mathbf{R}^n \to \mathbf{R}^n$ is continuous in (b, ξ) and measurable in x, coercive for some fixed $\alpha > 0$,

$$\left[a(x,b,\xi) - a(x,b,\hat{\xi})\right] \cdot (\xi - \hat{\xi}) \geq \alpha \, |\xi - \hat{\xi}|^2 \ , \tag{18}$$

it admits a potential $A = A(x, b, \xi)$, such that $\nabla_\xi A = a$, satisfying

$$|a(x,b,\xi)|^2 + |A(x,b,0)| + |\partial_b A(x,b,\xi)|^2 \leq C(1 + |B| + |\xi|^2) \ , \tag{19}$$

with $C > 0$ and $B(u) = \int_0^u [b(u) - b(v)] \, dv$, and a continuity property in b

$$\left|a(x, b(u), \xi) - a(x, b(\hat{u}), \xi)\right| \leq C(1 + |b(u)| + |b(\hat{u})| + |\xi|) \, \sigma(|u - \hat{u}|) \ , \tag{20}$$

$$\sigma(\tau) \searrow 0 \ \text{ as } \tau \searrow 0 \quad \text{and} \quad \int_{0+} \frac{d\tau}{\sigma^2(\tau)} = +\infty \ . \tag{21}$$

We note that the assumptions (18) and (19) are similar to the conditions of [3] and the continuity (20) with (21) is a slight generalization of the case $\sigma(\tau) = \sqrt{\tau}$ of Theorem 2.2 of [3], corresponding to a Hölder continuity of a with respect to b (recall (17)) of exponent $\lambda = 1/2$.

On the given data $\psi = \psi(x,t)$, $f = f(x,t)$, $g = g(x,t)$ and $b_0 = b_0(x)$, we suppose

$$\psi \in H^1(\Omega_T) = W_2^{1,1}(\Omega_T) \quad \text{such that} \quad \psi \leq 0 \ \text{ on } \Sigma_T \ , \tag{22}$$

$$f, \partial_t f \in L^2(\Omega_T) \quad \text{and} \quad g, \partial_t g \in L^2(S_T) \ , \tag{23}$$

$$\exists \, \xi \geq 0 : \quad \xi, \partial_t \xi \in L^2(\Omega_T) \ \text{ and } \ \xi \geq Q\psi - f \quad \text{in } \Omega_T \ , \tag{24}$$

$$\exists \, \eta \geq 0 : \quad \eta, \partial_t \eta \in L^2(S_T) \ \text{ and } \ \eta \geq N\psi - g \quad \text{on } S_T \ , \tag{25}$$

$$\exists \, u_0 \in V : \quad b_0 = b(u_0) \ \text{ and } \ u_0 \geq \psi(0) \quad \text{on } \Omega_0 \ . \tag{26}$$

Remark 1. The assumption (22) implies that $\mathbf{K}(\psi) \neq \emptyset$. On the other hand, since it does not give a sense to $N\psi = a(b(\psi), \nabla\psi) \cdot \vec{\nu}$ on S_T, we should regard the inequalities in (24) and (25) as the following condition for a.e. $t > 0$:

$$\int_\Omega \partial_t b(\psi) v + \int_\Omega a(b(\psi), \nabla\psi) \cdot \nabla v \leq \int_\Omega (f + \xi) v + \int_S (g + \eta) v \ , \qquad \forall v \in V, \ \ v \geq 0. \tag{27}$$

Clearly (27) is another form of (24)–(25), since it implies that $Q\psi$ is a measure in Ω_T and $N\psi$ is a measure on S_T with the corresponding inequalities taking place in the distributional sense. □

Theorem 1. *Under the assumptions (17)–(26) there exists a unique solution to the elliptic–parabolic variational inequality (15)–(16), which satisfies* $u \in L^\infty(0,T;V)$ *and*

$$f \leq Qu \leq f + \xi \quad \text{a.e. in } \Omega_T , \tag{28}$$

$$g \leq Nu \leq g + \eta \quad \text{a.e. on } S_T . \blacksquare \tag{29}$$

Remark 2. The above result extends the existence of strong solutions to the elliptic-parabolic case and (28)–(29) are the corresponding Lewy–Stampacchia inequalities. Note that if we can take

$$\xi = (Q\psi - f)^+ \quad \text{and} \quad \eta = (N\psi - g)^+ ,$$

we obtain the dual estimates (11) and (12), respectively, from (28) and (29) (recall $a \vee b = a + (b-a)^+$, $a^+ = a \vee 0$). □

Remark 3. Contrary to the parabolic case of [13] and [19], we have assumed in (23)–(25) the existence in L^2 of the time derivatives of f, g, ξ and η, in order to obtain the regularity $\partial_t b(u) \in L^2(\Omega_T)$. These conditions are perhaps not necessary, and it should be interesting to avoid them and to obtain (28) and (29) in the distributional sense only. □

Remark 4. From (29), if we know that $\eta \equiv 0$ on some subset Γ_T of S_T it is easily seen that the solution u verifies a Neumann boundary condition (2) on that part of the boundary. Analogously, if $\xi \equiv 0$ in an open subset $\mathcal{O} \subset \Omega_T$, from (28) it follows that u solves the equation (1) in $\mathcal{O}$. In particular, if $\xi \equiv 0$ everywhere in Ω_T, u solves a boundary obstacle problem of Signorini type. □

Theorem 2. *Under the assumptions (18) and (20)–(21), let u and $\hat{u}$ denote two solutions of (15)–(16) corresponding to admissible data f, g, ψ, b_0 and $\hat{f}$, $\hat{g}$, $\hat{\psi}$, $\hat{b}_0$, respectively. Then if $f \geq \hat{f}$, $g \geq \hat{g}$, $b_0 \geq \hat{b}_0$ and $\psi \geq \hat{\psi}$ we have*

$$u \geq \hat{u} \quad \text{in } \Omega_T . \tag{30}$$

Remark 5. This comparison result, which in particular gives the uniqueness in Theorem 1, extends a previous result of [3] for elliptic-parabolic equations and is similar to the results of [5], [15] and [14] for elliptic and parabolic variational inequalities. □

Remark 6. The results of this section can be extended to more general elliptic-parabolic unilateral problems, related to doubly nonlinear operators of the type considered in [48] or [16]. It would be also interesting to extend them to the framework of systems of elliptic-parabolic type in Orlicz spaces (see [27] for systems of equations). □

Remark 7. We have considered the homogeneous boundary condition on Γ_D only for simplicity. Actually, by replacing in (15) and (16) $\mathbf{K}$ by

$$\mathbf{K}(\psi, h) = \left\{ v \in L^2(0,T;H^1(\Omega)) : v \geq \psi \ \text{ in } \Omega_T, \ v = h \ \text{ on } \Sigma_T \right\}$$

for a sufficiently smooth Dirichlet data h, such that $h \geq \psi$ on Σ_T, we can easily extend Theorems 1 and 2. □

3 – Proof of the existence and comparison results

Proof of the Comparison Theorem 2: As in [15], we introduce the following approximation of the Heaviside function as $\varepsilon \to 0^+$

$$H_\varepsilon(s) = \begin{cases} 0 & \text{if } s \le \delta \\ \displaystyle\int_\delta^s \frac{d\tau}{\sigma^2(\tau)} & \text{if } \delta \le s \le \varepsilon \ , \\ 1 & \text{if } s \ge \varepsilon \end{cases} \tag{31}$$

which by the condition (21) is well defined for each $\varepsilon > 0$ and for some $\delta = \delta(\varepsilon) \in]0, \varepsilon[$. Then we observe that $H_\varepsilon(\hat{u} - u) \in L^2(0,T;V)$ and we can take

$$v = u + \delta H_\varepsilon(\hat{u} - u) \in \mathbf{K}(\psi) \quad \text{in the inequality (16) for } u \text{ and}$$

$$v = \hat{u} - \delta H_\varepsilon(\hat{u} - u) \in \mathbf{K}(\hat{\psi}) \quad \text{in the inequality (16) for } \hat{u}.$$

Since we can consider, in an equivalent way (see [40], [6] or [15]), the variational inequality (16) in each Ω_t for every $0 < t < T$, by subtraction we obtain

$$\begin{aligned} X_\varepsilon &\equiv \int_{\Omega_t} \partial_t \big[b(\hat{u}) - b(u)\big] H_\varepsilon(\hat{u} - u) \\ &\le \int_{\Omega_t} \big[a(b(u), \nabla u) - a(b(\hat{u}), \nabla\hat{u})\big] \cdot \nabla H_\varepsilon(\hat{u} - u) \equiv Y_\varepsilon \ . \end{aligned} \tag{32}$$

Letting $w = \hat{u} - u$ we have $\nabla H_\varepsilon(w) = \frac{\nabla w}{\sigma^2(w)} \chi_{\{\delta < w < \varepsilon\}}$ and using (18) and (20) we find in $O_\varepsilon \equiv \{\delta < w < \varepsilon\}$

$$\begin{aligned} Y_\varepsilon &\le -\alpha \int_{O_\varepsilon} \frac{|\nabla w|^2}{\sigma^2(w)} + \int_{O_\varepsilon} \big[a(b(u), \nabla\hat{u}) - a(b(\hat{u}), \nabla\hat{u})\big] \cdot \frac{\nabla w}{\sigma^2(w)} \\ &\le -\alpha \int_{O_\varepsilon} \frac{|\nabla w|^2}{\sigma^2(w)} + C \int_{O_\varepsilon} \big(1 + |b(u)| + |b(\hat{u})| + |\nabla\hat{u}|\big) \frac{|\nabla w|}{\sigma(w)} \\ &\le \frac{C^2}{4\,\alpha} \int_{O_\varepsilon} \big(1 + |b(u)| + |b(\hat{u})| + |\nabla\hat{u}|\big)^2 \to 0 \ \text{ as } \varepsilon \to 0. \end{aligned} \tag{33}$$

Taking $\varepsilon \to 0$ in (32) we include, by monotonicity of b, that

$$\int_\Omega \big[b(\hat{u}) - b(u)\big]^+(t) = \int_{\Omega_t} \partial_t \big[b(\hat{u}) - b(u)\big]^+ = \lim_\varepsilon X_\varepsilon \le 0 \ ,$$

which implies $b(\hat{u}) \le b(u)$. In particular, $b(u) = b(\hat{u})$ in the set $\{\hat{u} - u \ge \delta\} \supset O_\varepsilon$ and we get $X_\varepsilon = 0$ in (32). Consequently, combining (32) with (33) we find for $w_\delta^\varepsilon \equiv \delta \vee (w \wedge \varepsilon)$

$$\frac{\alpha}{\sigma^2(\varepsilon)} \int_{\Omega_t} |\nabla w_\delta^\varepsilon|^2 = \frac{\alpha}{\sigma^2(\varepsilon)} \int_{O_\varepsilon} |\nabla w|^2 \le \alpha \int_{O_\varepsilon} \frac{|\nabla w|^2}{\sigma^2(w)} \le 0 \ ,$$

that is to say $w_\delta^\varepsilon \equiv \delta$, since $w = \hat{u} - u = 0$ on Σ_T. Therefore we have $w = \hat{u} - u \le \delta(\varepsilon)$ and letting $\varepsilon \to 0$ we conclude $u \ge \hat{u}$. ∎

Proof of Theorem 1: It consists of four steps: i) considering solutions u_ε of a suitable parabolic penalized problem; ii) obtaining sufficient '*a priori*' estimates on u_ε; iii) showing that $u_\varepsilon \geq \psi$, i.e. $u_\varepsilon \in \mathbf{K}(\psi)$ for each $\varepsilon > 0$, and iv) letting $\varepsilon \to 0$ and showing that the limit point u of the sequence u_ε solves (15)–(16). Using the Theorem 2, we know there is at most one solution to (15)–(16) and consequently we shall obtain $u_\varepsilon \to u$ for the whole sequence, $\varepsilon \to 0$.

i) *Parabolic regularization and bounded penalization* — For $0 < \varepsilon < 1$, consider H_ε defined by (31) and let

$$b_\varepsilon(v) = b(v) + \varepsilon\, v\,, \qquad Q_\varepsilon v = \varepsilon\, \partial_t v + Qv\;, \tag{34}$$

$$f_\varepsilon(v) = f + \varepsilon\, \partial_t \psi + \xi\,[1 - H_\varepsilon(v - \psi)]\,, \qquad g_\varepsilon(v) = g + \eta\,[1 - H_\varepsilon(v - \psi)]\;. \tag{35}$$

We define u_ε as being the solution to the parabolic problem

$$Q_\varepsilon u_\varepsilon = f_\varepsilon(u_\varepsilon) \;\text{ in } \Omega_T\,, \qquad N u_\varepsilon = g_\varepsilon(u_\varepsilon) \;\text{ on } S_T\;, \tag{36}$$

$$u_\varepsilon = 0 \;\text{ on } \Sigma_T\,, \qquad u_\varepsilon(0) = u_0 \;\text{ in } \Omega_0\;. \tag{37}$$

By the assumptions this is a monotone problem with a unique solution $u_\varepsilon \in H^1(\Omega_T) \cap L^2(0,T;V)$ (see [38] or [40]) satisfying, for every $t \in]0,T]$,

$$\int_{\Omega_t} \partial_t b_\varepsilon(u_\varepsilon) v + \int_{\Omega_t} a(b(u_\varepsilon), \nabla u_\varepsilon) \cdot \nabla v = \int_{\Omega_t} f_\varepsilon(u_\varepsilon) v + \int_{S_t} g_\varepsilon(u_\varepsilon) v, \quad \forall v \in L^2(0,T;V). \tag{38}$$

ii) *'A priori' estimates* — The function u_ε satisfies the following estimates independently of $\varepsilon \in]0,1[$ and $t \in [0,T]$:

$$\int_\Omega B_\varepsilon(u_\varepsilon(t)) + \int_{\Omega_t} |\nabla u_\varepsilon|^2 \leq C\;, \tag{39}$$

$$\int_{\Omega_t} |\partial_t b(u_\varepsilon)|^2 + \varepsilon \int_{\Omega_t} |\partial_t u_\varepsilon|^2 + \int_\Omega |\nabla u_\varepsilon(t)|^2 \leq C\;. \tag{40}$$

The estimate (39) follows easily by taking $v = u_\varepsilon$ in (38), using the formula of integration by parts (7) and the assumptions (18) and (19). Recall that by Poincaré inequality for $u_\varepsilon(t) \in V$ we also control $\int_{\Omega_t} |u_\varepsilon|^2 \leq C$ and $|f_\varepsilon| \leq |f| + |\xi| + |\partial_t \psi|$, $|g_\varepsilon| \leq |g| + |\eta|$, since we have $0 \leq \vartheta_\varepsilon = 1 - H_\varepsilon \leq 1$.

The estimate (40) follows formally by taking $v = \partial_t u_\varepsilon$ in (38) and using (39), as in Theorem 2.3 of [3]. Actually, we need only to control the additional terms in the right-hand side of (38) containing f_ε and g_ε, since following [3] we obtain

$$\frac{1}{\beta} \int_{\Omega_t} |\partial_t b(u_\varepsilon)|^2 + \varepsilon \int_{\Omega_t} |\partial_t u_\varepsilon|^2 + \frac{\alpha}{2} \int_\Omega |\nabla u_\varepsilon(t)|^2 \leq$$

$$\leq C + \int_{\Omega_t} f_\varepsilon(u_\varepsilon)\, \partial_t u_\varepsilon + \int_{S_t} g_\varepsilon(u_\varepsilon)\, \partial_t u_\varepsilon\;. \tag{41}$$

If we set $\Theta_\varepsilon(s) = \int_0^s \vartheta_\varepsilon(v)\,dv = \int_0^s[1 - H_\varepsilon(v)]\,dv$, we have $|\Theta_\varepsilon(s)| \leq |s|$ and we can write

$$\int_{\Omega_t} f_\varepsilon(u_\varepsilon)\partial_t u_\varepsilon = \int_{\Omega_t} f\,\partial_t u_\varepsilon + \varepsilon\int_{\Omega_t} \partial_t\psi\,\partial_t u_\varepsilon + \int_{\Omega_t} \xi\,\partial_t\Theta_\varepsilon(u_\varepsilon - \psi) + \int_{\Omega_t} \xi\,\vartheta_\varepsilon\,\partial_t\psi.$$

Integrating by parts in time the first and third integrals of this right-hand side, with the assumptions (22)–(25), we can estimate this term by

$$\frac{\varepsilon}{2}\int_{\Omega_t} |\partial_t u_\varepsilon|^2 + \frac{\alpha}{4}\int_\Omega |\nabla u_\varepsilon(t)|^2 + C\ ,$$

which can be absorbed by the left-hand side of (41). Analogously, we can estimate the last integral of (41), by charging the time derivative on g and on η and using the trace theorem for $u_\varepsilon \in L^2(0,T;V)$:

$$\int_S |u_\varepsilon(t)|^2 \leq C\int_\Omega |\nabla u_\varepsilon(t)|^2 \quad \text{for each } t > 0\ .$$

iii) *$u_\varepsilon \geq \psi$ for each ε* — Observing that from (27) we have a.e. $t > 0$

$$\int_\Omega \partial_t b_\varepsilon(\psi)v + \int_\Omega a(b(\psi),\nabla\psi)\cdot\nabla v \leq \int_\Omega (f + \varepsilon\,\partial_t\psi + \xi)v + \int_S (g + \eta)v, \quad \forall v\in V,\ v\geq 0,$$

by choosing $v = H_\varepsilon(\psi - u_\varepsilon)$ in this inequality and in (38), we obtain, after subtraction, as in the proof of Theorem 2,

$$\int_{\Omega_t} \partial_t\Big[b_\varepsilon(\psi) - b_\varepsilon(u_\varepsilon)\Big]\,H_\varepsilon(\psi - u_\varepsilon) \leq Y_\varepsilon \to 0 \quad \text{as } \varepsilon \to 0\ , \tag{42}$$

where, remarking that $H_\varepsilon(\psi - u_\varepsilon)\,H_\varepsilon(u_\varepsilon - \psi) = 0$, we have

$$Y_\varepsilon = \int_{\Omega_t} \Big[a(b(u_\varepsilon),\nabla u_\varepsilon) - a(b(\psi),\nabla\psi)\Big]\cdot\nabla H_\varepsilon(\psi - u_\varepsilon)\ .$$

From (42), and recalling $\psi(0) \leq u_0$, we obtain $b_\varepsilon(\psi) \leq b_\varepsilon(u_\varepsilon)$, which by the strict monotonicity of b_ε implies $\psi \leq u_\varepsilon$.

iv) *Passage to the limit $\varepsilon \to 0$* — By the estimate (40), u_ε and also $b_\varepsilon(u_\varepsilon)$ belong to a bounded subset of $L^\infty(0,T;V)$, and $\partial_t b_\varepsilon(u_\varepsilon)$ is bounded in $L^2(\Omega_T)$. Therefore, there exists a subsequence $\varepsilon \to 0$ and limits u and b_* such that

$$u_\varepsilon \rightharpoonup u \ \text{ in } L^\infty(0,T;V)\text{-weak}^*\ , \tag{43}$$

$$b_\varepsilon(u_\varepsilon) \to b_* \ \text{ in } L^\infty(0,T;V)\text{-weak}^* \ \text{ and in } C^0([0,T];L^2(\Omega))\text{-strong}. \tag{44}$$

By monotonicity of b_ε, we have

$$\int_{\Omega_T} \Big[b_\varepsilon(u_\varepsilon) - b_\varepsilon(u + \lambda w)\Big]\,(u_\varepsilon - u - \lambda w) \ge 0\,, \qquad \forall w \in L^2(\Omega),\;\; \lambda \in \mathbf{R}\;;$$

letting $\varepsilon \to 0$, we find

$$\int_{\Omega_T} \Big[b_* - b(u + \lambda w)\Big]\,\lambda w \le 0\,, \qquad \forall w \in L^2(\Omega_T),\;\; \forall \lambda \in \mathbf{R}\,,$$

which implies $b_* = b(u)$ and, in particular, $b(u(0)) = b(u_0) = b_0$ due to (44). Since we have $b_\varepsilon(u_\varepsilon(t)) \to b(u(t))$ in $L^2(\Omega)$, for $\forall t \in [0,T]$, we obtain $\int_\Omega B_\varepsilon(u_\varepsilon(t)) \underset{\varepsilon\to 0}{\longrightarrow} \int_\Omega B(u(t))$, which by the formula (7) yields

$$\int_{\Omega_T} u_\varepsilon\,\partial_t b_\varepsilon(u_\varepsilon) \to \int_{\Omega_T} u\,\partial_t b(u) \qquad \text{as } \varepsilon \to 0\,. \tag{45}$$

Now we take an arbitrary $w \in \mathbf{K}(\psi)$ and we set $v = w - u_\varepsilon$ in (38) and we let $\varepsilon \to 0$. Observing that

$$\Big[1 - H_\varepsilon(u_\varepsilon - \psi)\Big]\,(w - u_\varepsilon) \ge \Big[1 - H_\varepsilon(u_\varepsilon - \psi)\Big]\,(\psi - u_\varepsilon) \ge -\varepsilon, \quad \text{and}$$
$$a(b(u_\varepsilon), \nabla w)\cdot\nabla(w - u_\varepsilon) \ge a(b(u_\varepsilon), \nabla u_\varepsilon)\cdot\nabla(w - u_\varepsilon)\,,$$

we obtain

$$\int_{\Omega_T} \partial_t b_\varepsilon(u_\varepsilon)\,(w - u_\varepsilon) + \int_{\Omega_T} a(b(u_\varepsilon), \nabla w)\cdot\nabla(w - u_\varepsilon) \ge \tag{46}$$
$$\ge \int_{\Omega_T} (f + \varepsilon\,\partial_t\psi)\,(w - u_\varepsilon) + \int_{S_T} g(w - u_\varepsilon) - \varepsilon\Big(\int_{\Omega_T} \xi + \int_{S_T} \eta\Big)\,.$$

Recalling (40), we know that $\partial_t b(u_\varepsilon) \rightharpoonup \partial_t b(u)$ in $L^2(\Omega_T)$-weak and $\varepsilon\,\partial_t u_\varepsilon \to 0$ in $L^2(\Omega_T)$. Since $a(b(u_\varepsilon), \nabla w) \to a(b(u), \nabla w)$ in $L^2(\Omega_T)$, using (45) and (43), we obtain from (46)

$$\int_{\Omega_T} \partial_t b(u)(w - u) + \int_{\Omega_T} a(b(u), \nabla w)\cdot\nabla(w - u) \ge \int_{\Omega_T} f(w - u) + \int_{S_T} g(w - u)\,,$$

for all $w \in \mathbf{K}(\psi)$, which is equivalent to (16), proving that u solves the variational inequality (15)–(16).

Finally, letting $v = u_\varepsilon \in \mathbf{K}(\psi)$ in (16) and $v = u - u_\varepsilon$ in (38) and combining (43), (45) and the coercivity (18) we obtain also

$$u_\varepsilon \to u \quad \text{in } L^2(0,T;V)\text{-strong.}$$

This property allows to pass to the limit in the following approximating estimates (which follow immediately from (36))

$$\varepsilon\,\partial_t\psi + f \le Q_\varepsilon u_\varepsilon \le f + \xi + \varepsilon\,\partial_t\psi \quad \text{in } \Omega_T\ ,$$
$$g \le N u_\varepsilon \le g + \eta \quad \text{on } S_T\ ,$$

showing (28) and (29) for u, and concluding the proof of Theorem 1. ∎

ACKNOWLEDGEMENT – The author wishes to thank fruitfull discussions with Professor D. Kröne on the occasion of a study visit to the University of Heidelberg.

REFERENCES

[1] AIKI T. – Two-Phase Stefan Problems for Parabolic-Elliptic Equations, *Proc. Japan Acad.* **64**, Ser. A (1988), 377–380.

[2] ALT H.W. – *Nonsteady fluid flow through porous media. Free Boundary Problems — Applications and Theory, III* (eds. A. Bossavit, A. Damlamian and M. Frémond), Research Notes Math. 120, Pitman, Boston (1985), 222–228.

[3] ALT H.W. & LUCKHAUS S. – Quasilinear elliptic-parabolic differential equations, *Math. Z.* **183** (1983), 311–341.

[4] ALT H.W., LUCKHAUS S. & VISINTIN A. – On nonstationary flow through porous media, *Ann. Mat. Pura Appl.* **136** (1984), 303–316.

[5] ARTOLA M. – Sur une classe de problèmes paraboliques quasi-linéaires, *Boll. Un. Mat. Ital.* (6), **5-B** (1986), 51–70.

[6] BENSOUSSAN A. & LIONS J.L. – *Applications des Inéquations Variationnelles en Contrôle Stochastique*, Dunod, Paris, 1978.

[7] BENSOUSSAN A. & LIONS J.L. – *Contrôle impulsionnel et inéquations quasivariationnelles*, Dunod, Paris, 1982.

[8] BLANCHARD D. & FRANCFORT G. – Study of a doubly nonlinear heat equation with no growth assumptions on the parabolic term, *SIAM J. Math. Anal.* **19** (1988), 1032–1056; *Erratum*, **20** (1989), 761–762.

[9] BLANCHARD D. & FRANCFORT G. – A few results on degenerate parabolic equations (to appear).

[10] BREZIS H. – On some degenerate non linear parabolic equation, *Proc. Symp. Pure Math.* **18**, Amer. Math. Soc. (1970), 28–38.

[11] BREZIS H., ROSENKRANZ W. & SINGER B. – On degenerate elliptic-parabolic equation occurring in the theory of probability, *Comm. Pure Appl. Math.* **24** (1971), 395–416.

[12] CARROL R.W. & SHOWALTER R.E. – *Singular and degenerate Cauchy Problems*, Academic Press, N.Y., 1976.

[13] CHARRIER P. & TROIANIELLO G.M. – Un résultat d'existence et de régularité pour les solutions fortes d'un problème unilatéral d'évolution avec obstacle dépendant du temps, *C.R. Acad. Sci. Paris* **281**, Sér. A (1975), 621–623; On strong solutions to parabolic unilateral problems with obstacle dependent on time, *J. Math. Anal. Appl.* **65** (1978), 110–125.

[14] CHIPOT M. & MICHAILLE G. – Uniqueness results and monotonicity properties for strongly nonlinear elliptic variational inequalities, *Ann. Scuola Norm. Sup. Pisa* **16** (1989), 137–166.

[15] CHIPOT M. & RODRIGUES J.F. – Comparison and stability of solutions to a class of quasilinear parabolic problems, *Proc. Royal Soc. Edinburgh* **110A** (1988), 275–285.

[16] DIAZ J.I. & THELIN F. – On doubly nonlinear parabolic equations arising in some models related to turbulent flows (to appear).

[17] DIBENEDETTO E. & FRIEDMAN A. – Periodic behaviour for the evolutionary dam problem and related free boundary problems, *Comm. Partial Diff. Equations* **11** (1986), 1297–1377.

[18] DIBENEDETTO E. & GARIEPY R. – Local behaviour of solutions of an elliptic parabolic problem, *Arch. Rational Mech. Anal.* **97** (1987), 1–18.

[19] DONATI F. & MATZEU M. – Solutions fortes et estimations duales pour des inéquations variationnelles paraboliques non linéaires, *C.R. Acad. Sci. Paris* **285**, Sér. A (1977), 347–350; On the strong solutions of some nonlinear evolution problems in ordered Banach spaces, *Boll. Un. Mat. Ital.* (5), **16-B** (1979), 54–73.

[20] VAN DUYN C.J. & PELETIER L.A. – Nonstationary filtration in partially saturated porous media, *Arch. Rational Mech. Anal.* **78** (1982), 173–198.

[21] FASANO A. & PRIMICERIO, M. – Liquid flow in partially saturated porous media, *J. Inst. Math. Appl.* **23** (1979), 503–517.

[22] GRANGE O. & MIGNOT E. – Sur la résolution d'une équation et d'une inéquation paraboliques non linéaires, *J. Funct. Anal.* **11** (1972), 77–92.

[23] HANOUZET B. & JOLY J.L. – Méthodes d'ordre dans l'interprétation de certaines inéquations variationnelles et applications, *C.R. Acad. Sci. Paris* **281**, Sér. A-B (1975), A373–A376; *J. Funct. Anal.* **34** (1979), 217–249.

[24] HORNUNG U. – A parabolic-elliptic variational inequality, *Manuscripta Math.* **39** (1982), 155–172.

[25] HULSHOF J. – Bounded weak solutions of an elliptic-parabolic Neumann problem, *Trans. Amer. Math. Soc.* **303** (1987), 211–227.

[26] HULSHOF J. & WOLANSKI N. – Monotone Flows in N-Dimensional Partially Saturated Porous Media: Lipschitz-Continuity of the Interface, *Arch. Rational. Mech. Anal.* **102** (1988), 287–305.

[27] KAČUR J. – On a solution of degenerate elliptic-parabollic systems in Orlicz–Sobolev spaces I–II, *Math. Z.* **203** (1990), 153–171 and 569–579.

[28] KENMOCHI N. – Neumann problems for a class of nonlinear degenerate parabolic equations, *Diff. Integral Eq.* **3** (1990), 253–273.

[29] KENMOCHI N. – Asymptotic Stability for Nonlinear Degenerate Parabolic Equations with Neumann Boundary Conditions, *Diff. Integral Eq.* **4** (1991), 803–816.

[30] KENMOCHI N., KRÖNER D. & KUBO M. – Periodic solutions to porous media equations of parabolic-elliptic type, *J. Partial Diff. Equations* **3** (1990), 63–77.

[31] KENMOCHI N. & KUBO M. – Periodic Stability of flow in partially saturated porous media, *Int. Series Numer. Maths* (Birkhäuser) **95** (1990), 127–152.

[32] KENMOCHI N. & PAWLOW I. – A class of nonlinear elliptic-parabolic equations with time-dependent constraints, *Nonlinear Anal. Th. Meth. Appl.* **10** (1986), 1181–1202.

[33] KENMOCHI N. & PAWLOW I. – Parabolic-elliptic free boundary problems with time-dependent obstacles, *Japan J. Appl. Math.* **5** (1988), 87–121.

[34] KENMOCHI N. & PAWLOW I. – Asymptotic Behaviour of Solutions to Parabolic-Elliptic Variational Inequalities, *Nonlinear Anal. Th. Meth. Appl.* **13** (1989), 1191–1213.

[35] KRÖNER D. – Parabolic regularization and behaviour of the free boundary for unsaturated flow in a porous media, *J. Reine Angew. Math.* **348** (1984), 180–196.

[36] KRÖNER D. & RODRIGUES J.F. – Global behaviour for bounded solutions of a porous media equation of elliptic parabolic type, *J. Math. Pures et Appl.* **64** (1985), 105–120.

[37] KUTTLER K.L. – Degenerate Variational Inequalities of Evolution, *Nonlinear Anal. Th. Meth. Appl.* **8** (1984), 837–850.

[38] LADYŽENSKAJA O.A., SOLONNIKOV V.A. & URAL'CEVA N.N. – *Linear and quasilinear equations of parabolic type*, Translation of Mathematical Monographs 23, American Mathematical Society, Providence, 1968.

[39] LEWY H. & STAMPACCHIA G. – On the smoothness of superharmonics which solve a minimum problem, *J. Anal. Math.* **23** (1970), 224–236.

[40] LIONS J.L. – *Quelques méthodes de résolution des problèmes aux limites non linéaires*, Dunod, Paris, 1969.

[41] MIGNOT F. & PUEL J.P. – Inequations d'évolution paraboliques avec convexes dépendant du temps. Applications aux inéquations quasi variationnelles d'évolution, *Arch. Ration. Mech. Analysis* **64** (1977), 59–91.

[42] RAVIART P.A. – Sur la résolution de certaines équations paraboliques non linéaires, *J. Funct. Anal.* **5** (1970), 299–328.

[43] RODRIGUES J.F. – Strong stability estimates for the evolution obstacle problem with Neumann condition, *Le Matematiche*, **XL** (1985), 29–44.

[44] RODRIGUES J.F. – *Obstacle Problems in Mathematical Physics*, North-Holland, Amsterdam, 1987.

[45] SCARPINI F. – Degenerate and Pseudoparabolic Variational Inequalities: Approximate Solutions, *Numer. Funct. Anal. and Optimiz.* **9** (1987), 859–879.

[46] SU N. – *The mathematical problems on the fluidal-solute-heat flow through porous media*, Ph. Thesis, Univ. Tsinghua, Beijing, 1987 (see also *Acta Math. Appl. Sinica* **6** (1990), 135–144; 145–157 and 224–237).

[47] VIVALDI M.A. – Existence of strong solutions for nonlinear parabolic variational inequalities, *Nonlinear Anal. Th. Meth. Appl.* **11** (1987), 285–295.

[48] XU X. – Existence and Convergence Theorems for Doubly Nonlinear Partial Differential Equations of Elliptic-Parabolic Type, *J. of Math. Anal. Appl.* **150** (1990), 205–223.

José-Francisco Rodrigues,
CMAF–INIC and Univ. of Lisbon,
Av. Prof. Gama Pinto, 2
1699 LISBOA Codex — PORTUGAL

G TALENTI AND F TONANI

Detecting subsurface gas sources

1 Introduction

Geologic applications, e.g. the surveillance of gas hazard in volcanic areas, may rise the following problem. Suppose gas flows out of some extended, plane, subsurface source — such as a water table — and travels towards the Earth's surface with parallel velocity. Suppose the pore structure of the involved medium is known, the bulk flow of gas is sampled at the surface over time, and measurements of gas concentration at and beneath the surface are available. Can one locate the gas source?

2 Problem statement

Under suitable assumptions, which will be detailed in a subsequent paper, a relevant mathematical model is the following:

$$u_t = Du_{xx} + F(t)u_x \qquad \text{for } 0 < x < L \text{ and } 0 < t < \infty, \tag{1}$$

$$u(0,t)=0 \text{ and } u(L,t)=1 \qquad \text{for } 0 < t < \infty \tag{2}$$

— a boundary value problem for a parabolic partial differential equation of evolution type with incomplete data. Here t and x denote time and depth, respectively. L is the depth of the gas source, D is the diffusion coefficient of the considerd gas, F stands for the velocity of gas in the pore space. F is proportional to flow density, i.e. velocity à la Darcy as measured outside the porous medium, by a factor depending on pore structure — for our present purpose, this factor can be made equal to the reciprocal coefficient of porosity. D is a constant, F is a function of time t only; D and F are given — in a typical situation D may have the value $0.15\ cm^2sec^{-1}$ and F, $cm\ sec^{-1}$, may be as in figure 1. u denotes the gas concentration, thus satisfies the constraint:

$$0 \leq u \leq 1\ . \tag{3}$$

Our problem amounts to estimating L, a positive number that causes some solution u to (1) & (2) & (3) to fulfill the following ovespecification

$$u_x(\text{at depth } 0 \text{ and a prescribed time } T) = \text{a prescribed quantity}\ . \tag{4}$$

3 Auxiliary theorems

Our method rests on theorems 1 and 2 below. Theorem 1 allows us to sandwich all solutions to (1) & (2) that satisfy (3); theorem 2 shows that all solutions to (1) & (2) lose memory of their initial profile in a comparatively short time.

Theorem 1. *Let u satisfy (1), (2) and (3); let m and M be constants such that*

$$m \leq F(t) \leq M \tag{5}$$

for $0 < t < \infty$. *Then the following inequalities*

$$h(t,L) \leq u_x(0,t;L) \leq H(t,L) \tag{6}$$

hold for every positive t and L.

Functions h and H are given by the following formulas:

$$h(t,L)= \tag{7}$$

$$\frac{1}{L}exp\Big(\frac{Lm}{2D}\Big)\left\{\frac{\frac{Lm}{2D}}{sinh\Big(\frac{Lm}{2D}\Big)} - 2\,exp\Big(-\frac{m^2t}{4D}\Big)\sum_{n=1}^{\infty}\frac{(-1)^{n-1}}{1+\Big(\frac{Lm}{2\pi D}\frac{1}{n}\Big)^2}\,exp\Big(-n^2\frac{\pi^2 Dt}{L^2}\Big)\right\},$$

$$H(t,L)= \tag{8}$$

$$\frac{\frac{M}{D}}{1-exp\Big(-\frac{LM}{D}\Big)} + \frac{2}{L}\,exp\Big(-\frac{M^2t}{4D}\Big)\sum_{n=1}^{\infty}\frac{1}{1+\Big(\frac{Lm}{2\pi D}\frac{1}{n}\Big)^2}\,exp\Big(-n^2\frac{\pi^2Dt}{L^2}\Big)\ ;$$

alternative, but more effective formulas are the following:

$$H(t,L)= \tag{9}$$

$$\frac{M}{2D}+\frac{1}{\sqrt{\pi Dt}}exp\Big(-\frac{M^2}{4\pi D^2}\pi Dt\Big)\,\varphi\Big(\frac{L^2}{\pi Dt}\Big)+\frac{M^2}{4\pi D^2}\,2\int_0^{\sqrt{\pi Dt}} exp\Big(-\frac{M^2}{4\pi D^2}s^2\Big)\,\varphi\Big(L^2s^{-2}\Big)ds\ ,$$

$$h(t,L)= \tag{10}$$

$$\frac{1}{L}exp\Big(\frac{Lm}{2D}\Big)\left\{exp\Big(-\frac{m^2t}{4D}\Big)\psi\Big(\frac{\pi Dt}{L^2}\Big)+\frac{m^2t}{4D}\,\frac{1}{\pi DL^{-2}t}\int_0^{\pi DL^{-2}t} exp\Big(-\frac{m^2t}{4D}\,\frac{s}{\pi DL^{-2}t}\Big)\psi(s)\,ds\right\}$$

— *here φ and ψ are (segments of) Jacobi theta functions, defined by*

$$\varphi(t)= 1+2\sum_{n=1}^{\infty} exp(-n^2\pi t) \quad , \quad \psi(t)= 1-2\sum_{n=1}^{\infty}(-1)^{n-1}exp(-n^2\pi t) \tag{11}$$

and satsfying

$$\varphi(t)= \frac{1}{\sqrt{t}}\,\varphi\Big(\frac{1}{\sqrt{t}}\Big) \quad , \quad \psi(t)= \frac{2}{\sqrt{t}}\sum_{n=1}^{\infty} exp\Big(-(n+\tfrac{1}{2})^2\,\frac{\pi}{t}\Big) \tag{12}$$

for every positive t.

Functions h and H can be efficiently computed via (9) and (10); their behavior is displayed in figures 2 and 3. Functions h and H enjoy the following properties:

(i) Let L be positive and fixed, and let t increase from 0 to $+\infty$. $h(t,L)$ has a zero of infinite order at t=0; then $h(t,L)$ **increases** *strictly from 0 to $(m/D)(1-exp(-Lm/D))^{-1}$. $H(t,L)$* **decreases** *strictly from $+\infty$ to $(M/D)(1-exp(-LM/D))^{-1}$.*

(ii) Let t be positive and fixed. Then

$$h(t,0+)=H(t,0+)=+\infty; \tag{13}$$

$h(t,L)$ and $H(t,L)$ **decrease** *strictly as L increases from 0 to $+\infty$;*

$$h(t,+\infty)=0, \tag{14}$$

$$H(t,+\infty)= max\Big(\frac{M}{D},0\Big)+\frac{1}{\sqrt{\pi Dt}}\,exp\Big(-\frac{M^2t}{4D}\Big)-\frac{|M|}{2D}\,erfc\Bigg(\sqrt{\frac{M^2t}{4D}}\Bigg) \tag{15}$$

— *here $erfc$ is the complementary error function, given by*

$$erfc(z)= 1-\frac{2}{\sqrt{\pi}}\int_0^z exp(-s^2)\,ds\ ; \tag{16}$$

$h(t,L)$ and $[H(t,L)-H(t,+\infty)]$ behave like $exp\Big(-\frac{L^2}{Dt}\Big)$ as L approaches $+\infty$.

Theorem 2 . *Let w be the* **difference** *of any two solutions to (1) & (2); let B be a constant such that*

$$|F(t)| \le B \tag{17}$$

for $0<t<\infty$. Then the following inequalities hold:

$$\Big(\int_0^L w^2(x,t)dx\Big)^{1/2} \le \Big(\int_0^L w^2(x,0)dx\Big)^{1/2} exp\Big(-\frac{\pi^2Dt}{L^2}\Big) \tag{18}$$

for $0<t<\infty$;

$$max\,\{\,|w(x,t)|:0\le x\le L\} \le 4.3402\,\Big(\frac{1}{L}\int_0^L w^2(x,0)dx\Big)^{1/2} exp\Big(\frac{BL}{4D}-\frac{\pi^2Dt}{L^2}\Big) \tag{19}$$

for $0.1430\times L^2/(\pi^2D)\le t<\infty$;

$$|w_x(0,t)| \le 65.2544\,\Big(\frac{1}{L}\int_0^L w^2(x,0)dx\Big)^{1/2} max\Big(\frac{B}{4D},\frac{1}{L}\Big)\,exp\Big(\frac{BL}{4D}-\frac{\pi^2Dt}{L^2}\Big) \tag{20}$$

for $0.2860\times L^2/(\pi^2D)\le t<\infty$.

Proof of theorem 1, outlined. Let v and V be the solutions to the problems:

$$v_t = Dv_{xx} + mv_x \qquad \text{for } 0 < x < L \text{ and } 0 < t < \infty, \tag{21a}$$
$$v(0,t) = 0 \text{ and } v(L,t) = 1 \qquad \text{for } 0 < t < \infty, \tag{21b}$$
$$v(x,0) = 0 \qquad \text{for } 0 < x < L; \tag{21c}$$

$$V_t = DV_{xx} + MV_x \qquad \text{for } 0 < x < L \text{ and } 0 < t < \infty, \tag{22a}$$
$$V(0,t) = 0 \text{ and } V(L,t) = 1 \qquad \text{for } 0 < t < \infty, \tag{22b}$$
$$V(x,0) = 1 \qquad \text{for } 0 < x < L. \tag{22c}$$

The maximum principle guarantees that v_x and V_x are positive; hence an easy comparison argument shows that $v \leq u \leq V$. Thus

$$v_x(0,t;L) \leq u_x(0,t;L) \leq V_x(0,t;L). \tag{23}$$

Integrations by separation of variables or Laplace transforms give

$$v_x(0,t;L) = \text{the r.h.s. of (7)} \quad , \quad V_x(0,t;L) = \text{the r.h.s of (8)}. \tag{24}$$

Inequality (6) follows.

Formulas (9) & (10) follow from (7) & (8) via manipulations.

Properties (i) & (ii) can be derived from (9) & (10). □

Proof of theorem 2, outlined. We have

$$w_t = Dw_{xx} + F(t)w_x \qquad \text{for } 0 < x < L \text{ and } 0 < t < \infty, \tag{25}$$
$$w(0,t) = w(L,t) = 0 \qquad \text{for } 0 < t < \infty. \tag{26}$$

Multiplying both sides of the equation in (25) by w, then integrating with respect to x between 0 and L give

$$\frac{1}{2}\frac{d}{dt}\int_0^L w^2(x,t)dx = -D\int_0^L w_x^2(x,t)dx \,.$$

In fact, F does not depend on x and

$$\int_0^L ww_{xx}dx = -\int_0^L w_x^2 dx \quad , \quad \int_0^L w_x w dx = 0$$

thanks to boundary conditions (26). On the other hand,

$$\int_0^L w_x^2 dx \geq \left(\frac{\pi}{L}\right)^2 \int_0^L w^2 dx$$

by virtue of (26). Hence

$$\frac{d}{dt}\int_0^L w^2(x,t)dx \leq -\frac{2\pi^2 D}{L^2}\int_0^L w^2(x,t)dx \,. \tag{27}$$

Inequality (18) follows.

Inequality (19) follows from (18) and the inequality:

$$max\{\,|\,w(x,t)\,|:0\le x\le L\}\le\left(\int_0^L w^2(x,s)dx\right)^{1/2}\times\sqrt{\frac{3}{L}}\,exp\left(\frac{BL}{4D}\right)\left(1+\frac{L}{4\sqrt{\pi D(t-s)}}\right), \tag{28}$$

where s is any positive number and $s<t<\infty$. Inequality (20) follows from (19) and the inequality:

$$|\,w_x(0,t)\,|\;\le max\{\,|\,w(x,s)\,|:0\le x\le L\}\times\frac{6}{L}\,max\left(\frac{BL}{4D},1\right)\left(1+\frac{L}{4\sqrt{\pi D(t-s)}}\right), \tag{29}$$

where s is any positive number and $s<t<\infty$.

Inequality (28) can be proved via techniques based on geometric measure theory and rearrangements — details are too complicated to be presented here. Inequality (29) can be proved via the following argument. Consider the solution W to the boundary value problem:

$$W_t=DW_{xx}+B\,|\,W_x\,| \qquad \text{for } 0<x<L \text{ and } 0<t<\infty, \tag{30a}$$

$$W(0,t)=W(L,t)=0 \qquad \text{for } 0<t<\infty, \tag{30b}$$

$$W(x,0)=1 \qquad \text{for } 0<x<L \tag{30c}$$

and let z be defined by:

$$z(x,t)=max\{\,|\,w(x',s)\,|:0\le x'\le L\}\times W(x,t-s)\pm w(x,t)\,.$$

It easy to check that

$$z_t-Dz_{xx}-F(t)z_x\ge 0 \qquad \text{for } 0<x<L \text{ and } s<t<\infty,$$

$$z(0,t)=z(L,t)=0 \qquad \text{for } s<t<\infty,$$

$$z(x,s)\ge 0 \qquad \text{for } 0<x<L.$$

The maximum principle tells us that

$$z(x,t)\ge 0$$

for $0<x<L$ and $s<t<\infty$. Consequently,

$$|\,w_x(0,t)\,|\;\le max\{\,|\,w(x,s)\,|:0\le x\le L\}\times W_x(0,t-s) \tag{31}$$

for $s<t<\infty$. Problem (30) can be solved via explicit formulas; these formulas give

$$|\,W_x(0,t)\,|\;\le\frac{6}{L}\,max\left(\frac{BL}{4D},1\right)\left(1+\frac{L}{4\sqrt{\pi Dt}}\right) \tag{32}$$

for $0<t<\infty$. Inequality (29) follows. □

4 Main results

Coupling condition (4) and theorem 1 results in estimates of L. L *is* ***bounded from below*** *by a positive quantity which depends only on the diffusion coefficient D, on a lower bound m for flow F, on time T and the right-hand side of equation (4). If the right-hand side of equation (4) is comparatively large* — i.e. a suitable datum is available at a suitable time, as made self-explanatory in fig.3 — *L can be* ***bounded from above*** *by a finite quantity which depends only on the diffusion coefficient D, on an upper bound M for flow F, on time T and the right-hand side of (4).* Both the lower and the upper bound are computable — graphically, they are abscissas of points where a horizontal straight line meets plots of h and H.

Observe that, due to the asymptotic behavior of h and H, the estimates in question may prove effective if L is small, but deteriorate if the ratio $L^2/(Dt)$ grows large, e.g. L is large and t is too small. In other words, a shallow gas source can be detected at a low cost; whilst locating deep gas sources requires collecting accurate data for a long period of time — figure 3 gives evidence to the situation.

A more precise estimate of L results from the following strategy. Let U be the solution to the boundary value problem:

$$U_t = DU_{xx} + F(t)U_x \quad \text{for } 0 < x < L \text{ and } 0 < t < \infty, \tag{33a}$$

$$U(0,t;L)=0 \text{ and } U(L,t;L)=1 \quad \text{for } 0 < t < \infty \tag{33b}$$

that satisfies the initial condition:

$$U(x,0;L)=x/L \quad \text{for } 0 < x < L, \tag{33c}$$

and let L' be determined by the equation:

$$U_x(0,T;L') = \text{the right-hand side of (4).} \tag{34}$$

An appropriate analysis of problem (33) points out that $U_x(0,T;L)$ decreases strictly from $+\infty$ to 0 as L increases from 0 to $+\infty$ and that the derivative of $U_x(0,T;L)$ with respect to L is negative and bounded away from zero — thus the solution L' of equation (34) is uniquely determined and stable. Importantly, theorem 2 tell us that

$$U_x(0,T;L) = \text{the right-hand side of (4) + an error,} \tag{35a}$$

$$|\,\text{error}\,| \le 65.2544\, max\left(\frac{B}{4D}, \frac{1}{L}\right) exp\left(\frac{BL}{4D} - \frac{\pi^2 Dt}{L^2}\right) \tag{35b}$$

— thus the error term approaches zero uniformly with respect to L as t approaches $+\infty$ and L stays bounded and bounded away from zero. *In conclusion, L' is an* ***estimate*** *of L which gets better and better as time T gets larger and larger.* Graphically, the estimate in question is obtained by drawing a plot of $U_x(0,T;L)$ versus L then singling out the point where such a plot meets a horizontal straight-line. Figure 4 shows plots of $U_x(0,t;L)$ versus L for different values of time t — the relevant flow F is displayed in figure 1.

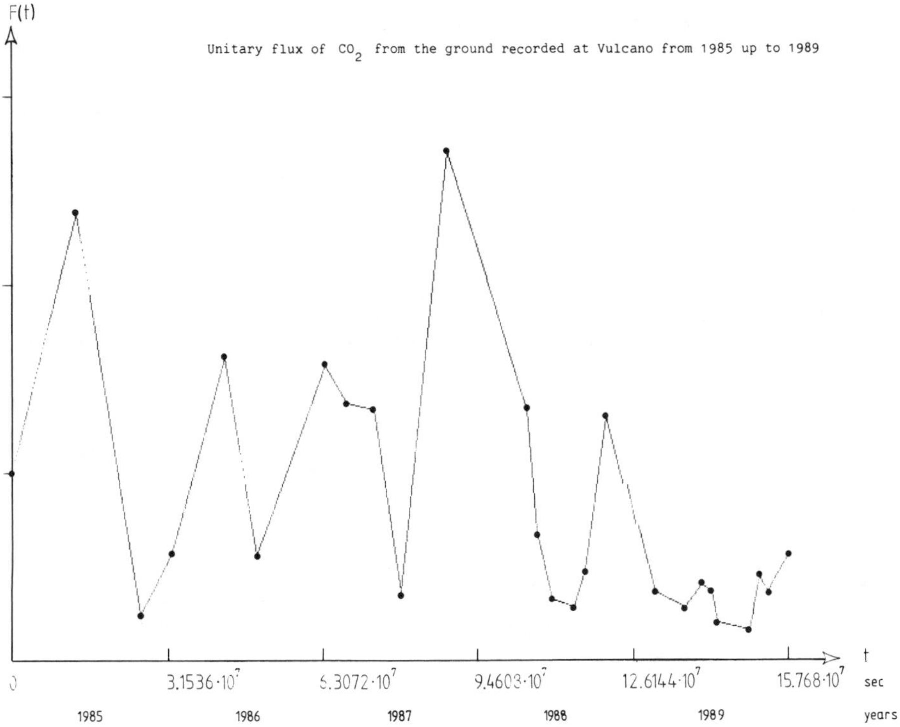
F(t)
Unitary flux of CO_2 from the ground recorded at Vulcano from 1985 up to 1989
t
0
3.1536·10^7
6.3072·10^7
9.4603·10^7
12.6144·10^7
15.768·10^7
sec
1985
1986
1987
1988
1989
years

fig.1

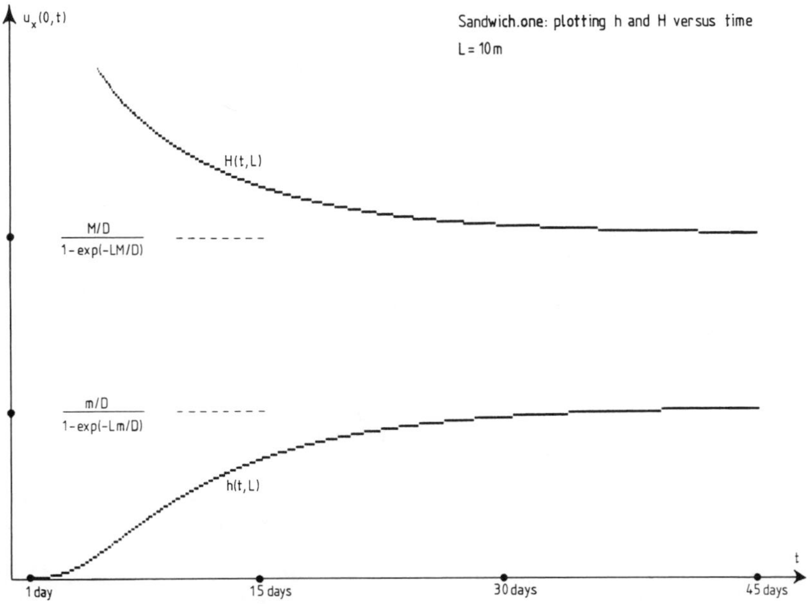
$u_x(0,t)$
Sandwich.one: plotting h and H versus time
L = 10m
H(t,L)
M/D
1 - exp(-LM/D)
m/D
1 - exp(-Lm/D)
h(t,L)
t
1 day
15 days
30 days
45 days

fig. 2

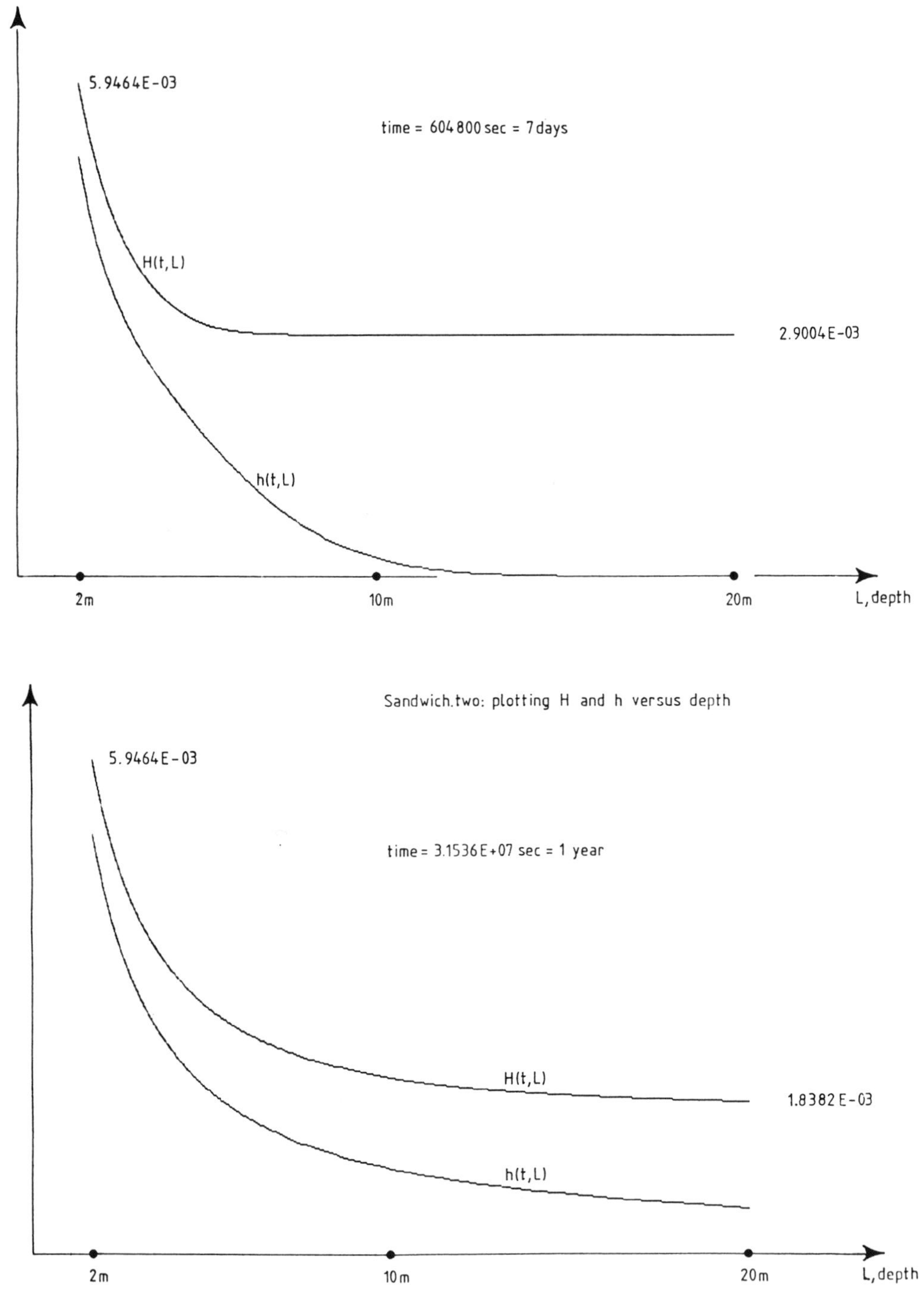
5.9464E-03
time = 604800 sec = 7days
H(t,L)
2.9004E-03
h(t,L)
2m
10m
20m
L,depth
Sandwich.two: plotting H and h versus depth
5.9464E-03
time = 3.1536E+07 sec = 1 year
H(t,L)
1.8382E-03
h(t,L)
2m
10m
20m
L,depth

fig. 3

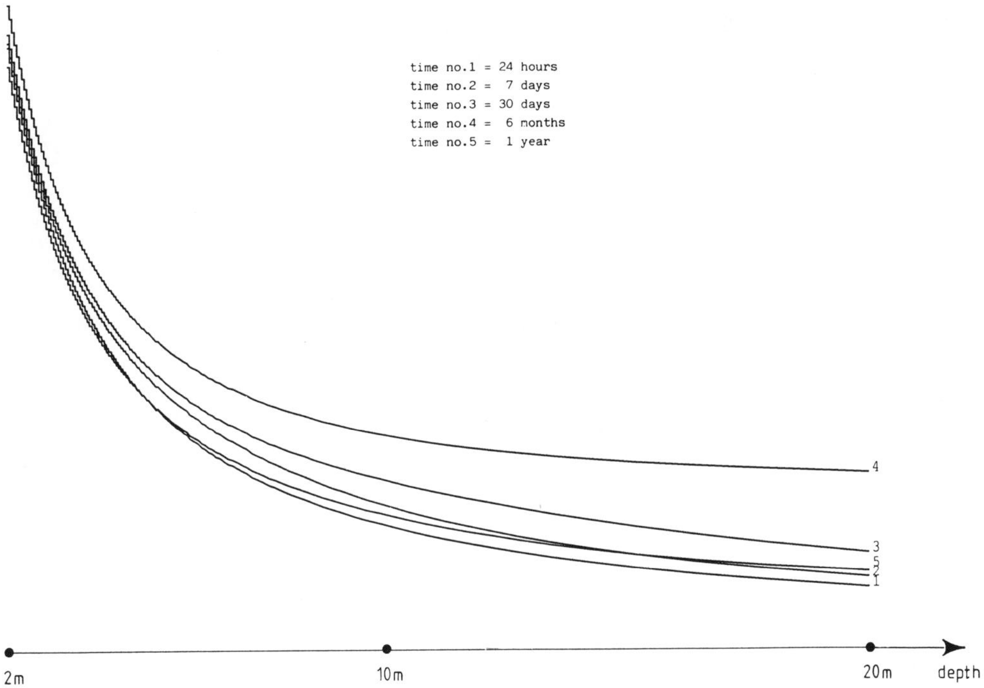

fig 4

Giorgio Talenti, Dipartimento di Matematica dell'Università, viale Morgagni 67A, 50134 Firenze, Italy

Franco Tonani, Istituto di Mineralogia, Petrografia e Geochimica dell'Università, via Archirafi 36, 90123 Palermo

A ABAKHTI-MCHACHTI AND J FLECKINGER-PELLÉ

Existence of positive solutions for non cooperative semilinear elliptic system defined on an unbounded domain

1. Introduction.

In the present work, we are interested in the study of the following variational problem :

$$(\mathcal{P}) \quad \begin{cases} (-\Delta + q(x))u = \lambda u - f(u) - v & \\ (-\Delta + q(x))v = \delta u - \gamma v & in \quad \Omega \\ u/_{\partial\Omega} = 0 \ , \ v/_{\partial\Omega} = 0 & \\ u(x) \xrightarrow[|x|\to+\infty]{} 0 \ , \ v(x) \xrightarrow[|x|\to+\infty]{} 0 & \end{cases}$$

where γ, δ are given positive numbers; λ is a real parameter, and Ω is an unbounded connected open set of R^n, with smooth boundary $\partial\Omega$.

We first notice that our system is non cooperative since $\delta > 0$. Non cooperative systems defined on bounded domains, when q is identically zero , have been intensively studied before, since they appear in many phenomena, for instance population dynamics; (see Rothe [16], Lazer-Mc Kenna [13], De Figueiredo-Mitidieri [8], Brown [2], Hernandez [10]). Under some assumptions on the nonlinearity f and the potential q, we prove here that there exists a unique pair (u,v) of positive solutions if and only if λ varies in an open interval (as when Ω is bounded and q is identically zero). Our result is an extension to unbounded open sets of earlier results obtained by De Figueiredo-Mitidieri and by Hernandez. Problem $(\mathcal{P})$ arises, for example, in gas lasers theory.

As in the above mentionned papers, we use a decoupling technique. Since Ω is unbounded, there does not exist an infinite sequence of positive eigenvalues for the Dirichlet Laplacian defined on Ω, but there does exist one for $-\Delta + q$, since we assume here that q tends to $+\infty$ at infinity. So that, after some technical preliminaries for such " Schrödinger operators", we are able to adapt proofs of [8], [10].

We make the following assumptions :

$$q \in C^o(R^n, R) \ ; \quad \exists k > 0 \quad q(x) \geq k \quad \forall x \in R^n; \quad q(x) \xrightarrow[|x|\to+\infty]{} +\infty \ . \qquad (1.1)$$

$$f \in C^2(R,R) \text{ is non decreasing; } \ f(0) = 0 \ ; \tag{1.2}$$

$$\frac{f(u)}{u} \text{ is strictly increasing for all } u > 0 \ ; \tag{1.3}$$

$$\frac{f(u)}{u} \longrightarrow f'(\infty) < +\infty \text{ as } u \longrightarrow +\infty \ . \tag{1.4}$$

Under Hypotheses (1.1) to (1.4), the nonlinearity f is strictly convex in R^+ and asymptotically linear. Problem $(\mathcal{P})$ admits the pair of trivial solutions (0,0). This enables us to give a complete description of the bifurcation diagram for the branch of positive solutions, using sub and supersolutions, and a local inversion and continuation method due to Crandall and Rabinowitz [4], as when Ω is bounded.

Notations : Throughout the paper, $(\ ,\ )_{L^2(\Omega)}$ denotes the usual scalar product in $L^2(\Omega)$; $H^1(\Omega)$ and $H^1_o(\Omega)$ are the usual Sobolev spaces.

2. Preliminaries : a maximum principle for a single equation.

Before dealing with the system, we first have to study the case of a single equation. Let us consider the Schrödinger operator $H_q := -\Delta + q$, considered as an operator in $L^2(\Omega)$, (see e.g. Kato [11], Reed and Simon [15], Edmunds and Evans [6]). We introduce, as in Fleckinger [9], the space $V^o_q(\Omega)$ defined as the completion of $C^\infty_o(\Omega)$ with respect to the norm $||.||_{q,\Omega}$, where

$$||u||_{q,\Omega} = \left(\int_\Omega |\nabla u|^2 + qu^2 \right)^{\frac{1}{2}} \qquad \forall u \in C^\infty_o(\Omega) \ .$$

Proposition 2.1 : $V^o_q(\Omega)$ is a Hilbert space which is dense in $L^2(\Omega)$; the embedding of $V^o_q(\Omega)$ into $L^2(\Omega)$ is continuous and compact.

Proof : The density and continuity of the embedding obviously follow from the definition of $V^o_q(\Omega)$. The compactness is established as in Fleckinger [9] :

Let $R > 0$, $\Omega'_R = \{x \in \Omega \, , |x| > R\}$ and $\Omega_R = B_R \cap \Omega$. Then,

$$\int_{\Omega'_R} u^2 \leq \sup_{x \in \Omega'_R} \left(\frac{1}{q(x)} \right) \int_{\Omega'_R} q(x) u^2 \leq \varepsilon(R) ||u||^2_{q,\Omega'_R}$$

where $\varepsilon(R) := \sup_{x \in \Omega'_R} \left(\frac{1}{q(x)} \right)$, which exists and tends to 0 as R tends to $+\infty$ by (1.1). So that :

$$||u||^2_{L^2(\Omega)} = \int_{\Omega_R} u^2 + \int_{\Omega'_R} u^2 \leq ||u||^2_{L^2(\Omega_R)} + \varepsilon(R) ||u||^2_{q,\Omega'_R} \ . \tag{2.1}$$

Let $\{u_k\}_{k\in N}$ be a bounded sequence in $V_q^o(\Omega)$, then u_{k/Ω_R} is bounded in $H^1(\Omega_R)$; since Ω_R is a bounded set, the embedding of $H^1(\Omega_R)$ into $L^2(\Omega_R)$ is continuous and compact, so that we can extract a subsequence which is strongly convergent in $L^2(\Omega_R)$. For R large enough, (2.1) implies that $\{u_k\}_{k\in N}$ is a Cauchy sequence in $L^2(\Omega)$, hence a convergent sequence in $L^2(\Omega)$. •

Let us denote by $a_q(\ ,\ )$ the inner product on $V_q^o(\Omega)$ associated with the norm $||\cdot||_{q,\Omega}$:

$$a_q(u,v) = \int_\Omega \nabla u \nabla v + quv \qquad \forall (u,v) \in V_q^o(\Omega) \times V_q^o(\Omega)$$

Considering the variational triple $(V_q^0(\Omega), L^2(\Omega), a_q)$, we define as usual by the Lax-Milgram lemma a unique operator H_q in $L^2(\Omega)$ by :

$$(H_q u, v)_{L^2(\Omega)} = a_q(u,v) \quad \forall u \in D(H_q) \quad \forall v \in V_q^o(\Omega)\ ,$$

with :

$$D(H_q) = \{u \in V_q^o(\Omega)\,,\ H_q u := (-\Delta + q)u \in L^2(\Omega)\}\ .$$

Since q is positive, the operator H_q is positive (in the sense that $(H_q u, u)_{L^2(\Omega)}$ is non negative for every u in $D(H_q)$) and it is selfadjoint. Its inverse H_q^{-1} is defined on $L^2(\Omega)$ with range $D(H_q)$; considered as an operator into $L^2(\Omega)$, it follows from Proposition 2.1 that it is compact; hence the spectrum of H_q consists of a countable sequence of eigenvalues $\{\lambda_k\}_{k\in N}$, each with finite multiplicity :

$$0 < \lambda_1 \le \lambda_2 \le \ldots \le \lambda_k \le \ldots \quad \text{with } \lambda_k \longrightarrow +\infty \text{ as } k \longrightarrow +\infty.$$

Remark 1 : According to the smoothing properties of the semigroup generated by H_q (see e.g. Simon [17]), every eigenfunction φ_k (associated with eigenvalue λ_k) is continuous and bounded on Ω, and there exist positive constants α and β such that $|\varphi_k(x)| \le \alpha e^{-\beta|x|}$ for $|x|$ large enough.

Proposition 2.2 : The first eigenvalue of H_q, denoted by λ_1 throughout the paper, is isolated and simple; moreover the associated eigenfunction φ_1 is (strictly) positive in Ω.

When $\Omega = R^n$ or $(R^n \setminus K)$, where K is a closed set of measure zero, this result can be found in Reed and Simon [15], or Faris and Simon [7]. When Ω satisfies our hypotheses, and q satisfies (1.1), this can be shown as in Manes-Micheletti [14].

Proof : We first note that λ_1 is with finite multiplicity; this holds since H_q^{-1} is compact. We then prove by contradiction that λ_1 is simple. If λ_1 is not simple, one associated eigenfunction (e.g. φ_1) changes sign; then φ_1^+ is also an eigenfunction associated with λ_1. Let us denote by $\Omega_+ = \{x \in \Omega\, ;\ \varphi_1(x) > 0\,\}$ and by $\lambda_1(H_q, \Omega_+)$ the first eigenvalue of H_q defined on Ω_+ with Dirichlet boundary conditions. Then

$\lambda_1 = \lambda_1(H_q, \Omega_+)$. Moreover, if we assume that $\Omega_+ \neq \Omega$, φ_1 and φ_1^+ are linearly independent. Hence, considering all the domains Ω_* such that $\Omega_+ \subset \Omega_* \subset \Omega$, we can construct an infinite number of linearly independent eigenfunctions associated with λ_1, which contradicts the fact that λ_1 is with finite multiplicity. Hence, λ_1 is simple, and therefore, φ_1 does not change sign. •

We now prove a (strong) maximum principle for a single equation.

Let us consider the following problem defined on Ω :

$$(\mathcal{E}) \quad \begin{cases} H_q u - \lambda u = g(x) & \text{in } \Omega \\ u/_{\partial\Omega} = 0 \ , \ u(x) \xrightarrow[|x|\to+\infty]{} 0 \ . \end{cases}$$

Theorem 2.3 : Assume that q satisfies (1.1) and that $\lambda < \lambda_1$. Then, for any g in $L^2(\Omega)$, there exists a unique u satisfying $(\mathcal{E})$; moreover, if g is non identically zero and non negative ae Ω, u is (stritly) positive ae Ω.

In the following, we make use of the next definition (see e.g. [15]) :

Definition : For a given real number $\lambda < \lambda_1$, we say that $(H_q - \lambda I)^{-1}$ is "positivity preserving" (resp. "positivity improving") if :

$$g \geq 0 \quad ae \quad \Omega \quad \Longrightarrow \quad (H_q - \lambda I)^{-1} g \geq 0 \quad (\text{resp. } u > 0\ ) \quad ae \quad \Omega$$

Consequently Theorem 2.3, which is a (strong) maximum principle for $(\mathcal{E})$, equivalently means that $(H_q - \lambda I)^{-1}$ is positivity improving for all $\lambda < \lambda_1$.

The proof of Theorem 2.3 relies on the spectral transform of the operator H_q. For convenience, let us recall here some abstract results which can be found, for example, in Reed and Simon [15]. In what follows, M denotes any measurable space with σ-finite measure μ, H is a selfadjoint operator defined on $L^2(M, d\mu)$ which is bounded below, and $\Sigma(H)$ is the spectrum of H. Then :

Proposition 2.4 : (Theorem XIII.44 Reed and Simon [15])

Assume that e^{-tH} is positivity preserving for all $t > 0$ and that $\inf \Sigma(H)$ is an eigenvalue. Then the following propositions are equivalent :

(i) $\inf \Sigma(H)$ is a simple eigenvalue, with strictly positive associated eigenvector

(ii) $(H - \lambda I)^{-1}$ is positivity improving for all $\lambda < \inf \Sigma(H)$

(iii) e^{-tH} is positivity improving for all $t > 0$.

Proposition 2.5 : (Trotter product formula (see [12],[15]))

Let A and B two selfadjoint operators, which are bounded below, defined on a Hilbert space $(\mathcal{H})$; assume that A + B is essentially selfadjoint on $D(A) \cap D(B)$, then : $\lim_{n\to+\infty} (e^{-\frac{tA}{n}} e^{-\frac{tB}{n}})^n = e^{-t(A+B)}$ in strong convergence sense.

Proof of Theorem 2.3 : We first prove the following lemma :

Lemma 2.6 : The operator e^{-tH_q} is positivity preserving for all $t > 0$.

Then Theorem 2.3 is a simple consequence of Lemma 2.6 and Propositions 2.2 and 2.4 combined with Proposition 2.5.

Proof of Lemma 2.6 : In our problem, $H = H_q$, defined as an operator in $L^2(\Omega)$, is selfadjoint, bounded below, and with compact inverse. We apply Proposition 2.5 with $\mathcal{H} = L^2(\Omega)$, $A = -\Delta$, $D(A) = H^2(\Omega) \cap H_o^1(\Omega)$, and B is the multiplication by q with domain $D(B) = \left\{ u \in L^2(\Omega) ; \int_\Omega q^2 u^2 < +\infty \right\}$. Under Hypothesis (1.1), $C_o^\infty(\Omega)$ is dense in $D(A) \cap D(B)$. Then, A + B is essentially selfadjoint on $D(A) \cap D(B)$ (see Edmunds and Evans [5], Theorem 2.5, p. 361) ; its unique selfadjoint extension is H_q. It follows from the first Beurling Deny criterion, (see e.g. Reed and Simon [15], Davies [5]), that $e^{t\Delta}$ is positivity preserving on $L^2(\Omega)$ for all $t > 0$. Moreover, for $t > 0$ and $\varphi \in L^2(\Omega)$, $e^{-\frac{tq}{n}}\varphi$ is non negative ae Ω whenever φ is non negative ae Ω. Hence, by Proposition 2.5 , we obtain Lemma 2.6. •

3. Linear decoupling and main results.

As in Rothe [16], Hernandez [10], De Figueiredo-Mitidieri [8], we reduce System $(\mathcal{P})$ to the study of a single nonlinear equation, using the following decoupling method :

Since $\gamma > 0$, $H_q + \gamma$ (with Dirichlet boundary conditions) is invertible. Hence, for each $u \in L^2(\Omega)$, the second equation $(-\Delta + q(x) + \gamma)v = \delta u$ has a unique solution $v = Bu$ in $D(H_q)$, where $B = \delta(-\Delta + q(x) + \gamma)^{-1}$. Replacing v by Bu in the first equation, System $(\mathcal{P})$ is equivalent to :

$$\begin{cases} Tu := (H_q + B)u = \lambda u - f(u) & \text{in } \Omega \\ u/_{\partial\Omega} = 0 \ , \quad u(x) \xrightarrow[|x| \to +\infty]{} 0 \end{cases} \tag{3.1}$$

We first notice that B : $L^2(\Omega) \longrightarrow L^2(\Omega)$ is selfadjoint and compact ; this follows from the results of the previous section. Thus $T := H_q + B$ may be consisered as a compact perturbation of H_q. The operator T is linear, selfadjoint and unbounded in $L^2(\Omega)$, with domain $D(T) = D(H_q)$. Moreover, it is positive (in the sense that $(T\varphi, \varphi)_{L^2(\Omega)} \geq 0$, $\forall \varphi \in D(T)$). Let us denote by $\rho(T)$ the resolvent set of T, by $\Sigma_p(T)$ the set of its eigenvalues, and by $\Sigma_{ess}(T)$ its essential spectrum (where

$\Sigma_{ess}(T) := \Sigma(T) \setminus \Sigma_p(T))$. Since B is compact, $\Sigma_{ess}(T) = \Sigma_{ess}(H_q)$; and since $\Sigma(H_q) = \Sigma_p(H_q)$, $\Sigma_{ess}(T) = \emptyset$.

If we denote by $\{\varphi_k\}_{k\in N}$ the eigenfunctions of H_q normalized by $||\varphi_k||_\infty = 1$ (in view of Remark 1), and by $\{\lambda_k\}_{k\in N}$ the associated eigenvalues, it is easy to compute that :

$$T\varphi_k = \widehat{\lambda}_k \varphi_k \quad \text{where} \quad \widehat{\lambda}_k = \lambda_k + \frac{\delta}{\gamma + \lambda_k} \qquad \forall k \in N \tag{3.2}$$

Since the system $\{\varphi_k\}_{k\in N}$ is a complete set in $L^2(\Omega)$, there are no other eigenfunctions, and $\Sigma(T) = \Sigma_p(T) := \{\widehat{\lambda}_k\}_{k\in N}$.

Besides, as the only eigenfunction which does not change sign in Ω is φ_1, necessarily $\widehat{\lambda}_1$, which is associated to φ_1, has to be smaller than $\widehat{\lambda}_k$ for any $k > 1$.

A sufficient condition for the function $x \longmapsto x + \dfrac{\delta}{\gamma + x}$ to be increasing in the halfline $(\lambda_1, +\infty)$ is :

$$\gamma + \lambda_1 > \sqrt{\delta}\ . \tag{3.3}$$

If (3.3) is satisfied, the sequence $\widehat{\lambda}_k$ is non decreasing; moreover, since λ_1 is simple, isolated and positive, so is $\widehat{\lambda}_1$.

Next, as in De Figueiredo-Mitidieri [8], we prove :

Proposition 3.1 : Assume (3.3) is satisfied; then, for any μ in $\rho(T)$, $T_\mu := (T - \mu I)^{-1}$ is compact.

Proof : We first prove this result for some real numbers by a factorization as in [8]. Let a and b be two real numbers such that :

$$a > -\lambda_1\ , \quad \gamma + b > -\lambda_1\ , \quad b \neq 0\ , \quad b\gamma + \delta = ab\ .$$

Then $(a + H_q)^{-1}$ and $(b + \gamma + H_q)^{-1}$ are compact. Hence $\lambda = -a - b$ is in $\rho(T)$ by (3.3); and one has : $T_\lambda = \Big(1 - b(\gamma + b + H_q)^{-1}\Big)(a + H_q)^{-1}$, so that, for this λ, T_λ is compact.

For any μ in $\rho(T)$, the same result holds, since :

$$\forall (\mu, \lambda) \in \rho(T) \times \rho(T) \quad , \quad T_\mu - T_\lambda = (\mu - \lambda) T_\mu T_\lambda\ . \bullet$$

It is now possible to deduce a (strong) maximum principle for (3.1).

Theorem 3.2 : For all λ such that (3.3) holds and

$$2\sqrt{\delta} - \gamma \leq \lambda < \widehat{\lambda}_1\ , \tag{3.4}$$

$T_\lambda := (T - \lambda I)^{-1}$ is positivity improving.

Proof : To a given λ satisfying (3.4), we associate b such that :

$$b^2 + (\gamma + \lambda)b + \delta = 0$$

Note that $b < 0$. Set $a = -b - \lambda$. We can factorize as above :

$$T_\lambda u = \Big(1 - b(\gamma + b + H_q)^{-1}\Big)(a + H_q)^{-1} u \ .$$

Let $u \geq 0$ ae ; by Theorem 2.3, $w = (a + H_q)^{-1} u$ is strictly positive since $a > -\lambda_1$. So is $(\gamma + b + H_q)^{-1} w$; and since $b < 0$, $T_\lambda u = w - b(\gamma + b + H_q)^{-1} w$ is strictly positive ae Ω and Theorem 3.2 holds. •

We establish now some necessary and sufficient conditions for the existence of positive solutions for System $(\mathcal{P})$.

Proposition 3.3 : (Necessary condition.) Suppose (1.1) to (1.4) hold; if $u \geq 0$ *ae* Ω is a non trivial solution of (3.1), then

$$\widehat{\lambda}_1 + f'(0) < \lambda < \widehat{\lambda}_1 + f'(\infty) \ . \tag{3.5}$$

Proof: The equation in (3.1) is equivalent to : $Tu + \dfrac{f(u)}{u} u = \lambda u$. Multiplying by φ_1 and integrating over Ω, we get :

$$\widehat{\lambda}_1 \int_\Omega u\varphi_1 + \int_\Omega \frac{f(u)}{u} u\varphi_1 = \lambda \int_\Omega u\varphi_1$$

As by hypothesis (1.4), $f'(0) < \dfrac{f(u)}{u} < f'(\infty)$, we therefore have :

$$f'(0) \int_\Omega u\varphi_1 < (\lambda - \widehat{\lambda}_1) \int_\Omega u\varphi_1 < f'(\infty) \int_\Omega u\varphi_1 \ .$$

Since $\displaystyle\int_\Omega u\varphi_1 > 0$, we deduce Lemma 3.3. •

Proposition 3.4 : Let $u \geq 0$ *ae* Ω be a non trivial solution of (3.1). Assume conditions (1.1) to (1.4), (3.3) and (3.5) are satisfied, and that :

$$f'(\infty) - f'(0) \leq \widehat{\lambda}_1 + \gamma - 2\sqrt{\delta} \tag{3.6}$$

Then $u > 0$ ae Ω.

Proof : The equation in (3.1) may be rewritten as follows : $Tu - \mu u = \lambda u - f(u) - \mu u$. If μ is in $\left[2\sqrt{\delta} - \gamma, \widehat{\lambda}_1\right)$, then $T - \mu I$ satisfies a maximum principle according to Theorem 3.2. Let us choose λ and μ so that the second member increases with u, which implies that $(\lambda - \mu)u - f(u)$ is non negative whenever u is non negative. Then Proposition 3.4 holds in view of Theorem 3.2.

Note that the existence of a μ in $\left[2\sqrt{\delta} - \gamma, \widehat{\lambda}_1\right)$ such that $\lambda - f'(u) - \mu > 0$ for all $u > 0$ and all λ in $\left(\widehat{\lambda}_1 + f'(0), \widehat{\lambda}_1 + f'(\infty)\right)$ necessarily implies :

$$2\sqrt{\delta} - \gamma \leq \widehat{\lambda}_1 + f'(0) - f'(\infty). \bullet$$

We are able now to prove the main theorem of this section.

Theorem 3.5 : (Sufficient conditions.) Suppose (1.1) to (1.4), (3.3) to (3.6) are satisfied. Then there exists a unique non trivial positive solution $u(\lambda)$ to (3.1).

Proof :

a). Existence : By Proposition 3.1, for all μ in $\rho(T)$, T_μ is compact. And Theorem 3.3 is a maximum principle for (3.1). So that we may apply the method of sub and supersolutions to show the existence of positive solutions, as in [10]. We seek a subsolution u_o which is proportional to the first eigenfunction of H_q : $u_o \equiv c\varphi_1$. We have :

$$Tu_o + f(u_o) - \lambda u_o = c\varphi_1 \left(\widehat{\lambda}_1 + f'(0) - \lambda + \frac{f(c\varphi_1)}{c\varphi_1} - f'(0) \right)$$

which is negative for c small enough; thus u_o is a subsolution.

We compute now u^o, supersolution of our problem; by (3.3), $\widehat{\lambda}_1$ is a strictly increasing function of λ_1. Moreover, by definition of :

$$\lambda_1 = \inf_{u \in V_q^o(\Omega)} \frac{\displaystyle\int_\Omega |\nabla u|^2 + qu^2}{\displaystyle\int_\Omega u^2},$$

λ_1 does not increase as Ω increases; this holds since $V_q^o(\Omega') \supset V_q^o(\Omega)$ whenever $\Omega' \supset \Omega$. Hence, for $\lambda < \widehat{\lambda}_1 + f'(\infty)$, there exists an open set $\Omega' \supset \Omega$ such that the first eigenvalue λ' of T (defined on Ω' with Dirichlet boundary conditions) satisfies $\lambda - f'(\infty) < \lambda' < \widehat{\lambda}_1$. If we denote by ψ_1 the associated eigenfunction which is (strictly) positive in Ω' and hence in Ω, we can choose $d > 0$ large enough so that $u^o = d\psi_1$ is a supersolution which satisfies $u^o > u_o$ ae Ω.

Besides, under Assumptions (1.3) and (1.4), f(u) is in $L^2(\Omega)$ for all u in $L^2(\Omega)$. Hence, for all u in $L^2(\Omega)$, there exists a unique w in $D(T)$ solution of the variational problem :

$$\begin{cases} (T-\mu I)w = \lambda u - f(u) - \mu u \\ w/_{\partial\Omega} = 0 \ , \ \ w(x) \xrightarrow[|x|\to+\infty]{} 0 \end{cases}$$

Thus, we define a mapping : $L : L^2(\Omega) \longrightarrow L^2(\Omega)$, $u \longmapsto w = Lu$.

Lemma 3.6 : L is a nonlinear, continuous, compact operator in $L^2(\Omega)$. Moreover, L is order preserving in the interval $[u_o, u^o] = \{u \in L^2(\Omega),\ u_o(x) \leq u(x) \leq u^o(x)\ ae\ \Omega\}$ and $L([u_o, u^o]) \subset [u_o, u^o]$.

Proof : $L := (T-\mu I)^{-1}((\lambda-\mu)I - f(.))$, so that it is continuous and compact by Hypothesis (1.2) and Proposition 3.1. Assume that u_1 and u_2, with $u_1 \not\equiv u_2$, satisfie $u_o \leq u_1 \leq u_2 \leq u^o$ ae Ω and $u_1 \not\equiv u_2$, with :

$$\begin{cases} (T-\mu I)Lu_1 = \lambda u_1 - f(u_1) - \mu u_1 \\ Lu_1/_{\partial\Omega} = 0 \end{cases} \text{and} \begin{cases} (T-\mu I)Lu_2 = \lambda u_2 - f(u_2) - \mu u_2 \\ Lu_2/_{\partial\Omega} = 0 \end{cases}$$

Substracting the two equalities, we get :

$$\begin{cases} (T-\mu I)(Lu_2 - Lu_1) = (\lambda-\mu)(u_2 - u_1) - (f(u_2) - f(u_1)) \\ (Lu_2 - Lu_1)/_{\partial\Omega} = 0 \end{cases}$$

By Theorem 3.2, $Lu_2 - Lu_1 > 0$ ae Ω. Similarly, we can prove that $u_o < Lu_o$ and $u^o > Lu^o$ ae Ω. Finally, we obtain that for all u in $[u_o, u^o]$, $u_o \leq Lu_o \leq Lu \leq Lu^o \leq u^o$ ae Ω, and $L([u_o, u^o]) \subset [u_o, u^o]$.

L is a compact operator of $L^2(\Omega)$; $[u_o, u^o]$ is a closed, convex, bounded set with respect to the $L^2(\Omega)$ norm. By Schauder's Fixed Point Theorem, there exists at least one fixed point for L in $[u_o, u^o]$, which is a solution to (3.1).

b). Uniqueness : Considering again u_o and u^o, we construct two sequences as follows : u_o , $u_1 = Lu_o$,..., $u_n = Lu_{n-1}$ and $v_o = u_o$, $v_1 = Lv_o$,..., $v_n = Lv_{n-1}$ such that :

$$u_o \leq u_1 \leq ... \leq u_n \leq ... \leq v_n \leq ... \leq v_1 \leq v_o = u^o \qquad ae \qquad \Omega \tag{3.7}$$

By use of Remark 1 applied to φ_1 and ψ_1, we deduce that u_o and u^o belong to $C(\Omega) \cap L^2(\Omega)$. By Lebesgue's Dominated Convergence Theorem, there exist two functions $\underline{u}$ and $\overline{u}$ in $L^\infty(\Omega) \cap L^2(\Omega)$ such that $\lim_{n\to+\infty} u_n = \underline{u}$ and $\lim_{n\to+\infty} v_n = \overline{u}$ in $L^2(\Omega)$ norm. Since L is continuous, $\lim_{n\to+\infty} Lu_{n-1} = \lim_{n\to+\infty} u_n = \underline{u} = L\underline{u}$, so that $\underline{u}$ is a fixed point of L. Similarly, $L\overline{u} = \overline{u}$. Using (3.7), we obtain : $\underline{u} \leq \overline{u}$. Now, it can be checked that every solution z of (3.1) is such that $u_o \leq \underline{u} \leq z \leq \overline{u} \leq u^o$ ae Ω, so that $\underline{u}$ and $\overline{u}$ are, respectively, the minimal and maximal solutions of (3.1).

Let us assume now that z and w are two positive non trivial solutions of (3.1). By Proposition 3.4, $z > 0$ and $w > 0$ ae Ω. We can always construct a supersolution $u^o = d\psi_1$ with d large enough so that $u^o > z$ and $u^o > w$ ae Ω. Consequently there exists a maximal solution constructed as above, again denoted by $\overline{u}$ such that $\overline{u} \geq z$ and $\overline{u} \geq w$ ae Ω. First, we have :

$$Tz + f(z) = \lambda z.$$

Multiplying by $\overline{u}$ and integrating over Ω, we obtain :

$$(Tz, \overline{u})_{L^2(\Omega)} + \int_\Omega \frac{f(z)}{z} z\overline{u} = \lambda \int_\Omega z\overline{u} \, .$$

Similarly, $$(T\overline{u}, z)_{L^2(\Omega)} + \int_\Omega \frac{f(\overline{u})}{\overline{u}} \overline{u}z = \lambda \int_\Omega \overline{u}z \, .$$

By substraction : $$\int_\Omega \left(\frac{f(z)}{z} - \frac{f(\overline{u})}{\overline{u}} \right) z\overline{u} = 0 \, ,$$

and it follows from (1.3) that $\overline{u} \equiv z$. An analogous calculation shows that $\overline{u} \equiv w$, so that $z \equiv w$ ae Ω. •

4. Local Inversion and Continuation.

In this section, we still consider the asymptotically linear and convex Problem $(\mathcal{P})$, reduced to (3.1), under the assumptions of Theorem 3.5. We wish to show, as in Hernandez [10], that it is possible to obtain the existence and uniqueness of a non trivial and non degenerate positive solution through a local inversion theorem , and a continuation argument. We somehow improve Theorem 3.5, since we show here that there exists a (strictly) positive solution which is non degenerate.

Theorem 4.1 : (Theorem 3.5 revisited)

Assume that the hypotheses in Theorem 3.5 are satisfied, together with condition (4.1) below. Then, there exists a unique non trivial and non degenerate positive solution $u(\lambda)$. Moreover, the map $\lambda \longmapsto u(\lambda)$ from $\left(\widehat{\lambda}_1 + f'(0), \widehat{\lambda}_1 + f'(\infty)\right)$ into $L^2(\Omega)$ is C^2, and $\lim_{\lambda \nearrow \lambda_1 + f'(\infty)} ||u(\lambda)||_{L^2(\Omega)} = +\infty$.

Proof : The proof of Theorem 4.1 requires several intermediate lemmas. First, let us define the functional $F : I \times D(T) \longrightarrow L^2(\Omega)$, where $I = \left(\widehat{\lambda}_1 + f'(0), \widehat{\lambda}_1 + f'(\infty)\right)$ and $F(\lambda, u) = Tu + f(u) - \lambda u$, so that $u \in D(T)$ is a solution of (3.1) if and only if $F(\lambda, u) = 0$. F is C^2 on $I \times D(T)$, $F(\lambda, 0) = 0$, $\forall \lambda \in I$ and :

$$F_u(\mu, v)w = Tw + f'(v)w - \mu w$$

$$F_{\lambda u}(\mu, v)w = -w$$

Then
$$F_u(\widehat{\lambda}_1 + f'(0), 0)w = Tw - \widehat{\lambda}_1 w$$

and
$$Ker(F_u(\widehat{\lambda}_1 + f'(0), 0) = Span\varphi_1 \ .$$

Let us take $Z = R(F_u(\widehat{\lambda}_1 + f'(0), 0)) = (Span\varphi_1)^\perp$ as a supplementary subspace of $Ker(F_u(\widehat{\lambda}_1 + f'(0), 0)$ in $L^2(\Omega)$.

Then, $F_{\lambda u}(\widehat{\lambda}_1 + f'(0), 0)\varphi_1 = -\varphi_1 \notin Z$.

By the Local Inversion Theorem due to Crandall and Rabinowitz [4] (see also [18], Theorem 13.5, p. 173) there exists an interval J containing the origin and two C^1 functions $\lambda : J \longrightarrow R$ and $\psi : J \longrightarrow Z$ such that $\lambda(0) = \widehat{\lambda}_1 + f'(0)$, $\psi(0) = 0$ and $u(s) = s\varphi_1 + s\psi(s)$ implies $F(\lambda(s), u(s)) = 0$. Moreover, $F^{-1}(0)$ (in a neighborhood of $(\widehat{\lambda}_1 + f'(0), 0)$) consists only in the curves $u = 0$ and $(\lambda(s), u(s))$, s in J. To extend to the right the branch of positive solutions given by the Local Inversion Theorem, we need supplementary results :

Lemma 4.2 : Suppose that hypotheses of Theorem 3.5 hold, and that :

$$f'(\infty) - f'(0) \leq \widehat{\lambda}_2 - \widehat{\lambda}_1 \tag{4.1}$$

If v is a solution of (3.1), $v > 0$ ae Ω, then v is non degenerate (the second partial derivative $F_u(\lambda, v)$ is an isomorphism).

To prove Lemma 4.2, we need the following result, which is an extension of the classical ones of spectral theory (see e.g. [3]) :

Lemma 4.3 : Let ρ be in $C(\overline{\Omega})$, bounded, and $\rho \geq 0$ in $\overline{\Omega}$. Then there exists an increasing sequence of real numbers $\alpha_k(\rho)$ such that $\lim\limits_{n \to +\infty} \alpha_k(\rho) = +\infty$ and :

$$Tw + \rho(x)w = \alpha w \ , \ w/_{\partial\Omega} = 0 \ , \ w(x) \xrightarrow[|x| \to +\infty]{} 0 \tag{4.2}$$

admits a non trivial solution if and only if there exists k in N such that $\alpha = \alpha_k(\rho)$. Moreover, if $\widehat{\rho}$ is in $C(\overline{\Omega})$ with $\rho(x) \leq \widehat{\rho}(x)$ and $\rho \not\equiv \widehat{\rho}$ in Ω, then $\alpha_k(\rho) < \alpha_k(\widehat{\rho})$, for all k.

Proof : Since ρ is bounded, $D((T+\rho)^{\frac{1}{2}}) = V_q^o(\Omega)$. Thus $T+\rho$ with Dirichlet boundary conditions defined into $L^2(\Omega)$ has compact inverse, and its eigenvalues are characterized by the usual Weyl-Courant Maximum-Minimum Principle :

$$\alpha_{k+1}(\rho) = \sup_{E_k \in \mathcal{G}_k} \inf_{u \in E_k^{\perp} \cap V_q^o(\Omega)} \frac{(Tu,u)_{L^2(\Omega)} + \int_\Omega \rho u^2}{\int_\Omega u^2}$$

where $\mathcal{G}_k$ is the set of all k-dimensional subspaces in $L^2(\Omega)$. The supremum and infinimum in the definition of $\alpha_{k+1}(\rho)$ are actually attained for $u = \phi_{k+1}$ and $E_k = Span\{\phi_i\}_{i=1,k}$, where $\{\phi_k\}_{k\in N}$ denotes the set of the eigenfunctions of problem (4.2). So that $\alpha_k(\rho)$ increases as ρ increases. •

Proof of Lemma 4.2 : Let ε be a given positive number such that $\varepsilon < f'(\infty) - f'(0)$. For λ in $\left(\widehat{\lambda}_1 + f'(0), \widehat{\lambda}_1 + f'(0) + \varepsilon\right]$, (3.5) is satisfied and we consider the equation :

$$F(\lambda, v) = 0 \tag{4.3}$$

which is nothing else than (3.1). Suppose $F_u(\lambda, v)$ is not an isomorphism, then there exists $w \not\equiv 0$ such that $w \in Ker\ F_u(\lambda, v)$, which can be written :

$$Tw + f'(v)w = \lambda w\ . \tag{4.4}$$

(4.4) and Lemma 4.3 implie :

$$\lambda \geq \widehat{\lambda}_1\left(f'(v)\right) > \widehat{\lambda}_1\left(\frac{f(v)}{v}\right) \tag{4.5}$$

since, by (1.3), $f'(v) > \left(\frac{f(v)}{v}\right) \quad \forall v > 0$.

By Lemma 4.3, there exists $j \geq 1$ such that $\lambda = \widehat{\lambda}_j\left(\frac{f(v)}{v}\right)$. If $j \geq 2$, by the same comparison argument, we have :

$$\lambda \geq \widehat{\lambda}_2\left(\frac{f(v)}{v}\right) > \widehat{\lambda}_2\left(f'(0)\right) = \widehat{\lambda}_2 + f'(0)\ .$$

This last inequality, combined with (3.5) implies that $\widehat{\lambda}_2 - \widehat{\lambda}_1 < f'(\infty) - f'(0)$, which contradicts (4.1). Thus $j = 1$ and $\lambda = \widehat{\lambda}_1\left(\frac{f(v)}{v}\right)$, which is again in contradiction with (4.5). Consequently, if v is a solution of (3.1), then $F_u(\lambda, v)$ is an isomorphism. •

By the Local Inversion Theorem, there exists a small branch of non trivial positive solutions bifurcating from $(\widehat{\lambda}_1 + f'(0), 0)$, defined for $\lambda > \widehat{\lambda}_1 + f'(0)$. This branch may be extended using Lemma 4.2. Indeed, if $F(\lambda, v) = 0$, where $v > 0$, the Implicit Function Theorem (see e.g. [18], Theorem 13.3, p. 170) yields a curve $(s, u(s))$ of solutions for an interval containing λ; the mapping $\lambda \longmapsto u(\lambda)$ is C^2.

We first notice that we can apply Corollary 18.4, p. 675 and Corollary 19.5, p. 677 by Amann in [1] in view of Proposition 2.2 :

Lemma 4.4 : Suppose that the hypotheses of Theorem 3.5 hold. Then $\widehat{\lambda}_1 + f'(0)$ (resp. $\widehat{\lambda}_1 + f'(\infty)$) is the only bifurcation point (resp. the only asymptotic bifurcation point) for the positive solutions of (3.1).

Lemma 4.5 : Let ε be a given positive number such that $\varepsilon < f'(\infty) - f'(0)$. For every λ in $\left(\widehat{\lambda}_1 + f'(0), \widehat{\lambda}_1 + f'(0) + \varepsilon\right]$, the solution $u(\lambda)$ of (3.1) is strictly positive ae Ω.

Proof : If $u(\lambda)$ is not strictly positive ae Ω, there exists a $\overline{\lambda}$, such that

$$\overline{\lambda} > \widehat{\lambda}_1 + f'(0) , \tag{4.6}$$

and a subset ω of Ω with meas$(\omega) \neq 0$ such that $u(\overline{\lambda})/_{\omega} \equiv 0$. By Proposition 3.4, $u(\overline{\lambda}) \equiv 0$ and $\lim\limits_{\lambda \nearrow \overline{\lambda}} u(\lambda) = 0$. By Lemma 4.4, $\overline{\lambda} = \widehat{\lambda}_1 + f'(0)$, which contradicts (4.6). •

We thus obtain the existence of a unique non trivial (strictly) positive solution $u(\lambda)$ for λ in $\left(\widehat{\lambda}_1 + f'(0), \widehat{\lambda}_1 + f'(\infty)\right)$. •

This local inversion method can be adapted to the study of $(\mathcal{P})$ when the strict convexity condition (1.3) is replaced by a strict concavity condition . In that case, as in the scalar case, the sub and supersolution method is no more available.

REFERENCES

[1] H. AMANN, *Fixed point equations and nonlinear eigenvalue problems in ordered Banach spaces*, SIAM Review, 18, (1976), pp. 620-709.

[2] K.J. BROWN, *Non trivial solutions of predator-prey systems with small diffusion*, Non Linear Analysis, 11, (1987), pp. 685-689.

[3] R. COURANT - D. HILBERT, *Methods of mathematical physics*, (1953), Interscience.

[4] M. CRANDALL - P.H. RABINOWITZ, *Bifurcation, perturbation of simple eigenvalues and linearized stability*, Archive for Rational Mechanics and Analysis, 52, (1973), pp. 161-180.

[5] E.B. DAVIES, *Heat kernels and spectral theory*, Cambridge University Press.

[6] D.E. EDMUNDS - W.D. EVANS, *Spectral theory and differential operators*, Oxford University Press, (1987).

[7] W. FARIS - B. SIMON, *Degenerate and non degenerate ground states for Schrödinger operators*, Duke Mathematical Journal, vol.42, 3, (1975), pp.559-567.

[8] D.G. DE FIGUEIREDO - E. MITIDIERI, *A maximum principle for an elliptic system and applications to semilinear problems*, SIAM Journal in Mathematical Analysis, 17, (1986), pp. 836-849.

[9] J. FLECKINGER, *Estimates of the number of eigenvalues for an operator of Schrödinger type*, Proceedings of the Royal Society of Edinburgh, 89A,(1981), pp. 355-361.

[10] J. HERNANDEZ, *Maximum principles and decoupling for positive solutions of reaction-diffusion systems*, To appear.

[11] T. KATO, *Perturbation theory for linear operators*, (1966), Springer Verlag.

[12] M. LAPIDUS, *Generalization of the Trotter-Lie formula*, Integral Equations and Operator Theory, 4/3, (1981), pp. 366-415.

[13] A. LAZER - P.J. MCKENNA, *On steady-state solutions of a system of reaction-diffusion equations from biology*, Non Linear Analysis, 6, (1982), pp. 523-530.

[14] M. REED - B. SIMON, *Methods of modern mathematical physics*, IV, Analysis of Operators, (1979), Academic Press.

[15] A. MANES - A.M. MICHELETTI, *Un 'estensione della teoria variazionale classica degli autovalori per operatori ellittici del secondo ordine*, Bolletino U.M.I., vol.7, (1973), pp. 285-301.

[16] F. ROTHE, *Global existence of branches of stationary solutions for a system of reaction-diffusion equations from biology*, Non Linear Analysis, 5, (1981), pp.487-498.

[17] B. SIMON, *Scrödinger semigroups*, Bulletin (New Series) of the AMS, vol. 7, 3, (1982), pp. 447-526.

[18] J. SMOLLER, *Shock waves and reaction diffusion equations*, (1983), Springer Verlag.

*A. Abakhti – Mchachti * , J. Fleckinger-Pellé*
Université Toulouse III.
Laboratoire d'Analyse Numérique.
UFR MIG. 118, route de Narbonne. 31062 Toulouse Cedex.

* : *Research supported by C.E.E. grant n° 038 644 H/1/400 11*

F ALABAU

A decoupling method for proving uniqueness theorems for electro-diffusion equations

1 Introduction.

We consider, in this paper, a generalized version of the electro-diffusion equations (see e.g. [16]), which models the transport of two families of charged "species" in a device under external forces. Assuming vanishing generation-recombination and constant mobilities, the scaled equations are ([9], [2])

$$
\begin{aligned}
\varepsilon\psi''(x) &= q_1 n_1(x) + q_2 n_2(x) - N(x) && (1)\\
n_1'(x) &= \alpha_1 q_1 n_1(x)\psi'(x) + q_1 J_1(x) \quad x \in \Omega && (2)\\
n_2'(x) &= \alpha_2 q_2 n_2(x)\psi'(x) + q_2 J_2(x) && (3)\\
J_1'(x) &= 0 && (4)\\
J_2'(x) &= 0 && (5)
\end{aligned}
$$

subject to the boundary conditions

$$\psi_{/\Gamma} = \psi_D(V)\ , \quad n_{i_{/\Gamma}} = n_D^i \text{ for } i = 1,2 \qquad (6)$$

where $\Omega = (a,b)$, Ω bounded, $\Gamma = \{a,b\}$, $\psi_D(V) = \{\psi_a, \psi_b\}$, $n_D^i = \{n_a^i, n_b^i\}$, and where $\psi_a - \psi_b$ depends linearly on the parameter V.

Here the scaled unknowns are ψ, the electrostatic potential, n_i the density of the ith species and J_i the current density of the ith species.

The parameter ε is a small, strictly positive number which depends in particular on the Debye length of the device. The numbers q_i can be positive, negative, or zero; while the α_i are strictly positive numbers.

The function N is the scaled doping profile (resp. fixed charge density) in the framework of semiconductor devices (resp. biological membranes). Nevertheless, for the sake of simplicity, we will refer to it as the doping profile and will specify its characteristics later in this paper.

The boundary conditions $\psi_D(V)$, n_D^i are given and may depend (as is the case for semiconductors) on the values of N on Γ. Since n_i (for $i = 1, 2$) is a density, we shall only consider strictly positive n_D^i. The external forces act on the device through the parameter V (the applied bias), which varies from "minus infinity" to "plus infinity", while the other parameters of this problem are kept fixed.

Remark 1.1: For the semiconductor devices $\alpha_1 = \alpha_2 = 1$, $q_1 = -q_2 = 1$ hold.

Many results which concern the existence of solutions for this system, even in the multi-dimensional case, have been proved and only a few questions concerning this point are still unsolved (see e.g [9] and the references therein).

By comparison there is a lack of theoretical results on the analysis of uniqueness or multiplicity of the solutions, or of the properties of the associated linearized system, even in the one-dimensional case. This is due to the complex dependence of the solutions on the applied bias and on the doping profile. Indeed, it is well-known that uniqueness, even in the one-dimensional case, does not hold under general hypotheses on N and V (see e.g [11], [14], [17]). But Mock [11] and Rubinstein [14] conjectured that uniqueness should hold for a doping profile which changes its sign less than three times through the device. We therefore focus our analysis on uniqueness problems and refer the reader to [9] and the references therein for existence results.

Mock [12] proved the first uniqueness theorem for the multi-dimensional semiconductor device equations, provided that the applied bias V is sufficiently small. More recently Brezzi et al. [6] used asymptotic methods to prove a uniqueness theorem for one-dimensional reverse-biased diodes, provided that a chosen singular perturbation parameter is sufficiently small. It is interesting to compare these two results. Mock's result cannot hold for large $|V|$ since it does not require any hypotheses on N. Brezzi's result is more specific, but it is the first result of the literature which requires hypotheses on N (N is assumed to have only one discontinuity and to be piecewise-constant), while it allows strong (but limited) reverse biases. Both results are based in a certain way on the "same philosophy", even if the involved techniques are very different from each other. Indeed, using the Slotboom variables (see [12]), one can separate out from Poisson's equation (1) the monotonic part with respect to ψ, while the nonmonotonic part is controlled by the "smallness" of the current densities (as V goes to 0 in [12] and as λ goes to 0 in [6], where λ is a small singular perturbation parameter). Unfortunately, since these methods are "quantitative" in essence they do not and indeed cannot provide uniqueness results which are uniform in V.

In this paper, we consider a completely different approach, which provides global and local uniqueness theorems (for different hypotheses on N) which are uniform in the bias V. It is based on a transformation, indeed a decoupling, of the linearized system, which has been introduced by the author in [3], [1], [5], and which is now generalized to the case of piecewise-constant doping profiles. It allows us to characterize the kernel of the linearized operator and to give an upper bound of its dimension, which depends on the number of junctions (i.e discontinuities) of the device.

The paper is organized as follows. In section 2 we give the decoupled linearized system and give an upper bound of the dimension of the kernel of the linearized operator. In section 3 we give the uniqueness theorems. The case of a vanishing doping profile is treated in subsection 3.1. In subsection 3.2 we prove a new local uniqueness theorem for sufficiently strong forward-biased symmetric diodes. Moreover, we prove asymptotic estimates for the total current as the applied bias V goes to plus infinity.

2 Decoupling of the linearized system

We set $X = H^2(\Omega) \times (H^1(\Omega))^2$, $Y = (L^2(\Omega))^2$, $W = (\psi, n_1, n_2, J_1, J_2)$.

We define a map $\mathcal{F}$ from $\mathbb{R} \times X \times \mathbb{R}^2$ to $Y \times \mathbb{R}^6$ by

$$\begin{aligned}\mathcal{F}(V,W) = \\ &\Big(\varepsilon\psi'' - (q_1 n_1 + q_2 n_2 - N), n_1' - \alpha_1 q_1 n_1 \psi' - q_1 J_1, n_2' - \alpha_2 q_2 n_2 \psi' - q_2 J_2, \psi(a) - \psi_a, \\ &\psi(b) - \psi_b, n_1(a) - n_a^1, n_1(b) - n_b^1, n_2(a) - n_a^2, n_2(b) - n_b^2\Big) .\end{aligned}$$

The nonlinear system (1)–(6) can be written in the abstract form

$$\mathcal{F}(V,W) = 0 . \tag{7}$$

$\mathcal{F}$ is clearly a continuously Fréchet differentiable map. Moreover it is easy to check that the linearized operator $D_W\mathcal{F}(V,W)$ is a Fredholm operator of index 0 (see [15] for the proof in the multi-dimensional case).

Let us assume now that N is piecewise-constant and has $r-1$ junctions, i.e.:

Hypothesis (H1)

$$N = N_k \text{ on } \overline{\Omega}_k = [x_k, x_{k+1}] \text{ for } k = 0, \ldots, r-1$$

$$\text{where } x_0 = a, x_r = b, x_k < x_{k+1}, \overline{\Omega} = \bigcup_{k=0}^{r-1} \overline{\Omega}_k .$$

We now consider the linearized homogeneous equation

$$D_W\mathcal{F}(V,W).(u, u_1, u_2, j_1, j_2) = 0 \;, \quad (u, u_1, u_2, j_1, j_2) \in X \times \mathbb{R}^2 \tag{8}$$

In [1], [5] the decoupling method is introduced under more restrictive assumptions on N. Theorem 2.1 gives an extension of this method to the case of piecewise-constant doping profiles.

Theorem 2.1 (Characterization of the kernel) .

Assume that $q_1\alpha_1 \neq q_2\alpha_2$, $q_1q_2 \neq 0$ *and that (H1) holds. Let* $(\psi, n_1, n_2, J_1, J_2)$ *and* (u, u_1, u_2, j_1, j_2) *be respectively solutions of (1)–(6) and (8). Then*

(i) *There exist continuous functions* α, β, γ *on* $\overline{\Omega}$, *which depend only on* $(\psi, n_1, n_2, J_1, J_2)$ *and* V, *such that solving (8) is equivalent to finding* u *and constants* $(i, E, F_0, \ldots, F_{r-1})$ *which satisfy*

$$\varepsilon u^{(3)}(x)+\varepsilon\alpha(x)u''(x)+\beta(x)u'(x)+N_k\gamma(x)u(x) = i+(Ex+F_k)\gamma(x)\ , \quad x \in \Omega_k, \quad (9)$$

for $k = 0, \ldots, r-1$

subject to the conditions

$$u(a) = u(b) = u''(a) = u''(b) = 0\ , \quad [u]_k = [u']_k = 0 \quad k = 1, \ldots, r-1 \quad (10)$$

$$\varepsilon\psi'(a)u'(a) = Ea + F_0\ , \quad \varepsilon\psi'(b)u'(b) = Eb + F_{r-1} \quad (11)$$

$$\alpha_i q_i \int_a^b n_i e^{-\alpha_i q_i \psi} u' dt + q_i j_i \int_a^b e^{-\alpha_i q_i \psi} dt = 0\ , \quad i = 1, 2 \quad (12)$$

$$\varepsilon\psi'(x_k)u'(x_k)+N_k u(x_k) = \sum_{i=1}^{2} \frac{e^{\alpha_i q_i \psi(x_k)}}{\alpha_i} \left(\alpha_i q_i \int_a^{x_k} n_i e^{-\alpha_i q_i \psi} u' dt + q_i j_i \int_a^{x_k} e^{-\alpha_i q_i \psi} dt\right) \quad (13)$$

$$+Ex + F_k\ , \quad k = 1, \ldots, r-2$$

$$\varepsilon u''(x_k^+) = \sum_{i=1}^{2} q_i e^{\alpha_i q_i \psi(x_k)} \left(\alpha_i q_i \int_a^{x_k} n_i e^{-\alpha_i q_i \psi} u' dt + q_i j_i \int_a^{x_k} e^{-\alpha_i q_i \psi} dt\right)\ , \quad k = 1, \ldots, r-2 \quad (14)$$

where $[u]_k = u(x_k^+) - u(x_k^-)$, *and* (j_1, j_2) *is solution of*

$$i = q_1^2 j_1 + q_2^2 j_2 \quad (15)$$

$$E = -\left(\frac{q_1 j_1}{\alpha_1} + \frac{q_2 j_2}{\alpha_2}\right) \quad (16)$$

(ii) $dim(Ker(D_W\mathcal{F}(V, W))) \leq 4r + 2$

Proof The abstract equation (8) reads as

$$\varepsilon u''(x) = q_1 u_1(x) + q_2 u_2(x) \quad (17)$$

$$u_1'(x) - \alpha_1 q_1 u_1(x)\psi'(x) = \alpha_1 q_1 n_1(x) u'(x) + q_1 j_1 \quad (18)$$

$$u_2'(x) - \alpha_2 q_2 u_2(x)\psi'(x) = \alpha_2 q_2 n_2(x) u'(x) + q_2 j_2 \quad (19)$$

$$u(a) = u(b) = 0 \quad (20)$$

$$u_i(a) = u_i(b) = 0 \qquad i = 1, 2 \; . \tag{21}$$

We multiply (17) by ψ' and use (18), (19) and (1) to derive

$$\varepsilon(u'\psi')' + Nu = \left(\frac{u_1}{\alpha_1} + \frac{u_2}{\alpha_2}\right)' + E \tag{22}$$

where E is given by (16).

We integrate (22) in each Ω_k. Hence we obtain

$$-\frac{\varepsilon}{q_1\alpha_1}u'' + \varepsilon\psi' u' + N_k u = \left(\frac{q_1\alpha_1 - q_2\alpha_2}{q_1\alpha_1}\right) u_2 + Ex + F_k \tag{23}$$

where the F_k are unknown constants.

We multiply this last equation by $e^{-\alpha_2 q_2 \psi}$ and differentiate the resulting equation. By using (19) we deduce that (9) holds with

$\alpha(x) = -(\alpha_1 q_1 + \alpha_2 q_2)\psi'(x)$,

$\beta(x) = -(\alpha_1 q_1^2 n_1(x) + \alpha_2 q_2^2 n_2(x) - \varepsilon\alpha_1 q_1 \alpha_2 q_2 \psi'^2(x))$,

$\gamma(x) = \alpha_1 q_1 \alpha_2 q_2 \psi'(x)$.

The conditions (12)–(14) are derived from (17)–(19) and (23) This concludes the proof of (i).

By considering a fundamental set of solutions (see [8]) of the third order homogeneous equation defined by the left-hand side of (9) we prove that solving (8) is equivalent to resolving a finite dimension system of order $4r + 2$. This proves (ii).

Remark 2.1 The use of the terminology *decoupled* is justified by the fact that the coupling between u and (u_1, u_2) has been suppressed, while u still satisfies an ordinary differential equation (namely (9)).

Remark 2.2 The same technique can be used to decouple the equations satisfied by the difference between two solutions of (7). We will see in section 3 how this provides global uniqueness theorems, under more restrictive assumptions on N.

Remark 2.3 From theorem 2.1 we conclude that bifurcation may only occur when the determinant of a certain finite system of order $4r + 2$ is zero.

Remark 2.4 The cases where $q_1\alpha_1 = q_2\alpha_2$ or $q_1 q_2 = 0$ are completely solved in [1]. For these cases the decoupling method is not necessary and global uniqueness theorems are proved for any bounded functions N, and any applied bias V.

3 Application of the decoupling method — Uniqueness theorems

Considering now the third order equation (9) obtained by the decoupling method, we remark that when N is identically zero, this equation becomes a second order equation in u' to which a maximum principle can be applied (see [1]). It seems therefore "natural" to distinguish the case of a vanishing doping profile (which has a physical meaning, at least for the case of biological membranes) from other cases.

We assume throughout this paper that the following holds: $q_1\alpha_1 \neq q_2\alpha_2$ and $q_1q_2 \neq 0$.

This assumption is not a restriction since the cases $q_1\alpha_1 = q_2\alpha_2$ or $q_1q_2 = 0$ are completely solved in [1] (see remark 2.4).

3.1 Global uniqueness theorems in case of a vanishing doping profile

We denote by

$$\Lambda(V) = (\psi_a, \psi_b, n_a^1, n_b^1, n_a^2, n_b^2) \tag{24}$$

the set of boundary conditions satisfied by ψ, n_1 and n_2 where

$$V = \psi_a - \psi_b \tag{25}$$

is the applied bias. Of course, the admissible boundary conditions satisfy

$$n_x^i > 0 \quad \text{for } x = a, b \text{ and } i = 1, 2\ . \tag{26}$$

Theorem 3.1 *Assume that $N \equiv 0$ and that $q_1q_2 < 0$. Then for any set of boundary conditions $\Lambda(V)$ in $\mathbb{R}^6$ such that there exists one solution $(\psi, n_1, n_2, J_1, J_2)$ of (1)–(6) for which*

$$\psi' \textit{ keeps a constant sign in } \Omega\ , \tag{27}$$

the system (1)–(6) has a unique solution in $(H^1(\Omega))^3 \times \mathbb{R}^2$.

The proof follows the same lines as that of theorems 2.1 and 2.2 of [1], where the results were stated in the case $q_1 = -q_2 = 1$, $\alpha_1 = \alpha_2 = 1$.

The existence of sets of boundary conditions such that (27) holds is guaranteed by the following two corollaries.

Corollary 3.2 *Assume that $N \equiv 0$ and that $q_1q_2 < 0$. Then if $\Lambda(V)$ is such that*

$$0 \le q_1n_a^1 + q_2n_a^2\ , \quad q_1n_b^1 + q_2n_b^2 \le 0\ . \tag{28}$$

Then for every bias V satisfying

$$\max\left(\frac{1}{\alpha_1 q_1}\ln\left(\frac{n_a^1}{n_b^1}\right), \frac{1}{\alpha_2 q_2}\ln\left(\frac{n_a^2}{n_b^2}\right)\right) \le V\ , \tag{29}$$

the system (1)–(6) has a unique solution in $(H^1(\Omega))^3 \times \mathbb{R}^2$.

Proof We differentiate (1) and obtain

$$\varepsilon\psi^{(3)} = \left(\alpha_1 q_1^2 n_1 + \alpha_2 q_2^2 n_2\right)\psi' + q_1^2 J_1 + q_2^2 J_2\ . \tag{30}$$

From (28) we deduce that $0 \le \psi''(a)$, $0 \ge \psi''(b)$. Moreover (29) implies that $0 \le J_1$, $0 \le J_2$.

Therefore the maximum principle [13] applied to (30) yields $\psi' \le 0$. We conclude the proof by using theorem (3.1).

Corollary 3.3 *Assume that* $N \equiv 0$ *and that* $q_1 q_2 < 0$. *Then if* $\Lambda(V)$ *is such that*

$$q_1 n_a^1 + q_2 n_a^2 \le 0\ , \quad 0 \le q_1 n_b^1 + q_2 n_b^2\ . \tag{31}$$

Then for every bias V satisfying

$$V \le \min\left(\frac{1}{\alpha_1 q_1}\ln\left(\frac{n_a^1}{n_b^1}\right), \frac{1}{\alpha_2 q_2}\ln\left(\frac{n_a^2}{n_b^2}\right)\right)\ , \tag{32}$$

the system (1)–(6) has a unique solution in $(H^1(\Omega))^3 \times \mathbb{R}^2$.

The proof is similar to that of Corollary (3.2), except that (31) and (32) imply $0 \le \psi'$.

Remark 3.1 These theorems are, as far as we know, the *first uniqueness results which are uniform in* V.

3.2 The case of a nonvanishing doping profile

It is well-known that uniqueness of the solutions of (1)–(6) does not hold under general assumptions on N (see e.g. [11] for a numerical example with $q_1 = -q_2 = 1$, $\alpha_1 = \alpha_2 = 1$).

In this subsection we will restrict our analysis to the case of semiconductor devices. Hence we assume that $q_1 = -q_2 = 1$, $\alpha_1 = \alpha_2 = 1$. In this case n_1 (resp. n_2) represents a scaled density of electrons (resp. holes). For the sake of clarity, we use the standard notations $n_1 = n$, $n_2 = p$, $J_1 = J_n$, $J_2 = J_p$.

We now consider the case of a symmetric diode represented by the interval $\Omega = (-1,1)$. This means that we assume

$$N(x) = -N(-x) = 1 \quad \text{for } 0 < x \leq 1 \; , \tag{33}$$

$$V(-1) = -V(1) = V \; , \tag{34}$$

where $V(x)$ stands for the applied bias at the contact x.

The boundary conditions associated to (1)–(5) are

$$\psi(x) = \ln\left(\frac{N(x) + \sqrt{N^2(x) + 4\delta^4}}{2\delta^2}\right) + V(x) = \psi_x(V) \; , \quad x = -1, 1 \; , \tag{35}$$

$$n(x) = \frac{N(x) + \sqrt{N^2(x) + 4\delta^4}}{2} = n_x \; , \quad x = -1, 1 \; , \tag{36}$$

$$p(x) = \frac{-N(x) + \sqrt{N^2(x) + 4\delta^4}}{2} = p_x \; , \quad x = -1, 1 \; , \tag{37}$$

where δ is a strictly positive number.

Remark 3.2 The assumption (34) is not a restriction, since the potential ψ is defined up to an additive constant. The assumption (33) is, of course, restrictive but it implies symmetry properties for the system (1)–(5) which have already been used in different frameworks (e.g. singular perturbation analysis, see e.g. [10], [4]) and which allow us to get uniqueness results in the present context.

Under these assumptions there exist symmetric solutions (ψ, n, p, J_n, J_p) of (1)–(5) subject to (35)–(37), i.e. solutions which satisfy

$$\psi(-x) = -\psi(x) \; , n(-x) = p(x) \; , p(-x) = n(x) \; , \quad 0 \leq x \leq 1 \; , \quad J_n = J_p \; . \tag{38}$$

In this case the system (1)–(5) subject to (35)–(37) reduces to the following nonlinear system, in Ω

$$\varepsilon\psi'' = n - p - 1 \tag{39}$$

$$n' = n\psi' + I/2 \tag{40}$$

$$p' = -p\psi' - I/2 \tag{41}$$

$$\psi(0) = 0, \; \psi(1) = \psi_1(V), \; n(0) = p(0), \; n(1) = n_1, \; p(1) = p_1 \; , \tag{42}$$

where I is the total current flowing through the device.

The purpose of this subsection is to show that our method reveals general properties of the electro-diffusion equations, even for the case of nonvanishing doping profiles. Nevertheless fundamental difficulties arise from the fact that the doping profile N is not zero.

An example of these difficulties can be easily illustrated by considering the third order equation (9):

One of the main step of our method (see [1]) is to apply the maximum principle to u' in (9). Can we use the same type of argument when N is not zero (so that (9) is no longer a second order equation in u')?

In [5] we consider (39)–(42) and prove that the local uniqueness of the solutions holds for any reverse bias V (i.e. for any $V \leq 0$)).

We now analyze the more complex case of forward biased diodes (i.e. $V > 0$), and prove a new uniqueness theorem.

Let (ψ, n, p, I) be a solution of (39)–(42) and let (u, w, z, i) be a solution of the corresponding homogeneous linearized system at (ψ, n, p, I). We apply the decoupling method introduced in section 2 and consider the third order linear differential operator $\mathcal{L}$ defined by the left-hand side of (9). Under the symmetry assumptions considered in this subsection, $\mathcal{L}$ reduces to

$$\mathcal{L}v = \varepsilon v^{(3)} + \beta v' + \gamma v \ , \tag{43}$$

where β and γ are functions which depend on V. Moreover the decoupling method leads to the "decoupled" system (see [5] for more details)

$$\mathcal{L}u = i - F\psi' \ , \tag{44}$$

$$w' - w\psi' = nu' + i/2 \ , \tag{45}$$

$$z' + z\psi' = -pu' - i/2 \ , \tag{46}$$

$$u(0) = u(1) = 0 \ , \quad w(0) = z(0) \ , \quad w(1) = z(1) = 0 \ , \tag{47}$$

$$\varepsilon u'(1)\psi'(1) = F \ , \quad 2z(0) = \varepsilon u'(0)\psi'(0) - \varepsilon u'(1)\psi'(1) \ . \tag{48}$$

We set

$$V_1(\delta) = \ln\left(\frac{1 + \sqrt{1 + 4\delta^4}}{2\delta^2}\right) . \tag{49}$$

Theorem 3.4 *Let $V_1(\delta)$ be defined by (49). There exists $V^\star$, $V^\star > V_1(\delta)$, such that for $V \geq V^\star$, every solution (ψ, n, p, I) of the system (39)–(42) is locally unique in $\left(H^1(\Omega)\right)^3 \times \mathbb{R}$.*

The proof requires the following two lemmas, whose proofs are deferred to the end of this subsection.

Lemma 3.5 *There exists $V^\star$, $V^\star > V_1(\delta)$, such that for $V \geq V^\star$ every solution (ψ, n, p, I) of the system (39)–(42) satisfies*

$$\psi' < 0 \ . \tag{50}$$

Lemma 3.6 *Let $\mathcal{L}$ be defined by (43) and let $V^\star$ be as in lemma 3.5. Then for $V \geq V^\star$, $\mathcal{L}$ satisfies the following property*

$$v(0) = v(1) = 0 \ , \quad v'(0) \geq 0 \ , \quad v'(1) \geq 0 \ , \quad \text{and } \mathcal{L}v \leq 0 \Rightarrow v = 0 \ .$$

Proof of theorem 3.4 We proceed as in [5]. Let $V \geq V^{\star}$. We assume, without loss of generality that $i \leq 0$.

Assume first that $u'(0) \geq 0$ and $u'(1) \geq 0$. Since (50) holds we deduce that $\mathcal{L}u \leq 0$. Lemma 3.6 implies that $u = 0$; $(w, z, i) = 0$ follows at once.

Hence $u'(0) < 0$ or $u'(1) < 0$ hold. Following now the proof of lemma 3.3 in [5] we conclude that this case is impossible. Thanks to the implicit function theorem ([7]) theorem 3.4 is proved.

The proof of lemma 3.5 requires the following proposition.

Proposition 3.7 *Assume that $V > V_1(\delta)$. Let (ψ, n, p, I) be any solution of (39)–(42). Then the following property holds:*

If

$$0 \leq \psi'(0) \tag{51}$$

then there exist constants $C_1 > 0$, $C_2 > 0$ which are independent on V such that

$$(n+p)(x) \leq C_1 \quad \forall x \in [0,1] \ , \quad 0 < I \leq C_2 \ . \tag{52}$$

Proof We differentiate twice (39) and obtain

$$\varepsilon\psi^{(4)} = S\psi'' + \psi'^2 \ , \tag{53}$$

where

$$S = n + p + \varepsilon\psi'^2 > 0 \ . \tag{54}$$

Thanks to the maximum principle applied to ψ'', we deduce easily that

$$\psi'' \leq 0 \ . \tag{55}$$

Therefore we obtain

$$\psi'(1) \leq \psi'(x) \leq \psi'(0) \quad \forall x \in [0,1] \ . \tag{56}$$

Since $V > V_1(\delta)$ and since (51) is assumed to hold, there exists a unique $x^{\star}$ in $[0,1)$ such that

$$\psi'(x^{\star}) = 0 \ . \tag{57}$$

Moreover a straightforward application of the maximum principle to $n - p$, together with (55) and (57) lead to

$$0 \leq n - p \leq 1 \ , \quad |\psi'| \leq 1/\varepsilon \ . \tag{58}$$

From equation

$$(n+p)' = (n-p)\psi' \tag{59}$$

we deduce that the first inequality in (52) holds with $C_1 = n_1 + p_1 + 1/\varepsilon$. We set $r = \sqrt{C_1/\varepsilon}$. We claim that the second inequality in (52) holds with $C_2 = C_1 cosh(r)/\varepsilon r sinh(r)$. Suppose, on the contrary, that we have

$$C_2 < I\ . \tag{60}$$

It is easy to check that ψ' is a solution of $\varepsilon w'' = (n+p)w + I$, $w'(0) = -1/\varepsilon$, $w'(1) = 0$ and that if (60) holds then $\psi' < 0$, which contradicts (51). Moreover $V > 0 \Rightarrow I > 0$. This concludes the proof.

Proof of lemma 3.5 Assume first that (51) holds with $\psi'(0) \neq 0$. Since (59) and (39) hold, there exists a constant $\alpha(V)$ such that

$$(n+p)(x) = \varepsilon\frac{\psi'^2(x)}{2} + \psi(x) + \alpha(V) \quad \forall x \in [0,1]\ . \tag{61}$$

We set $x = 1$ in (61) and deduce from the second inequality in (58) (see prop. 3.7) that

$$\lim_{V\to+\infty} \alpha(V) = \infty\ .$$

We now set $x = x^\star$ (where $x^\star$ is the solution of (57)) in (61). From the first inequality of (52) we get

$$\lim_{V\to+\infty} \psi(x^\star) = -\infty\ , \quad \text{and} \tag{62}$$

$$(n-p)' = (n+p)\psi' + I \leq C_1\psi' + I \quad \text{on } [0,x^\star]\ .$$

We integrate the last inequality from 0 to $x^\star$ and get $(n-p)(x^\star) - C_1\psi(x^\star) \leq I$, which, together with (62), implies that $\lim_{V\to+\infty} I = +\infty$. This contradicts the second inequality of (52) if V is sufficiently large.

Assume now that $\psi'(0) = 0$. We set successively $x = 0$, and $x = 1$ in (61), and get $\lim_{V\to+\infty} n(0)+p(0) = +\infty$ which contradicts the first inequality of (52) if V is sufficiently large. Thus lemma 3.5 is proved.

Proof of lemma 3.6 Assume that $V \geq V^\star$, where $V^\star$ is as in lemma 3.5. Let v be such that

$$v(0) = v(1) = 0\ , \quad v'(0) \geq 0\ , \quad v'(1) \geq 0\ , \quad \text{and } \mathcal{L}v \leq 0\ .$$

Equation (53) reads as $\mathcal{L}\psi' = 0$. Since for $V \geq V^\star$ (50) holds, we can set

$$y = (v/\psi')' \tag{63}$$

Therefore $y(0) \leq 0$, $y(1) \leq 0$ and $\mathcal{L}v = \mathcal{M}y \leq 0$, where $\mathcal{M}$ is the second order linear differential operator

$$\mathcal{M}y = \varepsilon\psi' y'' + 3\varepsilon\psi'' y' + (3\varepsilon\psi^{(3)} - S\psi')y \ . \tag{64}$$

As in lemma 3.2 of [5] we have

$$\mathcal{M}\left((\psi')^{-2}\right) = I(\psi')^{-2} - \varepsilon\psi' \ . \tag{65}$$

Since (50) holds for $V \geq V^\star$ and since $I > 0$, we deduce that $\mathcal{M}\left((\psi')^{-2}\right) > 0$. This proves that $\mathcal{M}$ satisfies the generalized maximum principle (see [13]). Therefore we deduce that $y \leq 0$; $v = 0$ follows at once. This concludes the proof.

We now prove, for the sake of completeness, asymptotic estimates for I.

Lemma 3.8 (Asymptotic estimates for the total current I) .

For $V \geq V^\star$ every solution (ψ, n, p, I) of the system (39)–(42) satisfies the following estimate

$$k_1(\delta) + Vk_2(\delta) \leq I \leq k_3(\delta) + Vk_4(\delta) + V^2 \tag{66}$$

where $k_i(\delta)$ are constants which depend only on δ, and where $k_2(\delta) > 0$.

Proof Equation (50) holds for $V \geq V^\star$. This, together with the first inequality in (58) imply $\psi' \leq (n-p)\psi'$. We integrate this inequality from 0 to 1 and get $(n+p)(0) \leq c(\delta)+V$, where $c(\delta)$ is a constant which depends on δ only.
Since (59) and (50) hold, we obtain $(n-p)' = (n+p)\psi' + I \leq (n_1+p_1)\psi' + I$. Integration of this last inequality from 0 to 1 proves the left-hand side of (66).

Moreover since $(n+p)(0)\psi' + I \leq (n-p)'$ holds, we deduce that $I \leq 1-(n+p)(0)\psi_1(V)$. The right-hand side of (66) follows at once. This concludes the proof.

Remark 3.3 Equation (66) proves that every solution (ψ, n, p, I) is such that the total current I is unbounded as V goes to plus infinity.

References

[1] F. ALABAU, *A method for proving uniqueness theorems for the stationary semiconductor device and electrochemistry equations.* To appear in JNA-TMA.

[2] ——, *Analyse asymptotique et simulation numérique des équations de base des semiconducteurs*, thèse de l'Université, Université Paris VI, Oct. 1987.

[3] ——, *Résultats d'unicité pour une classe d'équations stationnaires en physique des semi-conducteurs et des membranes biologiques*, C.R. Acad. Sc. Paris, Ser. I, 311 (1990), pp. 589–592.

[4] ——, *Uniform asymptotic error estimates for semiconductor device and electrochemistry equations*, JNA-TMA, 14 (1990), pp. 123–139.

[5] ——, *A uniqueness theorem for reverse biased diodes*, Preprint 9010, Université Bordeaux I, nov 1990. Submitted.

[6] F. BREZZI, A. CAPELO, AND L. GASTALDI, *A singular perturbation analysis of reverse-biased diodes*, Siam J. Math. Anal., 20 (1989), pp. 372–387.

[7] S. CHOW AND J. HALE, *Methods of bifurcation theory*, Springer, New York-Heidelberg-Berlin, 1982.

[8] E. CODDINGTON AND N. LEVINSON, *Theory of ordinary differential equations*, Mac Graw Hill, New York-Toronto-London, 1955.

[9] P. MARKOWICH, *The stationary semiconductor device equations*, Springer, Wien-New York, 1986.

[10] P. MARKOWICH AND C. SCHMEISER, *Uniform asymptotic representation of solutions of the basic semiconductor equations*, I.M.A. J. Appl. Math., 36 (1986), pp. 43–57.

[11] M. MOCK, *An example of nonuniqueness of stationary solutions in semiconductor device models*, Compel, 1 (1982), pp. 165–174.

[12] ——, *Analysis of mathematical models of semiconductor devices*, Boole Press, Dublin, 1983.

[13] M. PROTTER AND H. WEINBERGER, *Maximum principles in differential equations*, Prentice Hall, Englewood Cliffs, N.J., 1967.

[14] I. RUBINSTEIN, *Multiple steady states in one-dimensional electro-diffusion with electroneutrality*, Siam J. Math. Anal., 47 (1987), pp. 1076–1093.

[15] T. SEIDMAN, *Nonlinearly elliptic arising in semiconductor theory.* Séminaire IRIA Analyse et contrôle de systèmes, 1979.

[16] ——, *Steady-state solutions of diffusion-reaction systems with electrostatic convection*, JNA-TMA, 4 (1980), pp. 623–637.

[17] H. STEINRUCK, *A bifurcation analysis of the one-dimensional steady-state semiconductor device equations*, Siam J. Math. Anal., 49 (1989), pp. 1102–1121.

Author's address: CEREMAB, Université BORDEAUX I
351, cours de la Libération 33405 Talence Cedex (France)
e-mail: alabau@frbdx11.bitnet

S CARL

An existence result for discontinuous elliptic equations under discontinuous nonlinear flux conditions

1. Introduction

Let $\Omega \subset R^N$ be a bounded domain with Lipschitz boundary $\partial\Omega$. We consider the following boundary value problem (BVP)

$$- L u = f(u,u) \quad \text{in } \Omega , \quad \frac{\partial u}{\partial \nu} = g(u,u) \quad \text{on } \partial\Omega , \qquad (1)$$

where L is a uniformly elliptic differential operator of the form

$$L u = \frac{\partial}{\partial x_i} (a_{ij} \frac{\partial u}{\partial x_i}) - b_i \frac{\partial u}{\partial x_i}$$

with coefficients a_{ij}, $b_i \in L^\infty(\Omega)$, and $\partial/\partial\nu$ denoting the exterior conormal derivative on $\partial\Omega$.

The aim of this paper is to prove the existence of solutions of the BVP (1) provided the functions $(r,s) \to f(r,s)$ and $(r,s) \to g(r,s)$ on the right-hand sides of (1) have one of the following representations

$$f(r,s) = f_1(r) + f_2(s) , \quad f(r,s) = f_1(r) \cdot f_2(s) , \qquad (2)$$

and

$$g(r,s) = g_1(r) + g_2(s) , \quad g(r,s) = g_1(r) \cdot g_2(s) , \qquad (3)$$

respectively. The functions f, g are assumed to be continuous with respect to r whereas they may involve discontinuities with respect to s. Unlike recent works dealing with discontinuous problems cf. [2, 6, 8] in this paper we allow in addition the discontinuous nonlinearity to appear in the boundary condition. Upper and lower solution technique and monotone iteration are applied to obtain in a constructive way the existence of extremal solutions of the BVP (1) provided the discontinuous nonlinearities f and g satisfy a one-sided continuity condition with respect to s .

Boundary value problems with discontinuous nonlinearities arise in various fields of applications such as in chemical reaction theory to describe e.g. steady-state processes with zero order reaction (cf. [1, 5]) or in models of relay control systems (cf. [3]). Discontinuities in the boundary condition may be used to describe the dependence of the flux on a critical threshold value of the solution.

2. Hypotheses and the main result

Denote by $H^1(\Omega)$ the usual Sobolev space of square integrable functions. Define a partial ordering in the spaces $L^2(\Omega)$, $H^1(\Omega)$ by $u \leqq v$ if and only if v-u belongs to the set $L^2_+(\Omega)$ of all nonnegative elements of $L^2(\Omega)$.
Let a be the bilinear form associated with the elliptic operator L by

$$a(u, v) = \int_\Omega (a_{ij} \frac{\partial u}{\partial x_i} \frac{\partial v}{\partial x_j} + b_i \frac{\partial u}{\partial x_i} v)\, dx \quad .$$

Then the linear operator A mapping $H^1(\Omega)$ into its dual space $(H^1(\Omega))^*$ and defined by

$$\langle A\, u, v \rangle := a(u, v) + M \int_\Omega u\, v\, dx$$

is strongly monotone, coercive, continuous and bounded for M sufficiently large, cf. e.g. [11] .

<u>Definition 1.</u> A function $u \in H^1(\Omega)$ is called a solution

of the BVP (1) if

$$a(u, v) - \int_{\partial\Omega} g(u, u)\, v\, d\omega = \int_{\Omega} f(u, u)\, v\, dx$$

for all $v \in H^1(\Omega)$.

Definition 2. A function $\bar{u} \in H^1(\Omega)$ is said to be an upper solution to the BVP (1) if

$$a(\bar{u}, v) - \int_{\partial\Omega} g(\bar{u}, \bar{u})\, v\, d\omega \geq \int_{\Omega} f(\bar{u}, \bar{u})\, v\, dx$$

for all $v \in H^1(\Omega) \cap L^2_+(\Omega)$.

A lower solution $\underline{u}$ is defined similarly by reversing the inequality sign.

We shall impose the following hypotheses on the functions $f, g : R \times R \longrightarrow R$.

(H1) The BVP (1) has a lower solution $\underline{u}$ and an upper solution $\bar{u}$ such that $\underline{u} \leq \bar{u}$.

(H2) There exist functions $p \in L^2_+(\Omega)$ and $q \in L^2_+(\partial\Omega)$ such that

$|f(r, s)| \leq p(x)$ for $r, s \in [\underline{u}(x), \bar{u}(x)]$,

and a.e. $x \in \Omega$,

$|g(r, s)| \leq q(x)$ for $r, s \in [\underline{u}(x), \bar{u}(x)]$,

and a.e. $x \in \partial\Omega$.

(H3) There exists a constant $M \geq 0$ sufficiently large such that

$r \to f(r, s) - M r$ is continuous and nonincreasing, and
$s \to f(r, s) + M s$ is nondecreasing and either right continuous or left continuous for $r, s \in [\underline{u}(x), \bar{u}(x)]$. Analogous conditions are assumed to be satisfied for g.

Remark: Hypothesis (H2) means that the functions f and g are assumed to be locally $L^2(\Omega)$ and $L^2(\partial\Omega)$ bounded, respectively. By hypothesis (H3) we suppose that f and g are continuous in their first argument, while in their second argument they may be discontinuous. Further, a kind of weakened mixed monotonicity is imposed on f and g .

The main result of this paper is

Theorem 1. Let hypotheses (H1), (H2), (H3) be satisfied. Then there exists a greatest (resp. smallest) solution of the BVP (1) within the order interval $[\underline{u}, \bar{u}]$ provided the functions $s \to f(r,s)$, $s \to g(r, s)$ are right (resp. left) continuous.

In order to prove Theorem 1 some preliminaries are needed.

3. Preliminaries

In this section we shall assume that the hypotheses (H1), (H2), (H3) are satisfied.

Let $z \in H^1(\Omega)$ and $z \in [\underline{u}, \bar{u}]$ throughout this section. The monotonicity of f according to (H3) implies that the function $x \to f(r, z(x)) + M z(x)$ is measurable for each $r \in [\underline{u}(x), \bar{u}(x)]$.

Let $\bar{f}$ denote the 'truncated' function given by

$$\bar{f}(r, z(x)) = \begin{cases} f(\bar{u}(x), z(x)) & \text{for } r > \bar{u}(x) \\ f(r, z(x)) & \text{for } \underline{u}(x) \leq r \leq \bar{u}(x) \\ f(\underline{u}(x), z(x)) & \text{for } r < \underline{u}(x) , \end{cases} \tag{4}$$

then the function $(r, x) \longrightarrow \bar{f}(r, z(x)) + M z(x)$ satisfies the Caratheodorý conditions, and as a consequence of (H2) we obtain

Lemma 1. The Nemytskij operator $\bar{F}$ defined by

$(\bar{F}\, w)(x) := \bar{f}(w(x),\, z(x)) + M\, z(x)$

is continuous and bounded from $L^2(\Omega)$ to itself.

Analogous results hold for the function g but now defined for the traces of $w,\, z,\, \underline{u},\, \bar{u} \in H^1(\Omega)$.

For $z \in [\underline{u}, \bar{u}]$ fixed we define an operator $S : z \rightarrow w$ by the BVP

$$\begin{aligned} -Lw + 2Mw &= f(w, z) + 2Mz \quad \text{in } \Omega \\ \frac{\partial w}{\partial \nu} + Mw &= g(w, z) + Mz \quad \text{on } \partial\Omega \end{aligned} \tag{5}$$

Observe that each fixed point of S is a solution of the original BVP (1).

Lemma 2. The BVP (5) admits an uniquely defined solution $w = Sz$ satisfying $Sz \in [\underline{u}, \bar{u}]$ and $\|Sz\|_{H^1(\Omega)} \leq$ const. for each $z \in [\underline{u}, \bar{u}]$. Further, the operator S is increasing with respect to the order interval $[\underline{u}, \bar{u}]$.

Proof: a) Following an idea of Deuel and Hess in [4] we first show the existence of an uniquely defined solution of the 'truncated' problem associated with the BVP (5) replacing the nonlinearities f and g by $\bar{f}$ and $\bar{g}$, respectively. In its weak form this truncated problem reads as

$$\begin{aligned} a(w, v) + M \int_\Omega w\, v\, dx + \int_\Omega (Mw - \bar{f}(w,z))\, v\, dx \\ + \int_{\partial\Omega} (Mw - \bar{g}(w,z))\, v\, d\omega = \int_\Omega 2Mz\, v\, dx + \int_{\partial\Omega} Mz\, v\, d\omega\,, \end{aligned} \tag{6}$$

for all $v \in H^1(\Omega)$. For M sufficiently large by (H2), (H3) and lemma 1 one can ensure that the semilinear form on the left-hand side of (6) defines an operator $B: H^1(\Omega) \rightarrow (H^1(\Omega))^*$

which is strongly monotone, coercive, continuous and bounded. The right-hand side of (6) defines an element $b \in (H^1(\Omega))^*$. Thus (6) may be written as

$$\langle B\, w, v \rangle = \langle b, v \rangle , \tag{7}$$

and by the main theorem on monotone operators (cf. [11, p. 556]) there exists an uniquely defined solution w of (7). Now we show that this solution w belongs to $[\underline{u}, \bar{u}]$, and thus it must be a solution of the BVP (5).

By definition 2 the upper solution $\bar{u}$ satisfies

$$a(\bar{u}, v) - \int_{\partial\Omega} g(\bar{u}, \bar{u})\, v\, d\omega - \int_{\Omega} f(\bar{u}, \bar{u})\, v\, dx \geq 0 \tag{8}$$

for all $v \in H^1(\Omega) \cap L_+^2(\Omega)$. Subtracting (8) from (6) we get

$$a(w-\bar{u}, v) + 2\, M \int_{\Omega} (w-\bar{u})\, v\, dx + M \int_{\partial\Omega} (w-\bar{u})\, v\, d\omega$$

$$\leq \int_{\Omega} (\bar{f}(w, z) + 2\, M\, z - f(\bar{u}, \bar{u}))\, v\, dx - 2\, M \int_{\Omega} \bar{u}\, v\, dx$$

$$+ \int_{\partial\Omega} (\bar{g}(w, z) + M\, z - g(\bar{u}, \bar{u}))\, v\, d\omega - M \int_{\partial\Omega} \bar{u}\, v\, d\omega \tag{9}$$

for all $v \in H^1(\Omega) \cap L_+^2(\Omega)$. By (H3) the right-hand side of (9) becomes larger if we replace z by $\bar{u}$. Thus we obtain

$$a(w-\bar{u}, v) + 2\, M \int_{\Omega} (w - \bar{u})\, v\, dx + M \int_{\partial\Omega} (w - \bar{u})\, v\, d\omega$$

$$\leq \int_{\Omega} (\bar{f}(w, \bar{u}) - f(\bar{u}, \bar{u}))\, v\, dx + \int_{\partial\Omega} (\bar{g}(w, \bar{u}) - g(\bar{u}, \bar{u}))\, v\, d\omega\ .$$

Taking as special nonnegative test function $v = (w - \bar{u})^+$ we get

$$a(w-\bar{u},(w-\bar{u})^+) + 2\, M \int_{\Omega} (w-\bar{u})(w-\bar{u})^+ dx + M \int_{\partial\Omega} (w-\bar{u})(w-\bar{u})^+ d\omega \leq 0 \tag{10}$$

where $u^+ = \max(u, 0)$.

From (10) one readily follows that $(w - \bar{u})^+ = 0$, i.e. $w \leq \bar{u}$. Analogously one proves $w \geq \underline{u}$, and hence w is the uniquely defined solution of the BVP (5) within $[\underline{u}, \bar{u}]$.

b) By standard a priori estimate (cf. [7]) from (5) we obtain

$$\|w\|_{H^1(\Omega)} \leq C(\|f(w,z) + 2Mz\|_{L^2(\Omega)} + \|g(w,z) + Mz\|_{L^2(\partial\Omega)}).$$

Since $w \in [\underline{u}, \bar{u}]$ the right-hand side of the last inequality is uniformly bounded for each $z \in [\underline{u}, \bar{u}]$ by hypothesis (H2), i.e. $\|w\|_{H^1(\Omega)} \leq$ const.

c) Let $z_1, z_2 \in [\underline{u}, \bar{u}]$ with $z_1 \leq z_2$ and $w_i = S z_i$ be the corresponding solutions of (5). Then by subtracting the defining equations from each other and applying (H3) and special test function technique one finds similar to part a) that $w_1 \leq w_2$, and thus S is increasing. This completes the proof of lemma 2.

<u>Lemma 3.</u> Let the monotone sequence $\{u_n\}$ be norm bounded in $H^1(\Omega)$. Then this sequence possesses the following convergence properties:

(i) $u_n \to u$ strongly in $L^2(\Omega)$,

(ii) $u_n \to u$ weakly in $H^1(\Omega)$,

(iii) $u_n \to u$ strongly in $L^2(\partial\Omega)$.

<u>Proof:</u> Property (i) is a consequence of the compact imbedding $H^1(\Omega) \hookrightarrow L^2(\Omega)$ and the monotonicity of the sequence. The boundedness in $H^1(\Omega)$ implies first the weak convergence of a subsequence of u_n in $H^1(\Omega)$. However, according to property (i) all weakly convergent subsequences in $H^1(\Omega)$ must possess the same limit u, and thus the whole sequence is

weakly convergent to u in $H^1(\Omega)$. Finally property (iii) follows from (ii) by applying Theorem 5.1. of [10] and the compact imbedding $H^1(\Omega) \hookrightarrow H^{1-\varepsilon}(\Omega)$ as well as the continuous imbedding $H^{1-\varepsilon}(\Omega) \hookrightarrow L^2(\partial\Omega)$ for $0 < \varepsilon < 1/2$.

4. Proof of Theorem 1

We assume that the functions $s \longrightarrow f(r,s)$ and $s \longrightarrow g(r,s)$ are right continuous and prove the existence of a greatest solution of the BVP (1) within $[\underline{u}, \bar{u}]$.

For this purpose define the following iteration

$$u_{n+1} := S\, u_n \tag{11}$$

with $u_0 = \bar{u}$ and the operator S given by (5). By lemma 2 the iteration (11) yields a monotone nonincreasing sequence of iterates $u_n \in [\underline{u}, \bar{u}]$ which is norm bounded in $H^1(\Omega)$. Thus this sequence possesses the convergence properties of lemma 3. The monotonicity of the $u_n \in H^1(\Omega)$ implies also the monotonicity of their traces in $L^2(\partial\Omega)$ which by lemma 3 converge to the trace of u in $L^2(\partial\Omega)$.

Now we prove that the limit u is the greatest solution of the BVP (1) in $[\underline{u}, \bar{u}]$. In its weak formulation (11) reads as

$$a(u_{n+1},v) + 2\,M \int_{\Omega} u_{n+1}\, v\, dx + M \int_{\partial\Omega} u_{n+1}\, v\, d\omega$$

$$= \int_{\Omega} (f(u_{n+1}, u_n) + 2\,M\, u_n)\, v\, dx + \int_{\partial\Omega} (g(u_{n+1},u_n) + M\, u_n) v\, d\omega \tag{12}$$

From (12) we obtain by means of lemma 3

$$a(u,v) = \lim_{n \to \infty} \left[\int_{\Omega} f(u_{n+1},u_n)\, v\, dx + \int_{\partial\Omega} g(u_{n+1},u_n)\, v\, d\omega \right] \tag{13}$$

In order to pass to the limit under the integral sign on the right-hand side of (13) we argue as follows. The convergence of u_n in $L^2(\Omega)$ (resp. $L^2(\partial\Omega)$) and the monotonicity of u_n imply the convergence of u_n a.e. in Ω (resp. $\partial\Omega$). According to the special forms (2) and (3) of f(u, u) and g(u, u) the pointwise limits

$$\lim_{n\to\infty} f(u_{n+1}(x), u_n(x)) = f(u(x), u(x)) \quad \text{a.e. in } \Omega$$

exist, and analogously for g . Further, by (H2) the sequences $\{f(u_{n+1}, u_n)\}$ and $\{g(u_{n+1}, u_n)\}$ are $L^2(\Omega)$ and $L^2(\partial\Omega)$ bounded, respectively. Thus by means of Lebesgue's Theorem we can pass to the limit under the integral sign which shows that u is a solution of the BVP (1).

To prove that u is the greatest solution of (1) in $[\underline{u}, \bar{u}]$ take any other solution $\hat{u}$ of (1) lying in $[\underline{u}, \bar{u}]$. Then $\hat{u}$ is in particular a lower solution satisfying $\hat{u} \leq \bar{u}$. Taking $\hat{u}$ instead of $\underline{u}$ we can construct in the same way the sequence u_n of iterates which now satisfies $u_n \in [\hat{u}, \bar{u}]$ such that the limit u also belonges to $[\hat{u}, \bar{u}]$, i.e. $\hat{u} \leq u$.

Analogously one obtains by the iteration (11) the smallest solution of (1) within $[\underline{u}, \bar{u}]$ provided $s \to f(r,s)$, $s \to g(r,s)$ are left continuous, and starting the iteration with $u_0 = \underline{u}$.

<u>Remarks:</u> (i) By means of a general fixed point result in ordered Banach spaces which was recently obtained by Heikkilä and Kumpulainen in [9] it seems to be possible to prove Theorem 1 without assuming any one-sided continuity condition of f and g with respect to their second argument and without assuming the monotonicity of the functions $r \to f(r,s) - M\,r$ and $r \to g(r,s) - M\,r$, respectively. However, in this case the proof will be no longer constructive in the sense that the solution is obtained by monotone iteration.

(ii) The special representations of f(u, u) and g(u, u) according to (2) and (3), respectively, are only needed for passing to the limit under the integral sign on the right-hand side of (13). Thus other representations of f(u, u) resp. g(u, u) might be possible as well.

Example: Let us consider the BVP

$$- \Delta u = f(u) \quad \text{in } \Omega \ , \quad \frac{\partial u}{\partial n} = h(u - a) \quad \text{on } \partial\Omega \ , \tag{14}$$

where $\Omega := \{ x \in R^2 \mid \ |x| \leqq \varrho \}$, $\varrho \geqq 1/2$, h is the Heaviside step function, a = const. > 0 , and $f \in C^1$ with $f(0) = 0$, $f'(0) > 0$. Further we assume the existence of a constant $c \geqq a$ such that $f(c) \leqq -4$ and $f'(r) \leqq 0$ for $r \geqq c$.

Under these assumptions we can ensure by theorem 1 the existence of a nontrivial and nonnegative solution u of the BVP (14), since $\underline{u}(x) \equiv \varepsilon$ with $0 < \varepsilon < a$ is a lower solution and $\bar{u}(x) = c + |x|^2$ is an upper solution to the BVP (14).

References

[1] R. Aris, The Mathematical Theory of Diffusion and Reaction in Permeable Catalysts, Vol. 1 and 2, Clarendon Press, Oxford, 1975.

[2] S. Carl and S. Heikkilä, On a parabolic boundary value problem with discontinuous nonlinearity, Nonlinear Anal. 15 (1990), 1091-1095.

[3] P.W. Davis and B.A. Fleishman, A discontinuous nonlinear problem: stability and convergence of iterates in a finite number of steps, Appl.Analysis 23 (1986), 139-157.

[4] J. Deuel and P. Hess, A criterion for the existence of solutions of nonlinear elliptic boundary value problems, Proc. Royal. Soc. Edinburgh 74A (1974/75), 49-54.

[5] B.A. Fleishman and T.J. Mahar, A step-function model in chemical reactor theory: multiplicity and stability of solutions, Nonlinear Anal. 5 (1981), 645-654.

[6] L.S. Frank and W.D. Wendt, On an elliptic operator with discontinuous nonlinearity, J. Differential Equations 54 (1984), 1-18.

[7] D. Gilbarg and N.S. Trudinger, Elliptic Partial Differential Equations of Second Order, 2 nd edition, Springer, Berlin, 1983.

[8] S. Heikkilä, On an elliptic boundary value problem with discontinuous nonlinearity, Appl. Analysis 37 (1990), 183-189.

[9] S. Heikkilä and M. Kumpulainen, On a generalized iteration method with applications to fixed point theorems, Preprint, Mathematics, University of Oulu, 1991.

[10] J.L. Lions, Quelques Methodes de Résolution des Problémes aux Limites non Lineaires, Dunod Gauthier-Villars, Paris (1969). (Russian translation: Mir, Moscow, 1972)

[11] E. Zeidler, Nonlinear Functional Analysis and its Applications, Vol. II/B, Springer, Berlin, 1990.

Siegfried Carl
Fachbereich Mathematik/Informatik
TH Merseburg
Geusaer Straße
O-4200 Merseburg, Germany

M BARCELÓ-CONESA

A remark on quadratic growth of solutions for a class of fully nonlinear second order PDEs

1.0 Introduction. In this note we are interested with two equations: fully nonlinear parabolic equations

$$\begin{aligned} &u_t + F(x, Du, D^2u) = 0 \quad \text{in} \quad \Omega \\ &u(0,t) = \psi(x) \quad \text{for} \quad x \in \mathbb{R}^n \end{aligned} \tag{CP}$$

that is, the Cauchy problem. And, the elliptic equation

$$u + f(x, Du, D^2u) = f(x) \quad \text{in} \quad \Omega \tag{SP}$$

in which Ω is a bounded subset of $\mathbb{R}^n$, u represents a function of (t,x) on $(0,\infty) \times \overline{\Omega}$, Du, D^2u mean $D_x u(t,x)$ and $D_x^2 u(t,x)$, (gradient and Hessian matrix of u, respectively). F is a given real function on $\overline{\Omega} \times \mathbb{R}^n \times S^n$, where S^n denotes the space of $n \times n$ real symetric matrices with the usual ordering.

Moreover, F satisfies the degenerate ellipticity. We assume that

$$F(x,p,X) \leq F(x,p,Y) \quad \text{whenever} \quad Y \leq X \tag{1.1}$$

for $x \in \overline{\Omega}$, $p \in \mathbb{R}^n$ and $X, Y \in S^n$.

There is a great deal of literature about the existence and uniqueness theorems on fully nonlinear second order equations in $\mathbb{R}^n$ (see [**7**], [**8**],[**3**] and references therein). However, there seem to be few results relating solutions of quadratic growth with siutable initial data (see [**4**], [**5**], and [**7**]).

This note use the approach and the spirit of [**4**] requiring extra assumptions on F to achieve the same goals. Must be mentioned that equations of the form $u+G(Du, D^2u) = f(x)$ are not necessarily solvable in the case if f is a pure quadratic, noted in [**4**]; hence, the uniqueness of solutions with quadratic growth are not obtained from [**7**]. This remark, motives enlarge the basic paper of Crandall and Lions to other types of equations.

1.1 The concept of viscosity solutions. We begin by recalling the definition of viscosity solution of (CP), a good reference is [**3**]. For all this note we assume that F satisfies (1.1).

Suppose that u, φ (test function) are C^2 on $\mathbb{R}^n$ and $\widehat{x}$ is a local maximum of $u-\varphi$. By Taylor aproximation we have

$$u(x) \leq u(\widehat{x}) + (p, x-\widehat{x}) + \frac{(X(x-\widehat{x}), x-\widehat{x})}{2} + O(|x-\widehat{x}|^2) \tag{1.2}$$

as $x \to \widehat{x}$ and where $p = D\varphi(\widehat{x})$, $X = D^2\varphi(\widehat{x})$.

Let $\mathcal{O} \subset \mathbb{R}^n$ be a set on which $F \leq 0$ is to hold. If $u : \mathcal{O} \longrightarrow \mathbb{R}$ is function such that , for $\widehat{x} \in \mathcal{O}$, (1.2) holds as $x \longrightarrow \widehat{x}$, $x \in \mathcal{O}$. Then, we say $(p, X) \in J^{2,+}_{\mathcal{O}} u(\widehat{x})$ the second order "superjet"of u at $\widehat{x}$. Switching the inequality sign in (1.2) we can define the second order "subjet" $J^{2,-}_{\mathcal{O}} u(\widehat{x})$. If x is an interior point for all sets $\mathcal{O}$, we denote $J^{2,+}u$ or $J^{2,-}u$.

1.2 Definition. *Let F satisfy (1.1) and $\mathcal{O} \subset \mathbb{R}^n$. A viscosity subsolution (supersolution) of $F = 0$ on $\mathcal{O}$, is a function $u \in USC(\mathcal{O})$ $(u \in LSC(\mathcal{O}))$ such that $F(x,p,X) \leq 0$ $((F(x,p,X) \geq 0)$ for all $x \in \mathcal{O}$ and $(p,X) \in J^{2,+}_{\mathcal{O}} u(x)$ $(J^{2,-}_{\mathcal{O}} u(x))$.*

Where $USC(\mathcal{O})$ $(LSC(\mathcal{O}) = \{$upper (lower) - semicontinuous functions $u : \mathcal{O} \longrightarrow \mathbb{R}\}$ u is called a viscosity solution of $F = 0$ in $\mathcal{O}$ if it is both a viscosity subsolution and supersolution of $F = 0$ in $\mathcal{O}$.

1.3 Function spaces involved. Given functions $g : \mathbb{R}^n \longrightarrow \mathbb{R}$, we are concerned with the following type of function spaces

i) Uniformly continuous function spaces, noted $UC(\mathbb{R}^n)$.

$$UC(\mathbb{R}^n) = \{g : g \quad \text{is uniformly continuous}\}.$$

ii) Pure quadratic function spaces, noted $Q(\mathbb{R}^n)$.

If we denote the Euclidean inner-product of $x, y \in \mathbb{R}^n$ by $< x, y >$, then

$$Q(\mathbb{R}^n) = \{g : \quad \text{exists} \quad A \in S^n \quad \text{such that} \quad g(x) = < Ax, x > \quad \text{for} \quad x \in \mathbb{R}^n\}$$

iii) Quadratic like moduli of continuity funtion spaces, noted $QL(\mathbb{R}^n)$.

$$QL(\mathbb{R}^n) = \{g : \text{ exists a modulus } \omega \text{ such that}$$
$$|g(x) - g(y)| \leq \omega(1 + |x|^2 + |y|^2)^{1/2}|x-y|) \quad \text{for} \quad x, y \in \mathbb{R}^n\}$$

where a modulus is a function $\omega : [0,\infty) \longrightarrow [0,\infty)$ satisfying $\omega(0) = 0$ continuous, subadditive and nondecreasing.

iv) Quadratic growth function spaces, noted $QG(\mathbf{R}^n)$.

$$QG(\mathbf{R}^n) = \{g : \quad \text{exists } C \text{ such that } \quad |g(x)| \le C(1+|x|^2) \quad \text{for} \quad x \in \mathbf{R}^n\}$$

v) Quadratic uniformly continuous function spaces, noted $QUC(\mathbf{R}^n)$. In which , if $f \in QUC(\mathbf{R}^n)$ implies that f has a unique decomposition $f = g + h$ with $g \in Q(\mathbf{R}^n)$ and $h \in UC(\mathbf{R}^n)$, i.e.

$$QUC(\mathbf{R}^n) = Q(\mathbf{R}^n) + UC(\mathbf{R}^n).$$

Moreover, it is obvious that

$$QUC(\mathbf{R}^n) \subset QL(\mathbf{R}^n) \subset QG(\mathbf{R}^n).$$

1.4 Elliptic problem. We begin with the following stationary problem

$$u + F(x, Du, D^2u) = f(x) \tag{SP}$$

remarking that methods used in this note are adaptations of [**4**] to above equation.

To state the next theorem, we assume the following hypothesis.

Assumes that for each $R > 0$

$$\lim_{r \to 0} \sup\{|F(x,p,X+P) - F(x,p,X)| : X, P \in S^n, \quad x, p \in \mathbf{R}^n$$
$$\| X \| \le R, \| P \| \le r, \quad 0 \le r \le R\} = 0 \tag{H.0}$$
$$\lim_{r \to 0} \sup\{|F(x,p+q,,X) - F(x,p,X)| : x,p,q \in \mathbf{R}^n, \quad X \in S^n, \quad p \in B_R$$
$$q \in B_\lambda, \quad 0 \le \lambda \le R\} = 0 \tag{H.1}$$

where B_R (B_λ) will denote the ball of radius $R(\lambda)$ and center O in $\mathbf{R}^n$.

There is a modulus ω such that

$$|F(y, \lambda p, Y) - F(x, \lambda p, X)| \le \omega(\lambda \| X - Y \|^2 + \| X - Y \|) \quad \text{for} \quad x, y \in \mathbf{R}^n \tag{H.2}$$

and $X, Y \in S^n$, $\lambda \ge 0$.

Thus, we will prove:

Theorem 1.1. *Let F be continuous, (H.O), (H.1), (H.2) and (1.1) hold*

i) *If $f \in UC(\mathbb{R}^n)$, then (SP) has a unique solution $u \in QG(\mathbb{R}^n)$ and moreover satisfies $u \in UC(\mathbb{R}^n)$.*

ii) *If $f \in QUC(\mathbb{R}^n)$, then (SP) has a unique solution $u \in QG(\mathbb{R}^n)$ and moreover satisfies $u \in QUC(\mathbb{R}^n)$.*

iii) *If $f \in QL(\mathbb{R}^n)$, then (SP) has a unique solution $u \in QG(\mathbb{R}^n)$ and moreover satisfies $u \in QL(\mathbb{R}^n)$.*

Proof: The proof proceeds in two steps. First, we begin with the following lemma.

Lemma 1.2. *Let F be continuous, (H.0), (H.1), (H.2) and (1.1) hold $a, b, c \geq 0$ and f satisfy*

$$\sup_{\mathbb{R}^n \times \mathbb{R}^n} \{f(x) - f(y) - (b|x-y|^2 + c(|x|^2 + |y|^2))\} \leq a \tag{1.2}$$

If $u(v)$ is an upper (lower) semicontinuous subsolution (supersolution) of (SP), u^+, $u^- \in QG(\mathbb{R}^n)$, then there is a constant $\widehat{a}$ such that

$$\sup_{\mathbb{R}^n \times \mathbb{R}^n} \{u(x) - u(y) - (2b|x-y|^2 + 2c(|x|^2 + |y|^2))\} \leq \widehat{a}\,. \tag{1.3}$$

Proof: We abridge the proof making use of [4, Lemma 1.3]. By the linear growth, we have

$$u(x) - v(y) \leq k(1 + |x| + |y|) \quad \text{on} \quad \mathbb{R}^n \times \mathbb{R}^n \tag{1.4}$$

and

$$\Phi(x,y) = u(x) - v(y) - (2b|x-y|^2 + 2c(|x|^2 + |y|^2)) - \beta_R(x) - \beta_R(y) \tag{1.5}$$

attains its maximum at some point $(\widehat{x}, \widehat{y})$. Where $\beta_R : \mathbb{R}^n \longrightarrow [0, \infty)$ is a family of radial C^2 function with the properties

$$\begin{aligned}
&\text{i)} && \beta_R \geq 0 \\
&\text{ii)} && \liminf_{|x| \to \infty} \frac{\beta_R(x)}{|x|} \geq 2k \\
&\text{iii)} && |D\beta_R(x)| + \| D^2\beta_R(x) \| \leq C \quad \text{for } R \geq 1,\ x \in \mathbb{R}^n \\
&\text{iv)} && \lim_{R \to \infty} \beta_R(x) = 0 \quad \text{for } x \in \mathbb{R}^n
\end{aligned} \tag{1.6}$$

It follows that

$$2b|\widehat{x} - \widehat{y}|^2 + 2c(|\widehat{x}|^2 + |\widehat{y}|^2) \leq u(\widehat{x}) - v(\widehat{y})\,. \tag{1.7}$$

Noting that from [**1**, Theorem 1] there are $X, Y \in S^n$ such that

$$X \leq Y, \quad \| X \|, \quad \| Y \| \leq 16b$$

$$(4b(\widehat{x}-\widehat{y})+4c\widehat{x}+D\beta_R(\widehat{x}),\ X+4cI+D^2\beta_R(\widehat{x})) \in J^{2,+}u(\widehat{x})$$
$$(4b(\widehat{x}-\widehat{y})-4c\widehat{y}-D\beta_R(\widehat{y}),\ Y-4cI-D^2\beta_R(\widehat{y})) \in J^{2,-}v(\widehat{y}) \tag{1.8}$$

Moreover, u is subsolution and v is a supersolution of (SP) and using (1.6) we have

$$\begin{aligned} u(\widehat{x})-v(\widehat{y}) \leq {} & 2(f(\widehat{x})-f(\widehat{y})-(b|\widehat{x}-\widehat{y}|^2+c(|\widehat{x}|^2-|\widehat{y}|^2))+ \\ & +2(F(\widehat{y},4b(\widehat{x}-\widehat{y})-4c\widehat{y}-D\beta_R(\widehat{y}),Y-4cI-D^2\beta_R(\widehat{y}))- \\ & -F(\widehat{x},4b(\widehat{x}-\widehat{y})+4c\widehat{x}+D\beta_R(\widehat{x}),X+4cI+D^2\beta_R(\widehat{x}))) \end{aligned} \tag{1.9}$$

From (1.2) and (1.9) yields

$$u(\widehat{x})-v(\widehat{y}) \leq 2a+2d \tag{1.10}$$

where

$$\begin{aligned} d = {} & F(\widehat{y},4b(\widehat{x}-\widehat{y})-4c\widehat{y}-D\beta_R(\widehat{y}),Y-4cI-D^2\beta_R(\widehat{y}))- \\ & -F(\widehat{x},4b(\widehat{x}-\widehat{y})+4c\widehat{x}+D\beta_R(\widehat{x}),X+4cI+D^2\beta_R(\widehat{x})) \end{aligned} \tag{1.11}$$

so that will require a bound of d.

But d can be expressed by

$$\begin{aligned} d = {} & \frac{F(\widehat{y},4b(\widehat{x}-\widehat{y})-4c\widehat{y}-D\beta_R(\widehat{y}),Y-4cI-D^2\beta_R(\widehat{y}))}{\alpha}- \\ & -(1-\alpha)F(\widehat{x}-4b(\widehat{x}-\widehat{y})+4c\widehat{x}+D\beta_R(\widehat{x}),X+4cI+D^2\beta_R(\widehat{x})) \end{aligned} \tag{1.12}$$

where $\alpha \in (0,1)$ is arbitrary.

Using hypothesis (H.0) we only must bound the term

$$\begin{aligned} d_1 = {} & \frac{F(\widehat{y},4b(\widehat{x}-\widehat{y})-4c\widehat{y}-D\beta_R(\widehat{y}),Y)}{\beta}- \\ & -(1-\beta)F(\widehat{x},4b(\widehat{x}-\widehat{y})+4c\widehat{x}+D\beta_R(\widehat{x}),X) \end{aligned} \tag{1.13}$$

with $\beta \in (0,1)$ arbitrary.

By assumption (H.1) we will require a bound on

$$F(\widehat{y},4b(\widehat{x}-\widehat{y}),Y)-F(\widehat{x},4b(\widehat{x}-\widehat{y}),X)$$

but by hypothesis (H.2)

$$|F(\widehat{y}, 4b(\widehat{x}-\widehat{y}), Y) - F(\widehat{x}, 4b(\widehat{x}-\widehat{y}), X)| \leq \omega(4b \parallel X - Y \parallel^2 + \parallel X - Y \parallel). \quad (1.14)$$

Then we can take

$$d = \sup_{\|X\|,\|Y\| \leq 16b} \omega(4b \parallel X - Y \parallel^2 + \parallel X - Y \parallel) \quad (1.15)$$

and concludes $u(\widehat{x}) - v(\widehat{y}) \leq 2a + 2d$.

Taking into account this

$$\Phi(x, y) \leq \Phi(\widehat{x}, \widehat{y}) \leq u(\widehat{x}) - v(\widehat{y}) \quad (1.16)$$

the bound (1.15) and the definition of Φ, implies that (1.3) holds for $R \longrightarrow \infty$ with $\widehat{a} = 2a + 2d$ ∎

PROOF OF THEOREM 1.1:
Case i) If $f \in UC(\mathbb{R}^n)$ then for every $\gamma > 0$ and for $\varepsilon > 0$, there is an $M_\gamma \geq 0$ such that

$$|f(x) - f(y)| \leq \gamma + M_\gamma |x-y| \leq \gamma + \frac{M_\gamma \varepsilon}{2} + \frac{M_\gamma |x-y|^2}{2\varepsilon}. \quad (1.17)$$

From Lemma 1.2, if $u, v \in QG(\mathbb{R}^n)$ are subsolutions and supersolutions of (SP), we know that $u(x) - v(y) - \frac{M_\gamma |x-y|^2}{\varepsilon}$ is bounded (with $c = 0$, $b = \frac{M_\gamma}{2\varepsilon}$).

With this information, we repeat the proof of the lemma replacing β_R by $\lambda|x|^2$ where $\lambda > 0$ is arbitrary. It leads to the following estimation

$$u(x) - v(y) - \frac{M_\gamma |x-y|^2}{\varepsilon} - \lambda(|x|^2 + |y|^2) \leq 2\gamma + M_\gamma \varepsilon + 2d(\widehat{x}, \widehat{y}, \gamma, \varepsilon, \lambda) \quad (1.18)$$

where now (1.11) has become

$$\begin{aligned} d(\widehat{x}, \widehat{y}, \gamma, \varepsilon, \lambda) = {} & F(\widehat{y}, 4b(\widehat{x}-\widehat{y}) - 2\lambda\widehat{y}, Y - 2\lambda I) - \\ & - F(\widehat{x}, 4b(\widehat{x}-\widehat{y}) + 2\lambda\widehat{x}, X + 2\lambda I) \end{aligned} \quad (1.19)$$

with $\parallel X \parallel, \parallel Y \parallel \leq \frac{8M_\gamma}{\varepsilon}$.

Assuming (H.0) and (H.1), we deduce

$$d(\widehat{x}, \widehat{y}, \gamma, \varepsilon, \lambda) \leq \frac{F(\widehat{y}, 4b(\widehat{x}-\widehat{y}), Y) - F(\widehat{x}, 4b(\widehat{x}-\widehat{y}), X)}{\alpha} + \delta(\gamma, \varepsilon, \lambda) \quad (1.20)$$

where $\delta(\gamma, \varepsilon, \lambda) \longrightarrow 0$ as $\lambda \longrightarrow 0$ for fixed $b > 0$.

Now, we state a lemma analogous to [**4**, Lemma 1.6].

Lemma 1.3. *Let* $u, v : \mathbb{R}^n \longrightarrow \mathbb{R}$. *For* $b > 0$ *set*

$$M_b = \sup_{\substack{x,y \in \mathbb{R}^n \\ \|X\|,\|Y\| \leq 16b}} \{u(x) - v(y) - 2b|x-y|^2 - 2b \parallel X - Y \parallel^2\}. \tag{1.21}$$

If $M_b < \infty$ *for some* $b > 0$ *and* x_b, y_b *are such that*

$$M_b - (u(x_b) - v(y_b) - 2b|x_b - y_b|^2 - 2b \parallel X - Y \parallel^2) \longrightarrow 0 \tag{1.22}$$

then

$$b|x_b - y_b|^2 \longrightarrow 0 \quad \text{and} \quad b \parallel X - Y \parallel^2 \longrightarrow 0 \quad \text{as} \quad b \longrightarrow \infty. \tag{1.23}$$

Proof: It is clear from [**4**] by modification of M_b ∎

We observe from hypothesis (H.2) that

$$|F(\widehat{y}, 4b(\widehat{x} - \widehat{y}), Y) - F(\widehat{x}, 4b(\widehat{x} - \widehat{y}), X) \leq \omega(4b \parallel X - Y \parallel^2 + \parallel X - Y \parallel). \tag{1.24}$$

Applying the above lemma, we see that

$$4b \parallel \widehat{X} - \widehat{Y} \parallel^2 + \parallel \widehat{X} - \widehat{Y} \parallel \leq \alpha(\gamma, \varepsilon, \lambda) \tag{1.25}$$

where $\widehat{X} = X(\widehat{x})$, $\widehat{Y} = Y(\widehat{y})$ and $\alpha(\gamma, \varepsilon, \lambda) \longrightarrow 0$ if $\lambda \longrightarrow 0$, $b \longrightarrow \infty$.

From (1.18), (1.20) and (1.24) we can establish that

$$u(x) - v(y) - (2b|x-y|^2 + \lambda(|x|^2 + |y|^2)) \leq 2\gamma + M_\gamma \varepsilon + d(\widehat{x}, \widehat{y}, \gamma, \varepsilon, \lambda) \tag{1.26}$$

where

$$d(\widehat{x}, \widehat{y}, \gamma, \varepsilon, \lambda) \leq \omega(\alpha(\gamma, \varepsilon, \lambda)) + \delta(\gamma, \varepsilon, \lambda). \tag{1.27}$$

Finally, since $\omega(\alpha(\gamma, \varepsilon, \lambda) + \delta(\gamma, \varepsilon, \lambda) \longrightarrow 0$ if $\lambda \longrightarrow 0$ for fixed $b > 0$ then

$$d(\widehat{x}, \widehat{y}, \gamma, \varepsilon, \lambda) \longrightarrow 0 \quad \text{as} \quad \lambda \longrightarrow 0 \tag{1.28}$$

upon setting $x = y$ and taking the iterated limit $\lambda \longrightarrow 0$, $\gamma \longrightarrow 0$, $\varepsilon \longrightarrow 0$ we conclude $u \leq v$. Moreover, if $u = v$ is a solution implies if in the limit $\lambda \longrightarrow 0$ that $u \in UC(\mathbb{R}^n)$.

By arguments from Perron's method [**6**] can be proved that $u \in QG(\mathbb{R}^n)$ (see [**4**, Theorem 1.1]).

Case ii). It is easy to prove that u is a solution of (SP) if only if $w = u - q$ is a solution of

$$u + G(x, Du, D^2u) = h \tag{1.29}$$

where

$$G(x, r, B) = F(x, p + 2Ar, B + 2A). \tag{1.30}$$

It follows [4] that (SP) has a unique solution $u \in QG(\mathbf{R}^n)$ and likewise $u \in QUC(\mathbf{R}^n)$.

Case iii). If $f \in QL(\mathbf{R}^n)$ for every $\lambda > 0$ exists an M_γ such that

$$|f(x) - f(y)| \leq \gamma + M_\gamma |x - y|(1 + |x|^2 + |y|^2)^{1/2} \tag{1.31}$$

and for $\varepsilon > 0$

$$|f(x) - f(y)| \leq \gamma + \frac{\varepsilon M_\gamma}{2} + \frac{M_\gamma |x-y|^2}{2\varepsilon} + \frac{\varepsilon M_\gamma(|x|^2 + |y|^2)}{2}. \tag{1.32}$$

From lemma 1.2 $u(x) - v(y) - \left(\frac{M_\gamma |x-y|^2}{\varepsilon} + \varepsilon M_\gamma(|x|^2 + |y|^2)\right)$ is bounded, where $u(x)$, $v(y)$ are a subsolution and a supersolution.

Therefore $u(x) - v(y) - \left(\frac{M_\gamma |x-y|^2}{\varepsilon} + \varepsilon M_\gamma(|x|^2 + |y|^2)\right)$ attains its maximum $(\widehat{x}, \widehat{y})$ on $\mathbf{R}^n \times \mathbf{R}^n$.

Following the process as i) we have an estimate

$$u(x) - v(y) - \left(\frac{M_\gamma |x-y|^2}{\varepsilon} + \varepsilon M_\gamma(|x|^2 + |y|^2)\right) \leq 2\gamma + \varepsilon M_\gamma + 2d(\widehat{x}, \widehat{y}, \varepsilon, \gamma) \tag{1.33}$$

with

$$\begin{aligned} d(\widehat{x}, \widehat{y}, \varepsilon, \gamma) = {} & F(\widehat{y}, 4b(\widehat{x} - \widehat{y}) - 2\varepsilon M_\gamma \widehat{y}, Y - 2\varepsilon M_\gamma I) - \\ & - F(\widehat{x}, 4b(\widehat{x} - \widehat{y}) + 2\varepsilon M_\gamma \widehat{x}, X + 2\varepsilon M_\gamma I) \end{aligned} \tag{1.34}$$

where $\| X \|, \| Y \| \leq \frac{8M_\gamma}{\varepsilon}$.

Proceeding as before we have $\lim_{\varepsilon \to 0} d(\varepsilon, \gamma) = \omega(\alpha(\varepsilon, \gamma)) + \delta(\varepsilon, \gamma) = 0$.

The rest as [4] and we deduce that $u \in QL(\mathbf{R}^n)$. ∎

2. The Cauchy Problem. For the Cauchy Problem (CP) the function spaces appropiated are

$C([0, \infty) \times \mathbf{R}^n) = \{g : [0, \infty) \times \mathbf{R}^n \longrightarrow \mathbf{R},\ g \text{ is continuous}\}$

$UC_x([0, \infty) \times \mathbf{R}^n) = \{g \in C([0, \infty) \times \mathbf{R}^n : g(t, x)$ is uniformly continuous in x, uniformly for $t \geq 0\}$

$Q_x([0,\infty)\times \mathbb{R}^n) = \{g : \text{exists } A \in S^n \text{ such that } g(t,x) = < Ax, x >$
for $x \in \mathbb{R}^n\} = Q(\mathbb{R}^n)$
$QL_x([0,\infty)\times \mathbb{R}^n) = \{g \in C([0,\infty)\times \mathbb{R}^n) : \text{exists a modulus } \omega \text{ such that}$
$|g(t,x) - g(t,y)| \le \omega(1+|x|+|y|)|x-y|$ for $x, y \in \mathbb{R}^n$, $t \ge 0\}$
$QUC_x([0,\infty)\times \mathbb{R}^n) = Q_x([0,\infty)\times \mathbb{R}^n) + UC_x([0,\infty)\times \mathbb{R}^n)$.
and
$QG_x([0,\infty)\times \mathbb{R}^n) = \{g : \text{for all } T > 0 \text{ exists } C \text{ such that } |g(t,x)| \le C(1+|x|^2)$
for $(t,x) \in [0,T]\times \mathbb{R}^n\}$.

In order to prove the next theorem we need a condition (H.2') stronger than (H.2).

There is a modulus ω such that

$$|F(x,\lambda p, X)| \le \omega(\lambda \parallel X \parallel^2 + \parallel X \parallel) \quad \text{for all} \quad x, p \in \mathbb{R}^n \tag{H.2'}$$

and $X \in S^n$, $\lambda \ge 0$.

We will prove

Theorem 2.1. *Let F be continuous and (H.0), (H.1), (H.2) and (1.1) hold.*

i) *If $\psi \in UC(\mathbb{R}^n)$, then (CP) has a unique solution*

$$u \in QG_x([0,\infty)\times \mathbb{R}^n) \cap C([0,\infty)\times \mathbb{R}^n)$$

and moreover if F satisfies (H.2') then $u \in UC_x([0,\infty)\times \mathbb{R}^n)$.

ii) *If $\psi \in QUC(\mathbb{R}^n)$, then (CP) has a unique solution*

$$u \in QG_x([0,\infty)\times \mathbb{R}^n) \cap C([0,\infty)\times \mathbb{R}^n)$$

and moreover if F satisfies (H.2') then $u \in QUC_x([0,\infty)\times \mathbb{R}^n)$.

iii) *If $\psi \in QL(\mathbb{R}^n)$, then (CP) has a unique solution*

$$u \in QG_x([0,\infty)\times \mathbb{R}^n) \cap C([0,\infty)\times \mathbb{R}^n)$$

and moreover if F satisfies (H.2') then $u \in QL_x([0,\infty)\times \mathbb{R}^n)$.

As before, we need a lemma.

Lemma 2.2. *Let F be continuous (H.0), (H.1), (H.2) and (1.1) hold, $a, b, c \ge 0$ and ψ satify*

$$\sup_{\mathbb{R}^n\times\mathbb{R}^n} \{\psi(x) - \psi(y) - (b|x-y|^2 + c(|x|^2+|y|^2))\} \le a. \tag{2.1}$$

If u (v) is upper (lower) semicontinuous subsolution (supersolution) of (CP) on $[0,\infty)\times \mathbf{R}^n$

$$u(0,x)\le \psi(x)\le v(0,x) \quad \text{on} \quad \mathbf{R}^n \tag{2.2}$$

and $u^+, u^- \in QG_x([0,\infty)\times \mathbf{R}^n)$, *then there is a constant α such that*

$$\sup_{\substack{\mathbf{R}^n\times\mathbf{R}^n\\ t\ge 0}} \{u(t,x)-v(t,y)-(b|x-y|^2+c(|x|^2+|y|^2))\}\le a+\alpha t\,. \tag{2.3}$$

Proof: Let $T>0$ clearly we have

$$u(t,x)-v(t,y)\le C(1+|x|^2+|y|^2) \quad \text{for} \quad x,y\in \mathbf{R}^n\,, \quad 0\le t\le T\,. \tag{2.4}$$

Setting

$$\Phi(t,x,y)=u(t,x)-v(t,y)-(b|x-y|^2+c(|x|^2+|y|))-\beta_R(x)-\beta_R(y)-\alpha t \tag{2.5}$$

β_R satisfies (1.6) and $\alpha>0$.

As lemma 1.2 Φ will attain its maximum at some point $(\widehat{t},\widehat{x},\widehat{y})$ on $[0,T]\times\mathbf{R}^n\times\mathbf{R}^n$. In the case $\widehat{t}=0$, we have

$$\Phi(\widehat{t},\widehat{x},\widehat{y})_{\widehat{t}=0}=\Phi(0,\widehat{x},\widehat{y})\le \psi(\widehat{x})-\psi(\widehat{y})-(b|\widehat{x}-\widehat{y}|^2+c(|x|^2+|y|^2))\le a\,. \tag{2.6}$$

If $\widehat{t}>0$, we know from [**2**] that there are $X,Y\in S^n$ such that

$$X\le Y\,,\|X\|\,,\|Y\|\le 8b$$

$$\begin{aligned}&(2b(\widehat{x}-\widehat{y})+2c\widehat{x}+D\beta_R(\widehat{x}),X+2cI+D^2\beta_R(\widehat{x}))\in P^{2,+}u(\widehat{x})\\ &(2b(\widehat{x}-\widehat{y})-2c\widehat{y}-D\beta_R(\widehat{y}),Y-2cI-D^2\beta_R(\widehat{y}))\in P^{2,-}v(\widehat{y})\end{aligned} \tag{2.7}$$

where $P^{2,+}$, $P^{2,-}$ are the parabolic variants of $J^{2,+}$, $J^{2,-}$ see [**3**]. And

$$\begin{aligned}\alpha&=\alpha(\widehat{x},\widehat{y},b,c)\le\\ &\le F(\widehat{y},2b(\widehat{x}-\widehat{y})-2c\widehat{y}-D\beta_R(\widehat{y}),Y-2cI-D^2\beta_R(\widehat{y}))-\\ &-F(\widehat{x},2b(\widehat{x}-\widehat{y})+2c\widehat{x}+D\beta_R(\widehat{x}),X+2cI+D^2\beta_R(\widehat{x}))\,.\end{aligned} \tag{2.8}$$

Now, we can observe that

$$\Phi(t,x,y)\le\Phi(\widehat{t},\widehat{x},\widehat{y})\le a\,,\quad T>0 \quad \text{arbitrary}$$

and for $R \longrightarrow \infty$ we have

$$u(t,x) - v(t,y) - (b|x-y|^2 + c(|x|^2 + |y|^2) \leq a + \alpha(\widehat{x}, \widehat{y}, b, c)t\,. \tag{2.9}$$

Repeating the technics used in the case elliptic, we can take

$$\alpha(b,c) = \sup_{\|X\|,\|Y\| \leq 8b} \omega(2b \parallel X - Y \parallel^2 + \parallel X - Y \parallel) \quad \blacksquare \tag{2.10}$$

Proof of Theorem 2.1: With conditions (H.0), (H.1), (H.2), (1.1) and moreover (H.2') it is a simpler exercice from [**4** , Theorem 2.7] to prove this theorem and we left the reader for details.

REFERENCES

1. M.G. Crandall, *Quadratic forms, semidifferentials and viscosity solutions of fully nonlinear elliptic equations*, Ann. I.H.P. Anal.Non. Lin. **6** (1989), 419–435.
2. M.G. Crandall and H. Ishii The maximum principle for semicontinuous functions, preprint.
3. M.G. Crandall, H. Ishii, and P.L. Lions, *A user's guide to viscosity solutions to fully nonlinear equations of second order*, preprint.
4. M.G. Crandall and P.L. Lions, *Quadratic growth of solutions of fully nonlinear second order equations in* $\mathbb{R}^n$, Diff. Int. Equations. **3** (1990), 601–616.
5. M.G. Crandall, R. Newcomb and Y. Tomita, *Existence and uniqueness of viscosity solutions of degenerate quasilinear elliptic equations in* $\mathbb{R}^n$, Applicable Analysis **34** (1989), 1–24.
6. H. Ishii, *Perron's method for Hamilton-Jacobi equations*, Duke Math. J. **55** (1987), 369–384.
7. H. Ishii, *On uniqueness and existence of viscosity solutions of nonlinear secon-order elliptic PDE's*, Comm. Pure Appl. Math. **42** (1989), 14–45.
8. H. Ishii and P.L. Lions, *Viscosity solutions of fully nonlinear second-order elliptic partial differential equations*, J. Diff. Equations **83** (1990), 26–78.

Miquel Barceló Conesa
Dept. de Matemàtica Aplicada I, ETSEIB
Universitat Politècnica de Catalunya
Diagonal 647, 08028 Barcelona, Spain

V FERONE AND M ROSARIA POSTERARO

Symmetrization for a class of non-linear elliptic equations

Symmetrization methods have shown to be useful to obtain sharp estimates for solutions of elliptic equations. In the linear case, results for Dirichlet problem relative to classes of equations in general form are well known ([1], [2], [3], [4], [9], [10], [11], [15], [17]). These results have been extended to classes of non-linear equations ([2], [12], [13], [16]). In particular, in [2] the following Dirichlet problem is considered:

$$\begin{cases} -\left(a_{ij}(x)u_{x_j}\right)_{x_i} = H(x, \nabla u) & \text{in } \Omega \\ u \in H_0^1(\Omega) \cap L^\infty(\Omega), \end{cases} \tag{1}$$

where $a_{ij}(x)$, $H(x,\xi)$ are measurable functions such that:

$$a_{ij}(x)\xi_i\xi_j \geq |\xi|^2, \qquad \forall \xi \in \mathbf{R}^n$$

$$H(x,\xi) \leq f(x) + c_0|\xi|^p,$$

with $f \in L_+^\infty(\Omega)$, $c_0 > 0$, $p \in [1,2]$.

The following theorem holds (see [2]):

Theorem. *If there exists a solution $v(x) = v^\#(x)$ of the problem:*

$$\begin{cases} -\Delta v = f^\#(x) + c_0|\nabla v|^p, & \text{in } \Omega^\# \\ v \in H_0^1(\Omega^\#) \cap L^\infty(\Omega^\#), \end{cases}$$

where $\Omega^\#$ is the n-dimensional ball centered at the origin and with the same measure as Ω and $f^\#(x)$ is the spherically symmetric decreasing rearrangement of f, then (1) has a solution $u(x)$ and

$$u^\#(x) \leq v(x), \qquad \text{in } \Omega^\#.$$

For related results see also [12], [13].

As first step in studying (1) when $H(x,\xi) = (f_i)_{x_i} + c_0|\xi|^p$, we consider the following Dirichlet problem:

$$\begin{cases} -\Delta u = |\nabla u|^2 + (f_i)_{x_i}, & \text{in } \Omega \\ u \in H_0^1(\Omega) \cap L^\infty(\Omega). \end{cases} \tag{2}$$

To point out the kind of result we can expect for this problem *we suppose that* f_i*,* $i = 1, \ldots, n$*, and* Ω *are enoughly regular to assure* $|\nabla u| \in L^\infty(\Omega)$ (technical hypothesis).

Existence and regularity results for problems like (1), (2) (for more general classes of non-linear elliptic equations) can be found in [5], [6], [7].

Let us recall some definitions which will be useful in the following. If Ω is an open bounded set of $\mathbf{R}^n$, we will denote by $|\Omega|$ its measure and by $\Omega^\#$ the ball of $\mathbf{R}^n$ centered at the origin whose measure is $|\Omega|$. Moreover if φ is a measurable function,

$$\mu(t) = |\{x \in \Omega : |\varphi(x)| > t\}|, \qquad t \geq 0,$$

is the distribution function of φ and

$$\varphi^*(s) = \sup\{t \geq 0 : \mu(t) > s\}, \qquad s \in [0, |\Omega|],$$

is its decreasing rearrangement. If C_n is the measure of the n-dimensional unit ball,

$$\varphi^\#(x) = \varphi^*(C_n|x|^n), \qquad x \in \Omega^\#,$$

is the spherically symmetric decreasing rearrangement of $\varphi(x)$. For an exhaustive treatment of rearrangements we refer to [4], [8], [14].

Theorem. *Let* u *be a weak solution of (2); we suppose that* $f_i \in L^\infty(\Omega)$ *and* $|\nabla u| \in L^\infty(\Omega)$*. Let* $v(x) = v^\#(x)$ *be solution of the problem*

$$\begin{cases} -\Delta v = |\nabla v|^2 + F\left(\dfrac{x_i}{|x|}\right)_{x_i} \\ v \in H_0^1(\Omega^\#) \cap L^\infty(\Omega^\#), \end{cases} \tag{3}$$

where:

$$F = \max_i \|f_i\|_\infty. \tag{4}$$

Then we have:

$$u^{\#}(x) \leq v^{\#}(x), \qquad x \in \Omega^{\#}.$$

PROOF. If u is a weak solution of (2), we have:

$$\int_{\Omega} u_{x_i} \varphi_{x_i}\, dx = \int_{\Omega} |\nabla u|^2 \varphi\, dx + \int_{\Omega} f_i \varphi_{x_i}\, dx, \qquad \forall \varphi \in H_0^1(\Omega) \cap L^{\infty}(\Omega).$$

For $h > 0$, we choose the usual test function

$$\varphi_h(x) = \begin{cases} h \operatorname{sign} u & \text{if } |u(x)| > t + h \\ (|u(x)| - t) \operatorname{sign} u & \text{if } t + h \geq |u(x)| > t \\ 0 & \text{otherwise} \end{cases}$$

and in a standard way ([2], [9], [15]) we get:

$$-\frac{d}{dt} \int_{|u|>t} |\nabla u|^2\, dx \leq \int_{|u|>t} |\nabla u|^2\, dx + F\left(-\frac{d}{dt} \int_{|u|>t} |\nabla u|\, dx\right), \tag{5}$$

where F is defined in (4).

Using the Fleming-Rishel formula and the isoperimetric inequality, it is easy to obtain:

$$nC_n^{1/n} (\mu(t))^{1-1/n} \leq -\frac{d}{dt} \int_{|u|>t} |\nabla u|\, dx \leq (-\mu'(t))^{1/2} \left(-\frac{d}{dt} \int_{|u|>t} |\nabla u|^2\, dx\right)^{1/2},$$

where $\mu(t)$ is the distribution function of u. Then, from (5) we have:

$$\left(-\frac{d}{dt} \int_{|u|>t} |\nabla u|^2\, dx\right)^{1/2} \leq \frac{(-\mu'(t))^{1/2}}{nC_n^{1/n} (\mu(t))^{1-1/n}} \int_{|u|>t} |\nabla u|^2\, dx + F(-\mu'(t))^{1/2}.$$

Putting $t = u^*(s)$, we get:

$$\left(\frac{d}{ds} \int_{|u|>u^*(s)} |\nabla u|^2\, dx\right)^{1/2} \leq \frac{s^{-1+1/n}}{nC_n^{1/n}} \int_{|u|>u^*(s)} |\nabla u|^2\, dx + F. \tag{6}$$

Obviously, if $v(x) = v^{\#}(x)$ is solution of (3), we also have:

$$\left(\frac{d}{ds} \int_{v>v^*(s)} |\nabla v|^2\, dx\right)^{1/2} = \frac{s^{-1+1/n}}{nC_n^{1/n}} \int_{v>v^*(s)} |\nabla v|^2\, dx + F. \tag{7}$$

Now we put:

$$y(s) = \int_{|u|>u^*(s)} |\nabla u|^2\, dx, \qquad z(s) = \int_{v>v^*(s)} |\nabla v|^2\, dx,$$

and then (6) and (7) can be written as

$$(y')^{1/2} \leq \frac{s^{-1+1/n}}{nC_n^{1/n}}\, y + F \tag{8}$$

$$(z')^{1/2} = \frac{s^{-1+1/n}}{nC_n^{1/n}}\, z + F \tag{9}$$

First of all, taking into account the fact that $|\nabla u| \in L^\infty(\Omega)$, we have:

$$\int_{|u|>u^*(s)} |\nabla u|^2\, dx \leq \|\nabla u\|_\infty^2\, s.$$

Then (8) gives

$$y'(0) \leq F^2. \tag{10}$$

We can also prove that

$$z'(0) = F^2. \tag{11}$$

By assumption, v is solution of (3); if we define $w(x) = e^{v(x)}$, we get:

$$-\Delta w = w\, F \left(\frac{x_i}{|x|}\right)_{x_i}.$$

A straightforward calculation gives:

$$-\left(e^{-F|x|}\widetilde{w}_{x_i}\right)_{x_i} + F\, e^{-F|x|}\frac{x_i}{|x|}\widetilde{w}_{x_i} - F^2 \widetilde{w} e^{-F|x|} = 0,$$

where $\widetilde{w}(x) = w(x)e^{F|x|}$. So we have $\widetilde{w} \in W^{2,p}(\Omega^\#) \cap C^{1,\alpha}(\Omega^\#)$, $\forall p > 1$, $\alpha \in]0,1[$ and

$$0 = \widetilde{w}'(0) = w'(0) + Fw(0) = e^{v(0)}\big(v'(0) + F\big),$$

where the derivatives are with respect to $|x|$. This means $v'(0) = -F$ and then (11) holds.

Now we want to show that $U \equiv y - z \leq 0$, $\forall s \in [0, |\Omega|]$. Let us suppose that $U > 0$ in an interval $(\alpha, \beta] \subset [0, |\Omega|]$. Clearly, α can be chosen in such a way that $U(\alpha) = 0$. Squaring (8) and (9) we get:

$$U' \leq \gamma^2(s)\, U^2 + 2\gamma(s)\big(\gamma(s)\, z + F\big)U,$$

where $\gamma(s) = \dfrac{s^{-1+1/n}}{nC_n^{1/n}}$, and then, if $W = \dfrac{1}{U}$, we have:

$$-W' \leq \gamma^2(s) + 2\gamma(s)\big(\gamma(s)\, z + F\big) W.$$

Integrating between x and β, we obtain:

$$\begin{aligned} W(x) \leq & W(\beta) \exp\left(\int_x^\beta 2\gamma(t)\big(\gamma(t)\, z + F\big)\, dt \right) + \\ & + \int_x^\beta \gamma^2(s) \exp\left(\int_x^s 2\gamma(t)\big(\gamma(t)\, z + F\big)\, dt \right)\, ds. \end{aligned} \tag{12}$$

By assumption $\lim_{x \to \alpha+} W(x) = +\infty$, while (12) implies that $W(x) < +\infty$, $\forall x \neq 0$; this means $\alpha = 0$. Thus we have proved that, if $U(x) > 0$ in an interval, then such an interval must be $(0, \beta)$ with $\beta \leq |\Omega|$.

Furthermore (12) gives

$$\frac{1}{U(x)} = W(x) \leq \exp\left(\int_x^\beta 2\gamma(t)\big(\gamma(t)\, z + F\big)\, dt \right) \left(W(\beta) + \int_x^\beta \gamma^2(t)\, dt \right),$$

and then it is easy to realize that we should have:

$$\lim_{x \to 0+} U'(x) = +\infty.$$

On the other hand, using (10), (11), it is possible to get:

$$U'(0) \leq 0$$

and then the contradiction.

Thus we have obtained $y(s) \leq z(s)$; (8) and (9) give:

$$y'(s) \leq z'(s).$$

Recalling the definition of y and z, we have:

$$-\big(u^*(s)\big)' \leq -\big(v^*(s)\big)'$$

and the theorem follows. ∎

Remark. The hypothesis $|\nabla u| \in L^\infty(\Omega)$ has been used only to be sure that

$$\lim_{s \to 0} \frac{y(s)}{s^{1-1/n}} = 0. \tag{13}$$

So the arguments can be repeated as soon as (13) holds.

REFERENCES

[1] Alvino A.-Lions P.L.-Trombetti G., "Comparaison des solutions d'équations paraboliques et elliptiques par symétrisation. Le cas de la symétrisation de Schwarz", *C.R. Acad. Sc. Paris*, **303** (1986), 947-950.

[2] Alvino A.-Lions P.L.-Trombetti G., "Comparison results for elliptic and parabolic equations via Schwarz symmetrization", *Ann. Inst. Henri Poincaré*, **7** (1990), 37-65.

[3] Alvino A.-Trombetti G., "Equazioni ellittiche con termini di ordine inferiore e riordinamenti", *Atti Accad. Naz. Lincei Rend. Cl. Sci. Fis. Mat. Natur.*, (8) **54** (1976), 706-716.

[4] Bandle C., *Isoperimetric Inequalities and Applications*, Monographs and Studies in Math., No. 7, Pitman, London, 1980.

[5] Boccardo L.-Murat F.-Puel J.P., "Existence de solutions faibles pour équations elliptiques quasilinéaires à croissance quadratique", *Nonlinear Partial Differential Equations an their Applications*, College de France, 4, Pitman, London, 1983, 19-73.

[6] Boccardo L.-Murat F.-Puel J.P., "Résultats d'existence pour certaines problémes elliptiques quasilinéaires", *Ann. Scuola Norm. Pisa*, **11** (1984), 213-235.

[7] Boccardo L.-Murat F.-Puel J.P., "L^∞ estimate for some nonlinear elliptic partial differential equation and application to an existence result", to appear on *SIAM J. Math. Anal.*.

[8] Chong K.M.-Rice N.M., *Equimeasurable rearrangements of functions*, Queen's papers in pure and applied mathematics, No. 28, Queen's University, Ontario, 1971.

[9] Ferone V.-Posteraro M.R., "Symmetrization results for elliptic equations with lower-order terms", to appear on *Atti Sem. Mat. Fis. Univ. Modena*.

[10] Giarrusso E.-Nunziante D., "Regularity theorems in limit cases for solutions of linear and nonlinear elliptic equations", *Rend. Ist. Mat. Univ. Trieste*, **20** (1988), 39-58.

[11] Giarrusso E.-Trombetti G., "Estimates for solutions of elliptic equations in a limit case", *Bull. Austral. Math. Soc.*, **36** (1987), 425-434.

[12] Maderna C.-Pagani C.-Salsa S., "Quasilinear Elliptic Equations with quadratic growth in the gradient", *preprint*.

[13] Maderna C.-Salsa S., "Dirichlet problem for Elliptic Equations with Nonlinear First Order Terms: a Comparison Result", *Ann. Mat. Pura Appl.*, **148** (1987), 277-288.

[14] Mossino J., *Inégalités isopérimétriques et applications en physique*, Collection Travaux en Cours, Hermann, Paris, 1984.

[15] Talenti G., "Elliptic Equations and Rearrangements", *Ann. Scuola Norm. Sup. Pisa*, (4) **3** (1976), 697-718.

[16] Talenti G., "Nonlinear elliptic equations, Rearrangements of functions and Orlicz spaces", *Ann. Math. Pura Appl.*, **120** (1979), 159-184.

[17] Talenti G., "Linear Elliptic P.D.E.'s: Level Sets, Rearrangements and a priori Estimates of Solutions", *Boll. U.M.I.*, (6) **4**-B (1985), 917-949.

Vincenzo Ferone - M. Rosaria Posteraro

Dipartimento di Matematica ed Applicazioni "Renato Caccioppoli"

Università di Napoli

Via Mezzocannone, 8

80134 NAPOLI - ITALY

F ABERGEL, D HILHORST AND F ISSARD-ROCH

On a Stefan problem with surface tension in the neighborhood of a stationary solution

Introduction.

Consider a system composed of a solid phase of a single compound and an incompressible liquid phase which is a dilute solution of that compound and suppose that the evolution in time is governed by two processes : a diffusion process in a diffusion layer in the fluid and a dissolution-growth process, located at the interface between solid and fluid.

The problem can be described by three basic equations :

· A diffusion equation in the diffusion layer :

$$C_t = \Delta C \tag{1.1}$$

where $C = C(x, y, t)$ is the concentration in the liquid phase ;

. A Neumann boundary condition at the interface, which describes mass balance at the interface :

$$\left.\frac{\partial C}{\partial n}\right|_{\Gamma(t)} = V_n \tag{1.2}$$

where $\Gamma(t)$ denotes the solid/fluid interface, n is the normal unit vector to the interface directed towards the fluid and V_n is the normal velocity of the interface ;

. The kinetics of dissolution and growth of the solid at the interface which gives the evolution in time of this interface,

$$V_n = -\alpha e^{\gamma K} + C_{|\Gamma(t)} \tag{1.3}$$

where α and γ are positive constants, γ is proportional to the surface tension of the interface and K is the curvature of the interface.

Let us note that in view of (1.3), the boundary condition (1.2) can be rewritten as:

$$\left(C - \frac{\partial C}{\partial n}\right) \mid \Gamma(t) = \alpha e^{\gamma K}.$$

We suppose here that the interface is parametrized in the form $y = f(x, t)$ and denote by e the width of the diffusion layer ; then, f and C satisfy the following (rescaled) equations :

$$
(P_0) \quad \begin{cases}
C_t = \Delta C \ in \ Q_f := \left\{(x,y,t) \in \mathbb{R}^2 \times \mathbb{R}^+, f(x,t) < y < f(x,t) + e\right\} \\
\left(\begin{array}{l} \left(C - \dfrac{\partial C}{\partial n}\right)(x, f(x,t), t) = \alpha e^{-\gamma f_{xx}/(1+f_x^2)^{3/2}(x,t)} \\ (x,t) \in \mathbb{R} \times \mathbb{R}^+ \end{array} \right. \\
\dfrac{\partial C}{\partial n}(x, f(x,t) + e) = 0, \quad (x,t) \in \mathbb{R} \times \mathbb{R}^+ \\
C(x + 2L, y, t) = C(x,y,t), \quad (x,y,t) \in Q_f \\
\left(\begin{array}{l} \dfrac{f_t}{(1+f_x^2)^{1/2}} = -\alpha e^{-\gamma f_{xx}/(1+f_x^2)^{3/2}(x,t)} + C(x, f(x,t), t) \\ (x,t) \in \mathbb{R} \times \mathbb{R}^+ \end{array} \right. \\
f(x + 2L, t) = f(x,t), \quad (x,t) \in \mathbb{R} \times \mathbb{R}^+ \\
f(x,0) = f_0(x), \quad x \in \mathbb{R} \\
\left(\begin{array}{l} C(x,y,0) = C_0(x,y) \\ (x,y) \in \Sigma_f := \{(x,y) \in \mathbb{R}^2, f_0(x) < y < f_0(x) + e\} \end{array} \right.
\end{cases}
$$

We suppose that f_0 and C_0 satisfy the hypothesis :

$$H_0 : \ \left(f_0 \in C_{per}^{3+\lambda}(\mathbb{R}), \ C_0 \in C_{per}^{2+\lambda}(\bar{\Sigma}_f) \text{for some } \lambda \in (0,1) \right.$$

and C_0 satisfying the compatibility conditions :

$$
H_1 : \quad \left(\begin{array}{l} \dfrac{\partial C_0}{\partial n}(x, f_0(x) + e) = 0 \quad x \in \mathbb{R}, \\ \left(C_0 - \dfrac{\partial C_0}{\partial n}\right)(x, f_0(x)) = \alpha e^{-\gamma f_0''(x)/(1+f_0'^2(x))^{3/2}} \quad x \in \mathbb{R} \end{array} \right.
$$

(By the subscript per, we denote functions periodic in the variable x with period $2L$).

The aim of this note is to establish an existence and uniqueness result in the neighborhood of the stationary solution $(f, C) = (0, \alpha)$. More precisely we prove that

THEOREM 1. *There exists a positive constant $\rho_0 = \rho_0(T)$ such that if the initial data (f_0, C_0) satisfies the condition $\|f_0\|_{\mathbb{R}}^{(3+\lambda)} + \|C_0 - \alpha\|_{\Sigma_f}^{(2+\lambda)} \leq \rho_0$ then Problem (P_0) has a unique solution (f, C) in $C_{per}^{2+\lambda,(3+\lambda)/2}(\mathbb{R} \times [0,T]) \times C_{per}^{2+\lambda,1+\lambda/2}(\bar{Q}_f^T)$ where $Q_f^T := \{(x,y,t) \in Q_f, t \in (0,T)\}$.*

Let us notice that the well-posedness of a very similar problem in the case of spherical symmetry has been studied by F. Conrad, D. Hilhorst and T.I. Seidman on a bounded domain [4] and by D.Hilhorst, F. Issard-Roch and T.I. Seidman on an unbounded domain [5]. For local and uniqueness results for related problems, we refer to paper by X.Y. Chen [3] and Xinfu Chen [2] about a two phase problem.

The main steps of the proof of Theorem 1 are the following :
- First we transform Problem (P_0) into a problem on a fixed domain :

$$Set\ \hat{x} = x,\ \hat{y} = y - f(x,t),\ \hat{t} = t,$$
$$\hat{g}(\hat{x},\hat{y},\hat{t}) = C(x,y,t) - \alpha,\ \hat{f}(\hat{x},\hat{t}) = f(x,t)$$

then some easy computations imply that $\hat{g}$ and $\hat{f}$ satisfy the following equations (where the hats are omitted for simplicity) :

$$(P)\quad \begin{cases} \begin{cases} g_t = \Delta g + f_x^2 g_{yy} - 2f_x g_{xy} - (f_{xx} - f_t) g_y \\ \text{in } Q_0 := \{(x,y,t) \in \mathbb{R} \times (0,e) \times \mathbb{R}^+\} \end{cases} \\ \begin{cases} g(x,0,t) - (1+f_x^2)^{1/2} g_y(x,0,t) + \dfrac{f_x}{(1+f_x^2)^{1/2}} g_x(x,0,t) = \\ = \alpha(e^{-\gamma f_{xx}/(1+f_x^2)^{3/2}} - 1),\quad (x,t) \in \mathbb{R} \times \mathbb{R}^+ \end{cases} \\ (1+f_x^2)^{1/2} g_y(x,e,t) - \dfrac{f_x}{(1+f_x^2)^{1/2}} g_x(x,e,t) = 0,\quad (x,t) \in \mathbb{R} \times \mathbb{R}^+ \\ g(x+2L,y,t) = g(x,y,t) \text{ in } Q_0 \\ g(x,y,0) = g_0(x,y) = C_0(x,y+f_0(x)) - \alpha,\quad (x,y) \in \mathbb{R} \times (0,e) \\ f_t = -(1+f_x^2)^{1/2} (\alpha e^{-\gamma f_{xx}/(1+f_x^2)^{3/2}} - g(x,0,t) - \alpha),\quad (x,t) \in \mathbb{R} \times \mathbb{R}^+ \\ f(x,+2L,t) = f(x,t),\quad (x,t) \in \mathbb{R} \times \mathbb{R}^+ \\ f(x,0) = f_0(x),\quad x \in \mathbb{R}. \end{cases}$$

- Secondly, we consider a related linear problem and prove its well-posedness.
- Then, using suitable test functions, we prove the uniqueness of the solution of Problem (P).
- The existence is obtained by means of a fixed point argument. The iterative scheme used is as follows : we consider as known the nonlinear terms and solve the linear problem. Then we prove that it is a strict contraction from a ball of small enough radius into itself.

The paper is organized as follows :
- In Section 1, we study the linear problem.
- In Section 2, we give the main steps for the proof of uniqueness.
- In Section 3, we describe the iterative map and show that it is a strict contraction.

For the complete proofs, we refer to [1].

1. The linear problem

In this Section, we consider the following linear problem :

$$
(L)\begin{cases}
g_t - \Delta g = F_1 \text{ in } Q_T := \{(x,y,t) \in \mathbb{R} \times (0,e) \times (0,T)\} & (2.1)\\
\left(\begin{array}{l}(g - g_y)(x,0,t) + \alpha\gamma f_{xx}(x,t) = F_2(x,t)\\ (x,t) \in \Sigma_T := \mathbb{R} \times (0,T)\end{array}\right. & (2.2)\\
g_y(x,e,t) = F_3(x,t), \quad (x,t) \in \Sigma_T & (2.3)\\
g(x+2L,y,t) = g(x,y,t), \quad (x,y,t) \in Q_T & (2.4)\\
g(x,y,0) = g_0(x,y), \quad (x,y) \in \mathbb{R} \times (0,e) & (2.5)\\
(f_t - \alpha\gamma f_{xx})(x,t) - g(x,0,t) = F_4(x,t), \quad (x,t) \in \Sigma_T & (2.6)\\
f(x+2L,t) = f(x,t), \quad (x,t) \in \Sigma_T & (2.7)\\
f(x,0) = f_0(x), \quad x \in \mathbb{R} & (2.8)
\end{cases}
$$

where T is a fixed (arbitrary) positive number, $F_1 \in C_{per}^{\lambda,\lambda/2}(\bar{Q}_T)$, $F_i \in C_{per}^{1+\lambda,(1+\lambda)/2}(\bar{\Sigma}_T)$ $i \in \{2,...,4\}$ for some $\lambda \in (0,1)$, $f_0 \in C_{per}^{3+\lambda}(\mathbb{R})$, $g_0 \in C_{per}^{2+\lambda}(\mathbb{R} \times [0,e])$ and satisfy the compatibility conditions $g_0y(x,e) = F_3(x,0)$ and $(g_0 - g_0y)(x,0) + \alpha\gamma f_0''(x) = F_2(x,0)$ for all $x \in \mathbb{R}$.

THEOREM 2. *Problem (L) has a unique solution*

$$(f,g) \in C_{per}^{3+\lambda,(3+\lambda)/2}(\bar{\Sigma}_T) \times C_{per}^{2+\lambda,1+\lambda/2}(\bar{Q}_T).$$

Furthermore there exists a constant $C = C(T)$ such that

$$\|g\|_{Q_T}^{(2+\lambda)} + \|f\|_{\Sigma_T}^{(3+\lambda)} \leq C\left(\|g_0\|_{\mathbb{R}\times(0,e)}^{(2+\lambda)} + \|f_0\|_{\mathbb{R}}^{(3+\lambda)} + \|F_1\|_{Q_T}^{(\lambda)} + \sum_{i=2}^{4}\|F_i\|_{\Sigma_T}^{(1+\lambda)}\right).$$

Sketch of the proof :

We define an affine map $\mathcal{L} : f \to \mathcal{L}(f) = h$ as follows : given $f \in C_{per}^{3+\lambda,(3+\lambda)/2}(\bar{\Sigma}_T)$, $f(0) = f_0$, we compute $g \in C_{per}^{2+\lambda,1+\lambda/2}(\bar{Q}_T)$ as the solution of equations (2.1)...(2.5) and $h = \mathcal{L}(f) \in C_{per}^{3+\lambda,(3+\lambda)/2}(\bar{\Sigma}_T)$ as the solution of equations (2.6)...(2.8).

Denote by $\mathcal{M}$ the corresponding linear map, we have to prove that $I_d - \mathcal{M}$ is invertible on $\{f \in C_{per}^{3+\lambda,(3+\lambda)/2}(\bar{\Sigma}_T), f(0) = 0\}$. We show that $\mathcal{M}$ is compact and $I_d - \mathcal{M}$ is one to one, so the result follows from Fredhlom alternative. To prove the uniqueness, we suppose that f satisfies $\mathcal{M}(f) = f$, we multiply the differential equation for f, first by f then by $-\alpha\gamma f_{xx}$ and integrate by parts. We also multiply the equation for g by g and integrate. Finally we sum the three inequalities obtained, use Young's inequality and the Lemma 1 below, and conclude thanks to Gronwall's lemma that $f = 0$.

LEMMA 1. *Let Ω be a bounded lipschitzian domain in $\mathbb{R}^2$. For all $\varepsilon > 0$, there exists a positive constant C_ε which is such that*

$$\int_{\partial\Omega} w^2(s)ds \leq \varepsilon \int\int_\Omega (\text{grad } w)^2 dz + C_\varepsilon \int\int_\Omega w^2 dz.$$

for all $w \in H^1(\Omega)$.

2. Uniqueness of the solution of Problem (P)

In fact, Problem (P) coincides with Problem (L) with :

$$F_1(f,g) = f_x^2 g_{yy} - 2f_x g_{xy} - f_{xx} g_y + f_t g_y$$

$$F_2(f,g) = \alpha(e^{-\gamma f_{xx}/(1+f_x^2)^{3/2}} - 1 + \gamma f_{xx}) + \left((1+f_x^2)^{1/2} - 1\right) g_y(x,0,t) - \\ - \frac{f_x}{(1+f_x^2)^{1/2}} g_x(x,0,t)$$

$$F_3(f,g) = \left(1 - (1+f_x^2)^{1/2}\right) g_y(x,e,t) + \frac{f_x}{(1+f_x^2)^{1/2}} g_x(x,e,t)$$

$$F_4(f,g) = -\alpha\left(1+f_x^2\right)^{1/2}\left(e^{-\gamma f_{xx}/(1+f_x^2)^{3/2}} - 1 + \gamma\frac{f_{xx}}{(1+f_x^2)^{1/2}}\right) + \\ + \left((1+f_x^2)^{1/2} - 1\right) g(x,0,t).$$

THEOREM 3. *There exists at most one solution (f,g) of Problem (P) such that*

$$f \in L^\infty(0,T;W^{3,\infty}(\mathbb{R})), \quad f_t \in L^\infty(\Sigma_T)$$
$$\text{and } g \in L^\infty(0,T;W^{2,\infty}(\mathbb{R}\times(0,e))), \quad g_t \in L^\infty(Q_T).$$

The method of proof is similar to the one used in the previous section to establish the uniqueness of the solution of the linear Problem (L). Let (f_1,g_1) and (f_2,g_2) be two solutions of Problem (P). First we multiply the difference of the differential equations for g_1 and g_2 by $g_1 - g_2$ and integrate by parts. Next, we write a differential equation for $f_1 - f_2$ and multiply this equation first by $f_1 - f_2$ then by $-(f_1 - f_2)_{xx}$. Adding the three inequalities obtained, we conclude, thanks to Gronwall's Lemma that $f_1 = f_2$ and $g_1 = g_2$.

3. Existence of the solution of Problem (P).

We recall that Problem (P) coincides with Problem (L) with the expressions of F_i, $i \in \{1,...,4\}$ given in the previous section. Let $f_0 \in C^{3+\lambda}_{per}(\mathbb{R})$ and $g_0 \in C^{2+\lambda}_{per}(\mathbb{R}\times[0,e])$ such that g_0 satisfies the compatibility conditions : $g_{0y}(x,e) = F_3(f_0,g_0)(x,0)$ and $(g_0 - g_{0y})(x,0) + \alpha\gamma f_0''(x) = F_2(f_0,g_0)(x,0)$ for all $x \in \mathbb{R}$. We introduce the affine space

$$\mathcal{A} = \left\{(f,g) \in C^{3+\lambda,(3+\lambda)/2}_{per}(\bar{\Sigma}_T) \times C^{2+\lambda,1+\lambda/2}_{per}(\bar{Q}_T), f(0) = f_0, g(0) = g_0\right\}$$

and endow $\mathcal{A}$ with the following norm :

$$\|(f,g)\| = \|f\|^{(3+\lambda)}_{\Sigma_T} + \|g\|^{(2+\lambda)}_{\bar{Q}_T}.$$

The main ingredient in the proof is the following key lemma.

LEMMA 2. *Let $\delta_0 > 0$ be arbitrary.*
1. *There exists a positive constant $K_1 = K_1(\delta_0)$ such that for all $\delta \in (0, \delta_0)$ and for all $(f,g) \in \mathcal{A}$ satisfying $\|(f,g)\| \leq \delta$ there holds*

$$\|F_1(f,g)\|^{\lambda}_{Q_T} + \sum_{i=2}^{4} \|F_i(f,g)\|^{(1+\lambda)}_{\Sigma_T} \leq K_1 \|(f,g)\|^2.$$

2. *There exists a positive constant $K_2 = K_2(\delta_0)$ such that for all $\delta \in (0, \delta_0)$ and for all $(f,g) \in \mathcal{A}$, $(h,k) \in \mathcal{A}$ satisfying* $\max\left(\|(f,g)\|, \|(h,k)\|\right) \leq \delta$ *there holds*

$$\|F_1(f,g) - F_1(h,k)\|^{\lambda}_{Q_T} + \sum_{i=2}^{4} \|F_i(f,g) - F_i(h,k)\|^{(1+\lambda)}_{\Sigma_T} \leq K_2 \delta \|(f-h, g-k)\|.$$

Proving these results is the most technical part of this work. It requires to estimate in Hölder spaces the norms of expressions such as :
· $f_1, f_2, \frac{1}{f_1}$ when f_i $i = 1,2$ belong to $C^{m+\lambda,(m+\lambda)/2}(\bar{Q}_T)$;
· $\varphi \circ f$ when $f \in C^{1+\lambda,(1+\lambda)/2}(\bar{Q}_T)$; $-M \leq f \leq M$ and $\varphi \in C^1([-M,M])$ with Lipschitz continuous derivatives.
· $\mu(r_1)$ and $\mu(r_1) - \mu(r_2)$ with $\mu(s) = e^{\gamma s} - 1 - \gamma s$ and

$$r_i \in C^{1+\lambda,(1+\lambda)/2}(\bar{\Sigma}_T), \ \|r_i\|^{(1+\lambda)}_{\lambda} \leq \delta, \ i = 1, 2.$$

For more details, we refer to [6,1].

THEOREM 4. *There exists a positive constant $\rho_0 = \rho_0(T)$ such that if (f_0, g_0) satisfy the conditions $\|f_0\|^{(3+\lambda)}_{\mathbb{R}} + \|g_0\|^{(2+\lambda)}_{\mathbb{R}\times(0,e)} \leq \rho_0$ then there exists a constant $\rho > \rho_0$ such that Problem (P) has a unique solution (f,g) in $C^{3+\lambda,(3+\lambda)/2}_{per}(\bar{\Sigma}_T) \times C^{2+\lambda,1+\lambda/2}_{per}(\bar{Q}_T)$ satisfying $\|(f,g)\| \leq \rho$.*

Sketch of the proof :

Consider the mapping $\tau : (f,g) \in \mathcal{A} \to \tau(f,g) = (\tilde{f}, \tilde{g})$ where $(\tilde{f}, \tilde{g})$ is the solution of the linear Problem (L) with initial condition (f_0, g_0) and right-hand sides $F_i = F_i(f,g)$ $i \in \{1, ..., 4\}$. Let $B(0,\rho)$ be the ball of center 0 and radius ρ. Then using the estimate in Theorem 2 about the linear Problem (L) and the previous estimates in Lemma 2, one can prove that there exists ρ_0 and $\rho > \rho_0$ such that if $\|f_0\|^{(3+\lambda)}_{\mathbb{R}} + \|g_0\|^{(2+\lambda)}_{\mathbb{R}\times(0,e)} \leq \rho_0$ then τ is a strict contraction from $B(0,\rho) \cap \mathcal{A}$ into itself. Thus there exists a unique solution (f,g) of Problem (P) in that set and Theorem 4 is proved.

References

[1] Abergel F., D. Hilhorst and F. Issard-Roch, On a dissolution growth problem with surface tension in the neighborhood of a stationary solution, to appear.

[2] Chen Xinfu, Generation and propagation of interfaces in reaction-diffusion systems, IMA Preprint Series n° 708 (1990).
[3] Chen X.Y., Dynamics of interfaces in reaction-diffusion systems, *Hiroshima Math. J. 21* (1991).
[4] Conrad F., D. Hilhorst and T.I. Seidman, Well-posedness of a moving boundary problem arising in a dissolution-growth process, *Nonlinear Analysis TMA 15* (1990), 445-465.
[5] Hilhorst D., F. Issard-Roch and T.I. Seidman, On a reaction-diffusion equation with a free boundary : the case of an unbounded domain, to appear in the Proceedings of : Free Boundary Problems, Theory and Applications, Montreal, (1990).
[6] Ladyzenskaja O.A., V.A. Solonnikov and N.N. Ural'ceva, Linear and quasilinear equations of parabolic types, Translations of Mathematical Monographs, Volume 23, Providence R.I., American Mathematical Society (1968).

F. Abergel, D. Hilhorst and F. Issard-Roch
Laboratoire d'Analyse Numérique
CNRS et Université Paris-Sud
Bâtiment 425,
91405 Orsay, France

B KAWOHL

On a class of singular elliptic equations

Abstract: The paper contains existence, uniqueness, regularity and concavity results for some nonlinear singular elliptic boundary value problems.

Introduction.

Consider the following model problem

$$\Delta u + a(x)\, u^{-p}(x) =0 \qquad \text{in } \Omega, \tag{1}$$

$$u = 0 \qquad \text{on } \partial\Omega, \tag{2}$$

where Ω is a bounded smooth domain in $I\!R^n$, the coefficient $a(x)$ is positive and bounded in Ω and the exponent p is positive. The results of the paper are actually derived for more general equations and boundary conditions. In [4,6] it was shown that if Ω is a ball and if a depends on the radial variable only, then for $0 < p < 1$ there exists a positive radial solution $u \in C^2(\Omega) \cap C^1(\overline{\Omega})$ of problem (1) (2). Moreover this solution is unique among all radial solutions. In the present paper we derive existence and uniqueness for general domains Ω in an elementary way. This implies that solutions of (1) (2) on a ball must be radial and that the uniqueness result in [4,6] holds in the wider class of general (possibly nonradial) solutions.

There are many results on semilinear equations in the mathematical literature, and the reader might wonder if the results of this paper are not already contained in some of the standard references. Therefore I wish to point out that in [2] the singular term has a sign opposite to the one in (1), and that [9] contains only results for regular nonlinearities. For autonomous nonlinearities, however, i.e. for constant $a(x)$, one can derive the existence theorem given here by monotone operator methods from [1,10]. In view of these methods it is not surprising that iterative methods have worked for regularizations of the problem, see [6].

After writing up the variational proof of the present paper I have learned from P.Waltman that A.C.Lazer and P.J.McKenna have found similar results on nonradial domains by the method of sub- and supersolutions. Moreover, their results (as well as some of mine) are already contained in a paper of M.Crandall, P.Rabinowitz and L.Tartar. Nevertheless I believe that the proofs presented here are of interest in view of their simplicity. Finally the present contribution contains some concavity results that are apparently new.

Results

Let us now consider the more general problem

$$\Delta u + f(x,u) = 0 \qquad \text{on } \Omega, \tag{3}$$

$$Bu = 0 \qquad \text{on } \partial\Omega, \tag{4}$$

where f satisfies some or all of the assumptions H1 – H5 below, and where $BU = \alpha\frac{\partial u}{\partial n} + \beta u$ for some numbers $\alpha \geq 0$ and $\beta > 0$. This includes the Dirichlet condition and the so-called third (or Robin) boundary condition. Here are the assumptions on f:

H1 $f : \Omega \times (0,\infty) \to (0,\infty)$ is Hölder-continuous.

H2 For each fixed $x \in \Omega$ the function f is decreasing in y. Moreover, $f(x,1) \in L^1(\Omega)$.

H3 $\lim\limits_{y\to 0+} f(x,y) = \infty$ and $\lim\limits_{y\to\infty} f(x,y) = 0$, with both limits being uniform on compact subsets of Ω.

H4 For every small $\theta > 0$ the integral $\int\limits_{\Omega} f(x, \theta\, d(x,\partial\Omega))\,dx$ exists. Here $d(x,\partial\Omega)$ is the distance of x to $\partial\Omega$.

H5 If $\int\limits_0^1 f(x,s)\,ds$ is unbounded for some $x \in \Omega$, and if $\alpha = 0$, then $f(x,u) = a(x)u^{-p}$ with $a(x)$ uniformly bounded on Ω and $p < 3$.

Remark 1: Assumption H4 is only needed to prove C^1-continuity of u up to the boundary, see Theorem 2, and Theorem 5 as one of its consequences. Assumption H5 appears to be technical and is only used to construct a function v for which the corresponding energy is finite. In [11] it is shown that one can remove the restriction $p < 3$. If $p \geq 3$ then u is no longer in $H^1(\Omega)$. There appears to be a way around this problem by working with functionals on $W^{1,q}(\Omega)$ with $q > 1$ sufficiently close to 1, but since it gets a little complicated it does not seem to be worth it in view of [11,12]. In [3] one can find Theorem 1 for the radial case without the structure assumption $f(x,u) = a(x)u^{-p}$ in H5. Therefore some of the results here exceed the ones given in [3,11,12], while others do not seem to be previously known.

Theorem 1. Existence and Uniqueness

Suppose that H1, H2, H3 *and* H5 *hold. Then Problem* (1.3) (1.4) *has a unique positive solution* $u \in C^2(\Omega)$. *It satisfies the boundary condition (1.4) in the sense of traces of* $H^1(\Omega)$ *functions.*

For the proof we introduce the function $F : \Omega \times I\!R \to I\!R \cup \{\infty\}$ defined by

$$F(x,u) = \begin{cases} \int_1^u f(x,s)\, ds & \text{if } u \geq 0 \text{ and if this integral exists,} \\ \infty, & \text{else,} \end{cases}$$

and the functional $E : H^1(\Omega) \to I\!R \cup \{\infty\}$ by

$$E(v) = \begin{cases} \int_\Omega \{ \tfrac{1}{2}|\nabla v|^2 - F(x,v)\}\, dx + \tfrac{\beta}{\alpha} \int_{\partial\Omega} v^2\, ds, & \text{if these exist,} \\ +\infty, & \text{else.} \end{cases}$$

Here I have abused notation. If $\alpha = 0$, then it is tacitly understood that E is only defined on $H_0^1(\Omega)$ so that the product which contains the boundary integral vanishes.

The functional E is strictly convex and its Euler equations are given by the weak version of (3) (4), i.e.

$$\int_\Omega \nabla u \nabla \phi - f(x,u)\, \phi\, dx + \int_{\partial\Omega} \frac{\beta}{\alpha}\, u\, \phi\, ds = 0. \tag{5}$$

Notice that upon integration by parts, this turns into

$$-\int_\Omega \{\Delta u + f(x,u)\}\, \phi\, dx + \int_{\partial\Omega} \left\{ \frac{\partial u}{\partial n} + \frac{\beta}{\alpha}\, u \right\} \phi\, ds = 0. \tag{6}$$

In order to get existence of a minimum for E we have to show that $E(v)$ is finite for some v and that E is coercive, i.e. $E(v) \to \infty$ whenever $v \to \infty$ in $H^1(\Omega)$.

To prove that E is finite for some v let us distinguish the cases $\alpha > 0$ and $\alpha = 0$. If $\alpha > 0$, then we can set $v = 1$ and see that $E(1)$ is finite. If $\alpha = 0$ and $\underline{\lim}_{s \to 0+} F(x, s) > -\infty$ for a.e. $x \in \Omega$ then $E(0)$ is finite. Finally suppose that H5 is nonvoid. Then for a suitably small, but positive constant γ we can pick $v = [d(x, \partial\Omega)]^\gamma$ and see that $E(v)$ is finite.

To prove the coerciveness of E take any function $v_0 \in H^1(\Omega)$ and consider $E(tv)$ for $t \to \infty$. Then

$$E(tv_0) = \int_\Omega \left\{ \frac{1}{2}|\nabla v_0|^2 t^2 - F(x, tv_0) \right\} dx + \frac{\beta}{\alpha} \int_{\partial\Omega} v_0^2 t^2 \, ds$$

is the difference of a quadratic function in t and the term

$$I(t) = \int_\Omega F(x, tv_0) \, dx.$$

We have to show that $I(t)$ grows less than quadratically in t. But this is easy once we recall that $I(1) = 0$, that $I'(1) = \int_\Omega f(x, 1) \, dx$ is finite and that $I(t)$ is concave.

This settles the existence question. Uniqueness follows from strict convexity of E. Notice that no strict monotonicity of f with respect to y as in [4,6] is needed.

∎

The following results on regularity for radial domains (Theorems 2 and 3) or for more general domains (Corollary 4) were already derived in [6,3] or [11,12] by different methods. Nevertheless the proofs presented here appear to be considerably shorter.

Theorem 2. Regularity near the boundary, the regular case

Suppose that hypotheses H1, H2 *and* H3 *hold. Let* $\Omega = B_R(0)$ *be a ball in* $\mathbb{R}^n$ *and suppose that* f *depends only on* r *and* u. *Then the solution* u *of problem* (3) (4) *is in* $C^0(\overline{\Omega})$ *and* $\underline{\lim}_{r \to R} u_r(r) < 0$.

Moreover, if hypothesis H4 *holds or if* $\alpha > 0$*, then* $u_r(R)$ *exists and is finite.*

The continuity statement follows from the fact that an H^1 function of a single (radial) variable is absolutely continuous. The negativity of u_r is an immediate consequence of Hopf's Lemma. For the last statement one observes that the case $\alpha > 0$ is covered by the boundary condition. If $\alpha = 0$, then we interpret u as a solution to a Neumann problem and see from the Fredholm alternative that the radial derivative of u is finite if and only if $f(r, u) \in L^1(\Omega)$. But u is bounded below by a small and positive multiple of $d(x, \partial\Omega)$. This and property H2 imply that

$$\int_\Omega f(r,u)\; dx \le \int_\Omega f(r, \theta\; d(x, \partial\Omega))\; dx < \infty,$$

which completes the proof of Theorem 2. ∎

Remark 2: Notice that condition H4 is satisfied e.g. for $f(r, u) = a(r)u^{-p}$ with $0 < p < 1$ and with $a(r) = (R - r)^{-q}$ ($q \in (0, 1 - p)$) degenerating on the boundary. This observation was kindly pointed out to me by V.Oliker.

Theorem 3. Regularity near the boundary, the singular case

Let $\Omega = B_R(0)$ *be a ball in* $\mathbb{R}^n$ *and suppose that* $f = a(r)u^{-p}$ *with* a *uniformly positive and bounded on* $[0, R]$ *and with* $p \geq 1$*. Finally suppose that* $\alpha = 0$*, i.e. the Dirichlet condition holds on* $\partial\Omega$*. Then there are two positive constants* c *and* C *such that* (7) *and* (8) *hold:*

$$c\,(R - r)^{2/(1+p)} \le\; u(r)\; \le C\,(R - r)^{2/(1+p)} \qquad \text{for } p > 1, \tag{7}$$

$$c\,(R - r) \le u(r)\; \le C(q)\,(R - r)^{2/(1+q)} \qquad \text{for } p = 1 \text{ and some } q > 1. \tag{8}$$

By a simple comparison argument it can be seen that if there are two positive constants a_0 and a_1 with $a_0 \le a(r) \le a_1$, then u is bounded by two functions $u_0 \le u \le u_1$, where u_i satisfies the equation

$$\Delta u_i + a_i\; u_i^{-p} = 0 \qquad \text{in } \Omega \tag{9}$$

under Dirichlet's boundary condition. Therefore it suffices to prove the Theorem for constant a.

Let me introduce the function $P(r,\alpha) = u_r^2 + \alpha\frac{a}{1-p}u^{1-p}$. Then for $\alpha = 2$ one sees immediately that

$$\frac{\partial P}{\partial r}(r,2) = -2\frac{n-1}{r}u_r^2 \leq 0$$

so that this and the observation $P(0,2) \leq 0$ implies $P(r,2) \leq 0$. An integration of the last inequality gives us the upper bound for u in (7).

To get the lower bound we pick $\alpha = 2/n$ and check that $\Delta P \geq 0$ in Ω, so that $P(r, 2/n)$ attains its maximum at $r = R$. In fact for every ball $B_r \subset B_R$ the function $P(\rho, 2/n)$ with $\rho \in [0,r]$ attains its maximum at r. This implies $P(r,2/n) \geq P(0,2/n)$ and thus the existence of a constant $\tilde{C}$ such that $-u_r \geq \tilde{C}\ u^{(1-p)/2}$, which in turn provides the lower bound for u in (7).

To prove (8) let us recall again that the lower bound follows from Hopf's Lemma. The upper bound follows from the observation that we can compare $f(r,u)$ with $\tilde{f}(r,u) = a_1\ \max\{u^{-q}, u^{-1}\}$ and thus reduce the proof to the case $p > 1$. ∎

Corollary 4.

Suppose that $\alpha = 0$, that Ω is convex and that $f(x,u) = u^{-p}$. Then the behaviour of u near $\partial\Omega$ is described by

$$\begin{aligned} cd(x\partial\Omega) \leq u &\leq Cd(x,\partial\Omega) && \text{if } 0 < p < 1, \\ cd(x\partial\Omega) \leq u &\leq Cd(x,\partial\Omega)^{2/(q+1)} \quad \text{with } q > 1 && \text{if } p = 1, \\ cd(x\partial\Omega)^{2/(q+1)} \leq u &\leq Cd(x,\partial\Omega)^{2/(q+1)} && \text{if } p > 1. \end{aligned}$$

This follows from Theorems 2 and 3 by comparison of Ω with inscribed and circumscribed balls. ∎

Numerical calculations in [3] suggest that solutions of $-\Delta u + u^{-p} = 0$ with Dirichlet boundary conditions should be concave at least if $p \in (0,1)$ and if Ω is a ball. Therefore let us give some concavity results first for balls, then for convex domains.

Theorem 5.

Suppose that $f(x,u) = f(u)$, *that* H1, H2 ,H3 *and* H4 *hold, that* $\alpha = 0$ *and that* $\Omega = B_R(0)$. *Then the function* u *is concave.*

We use the ordinary differential equation

$$u_{rr} + \frac{n-1}{r}\, u_r \; + f(u) = 0 \qquad r \in (0,R). \tag{10}$$

and see from (H1), Theorem 2, (10) and the boundary conditions $u_r(0) = 0$ and $u(R) = 0$ that $u_{rr}(0) \leq 0$ and $u_{rr}(R) \leq 0$. It remains to show that u_{rr} cannot attain a positive maximum in $(0,R)$. But if it did so in r_0, say, then

$$u_{rrr}(r_0) = 0 = \frac{n-1}{r^2}\, u_r + \frac{1-n}{r}\, u_{rr} - f'(u)u_r$$

and consequently (11) would hold:

$$u_{rr}(r_0) = \frac{1}{r}u_r - \frac{r}{n-1}f'(u)u_r. \tag{11}$$

If we multiply (11) by $n-1$ and add it to (10) we arrive at the desired contradiction

$$0 < n\; u_{rr}(r_0) = -rf'(u)u_r - f(u) \leq 0.$$

∎

Theorem 6.

Suppose that $f(x,u) = u^{-p}$, *that* $\alpha = 0$ *and that* Ω *is convex. Then for* $p \in (0,1)$ *the function* $u^{(1+p)/2}$ *is concave, while for* $p > 1$ *the stronger statement that the function* u *is concave holds.*

To prove these statements, one can use arguments from [7, Example 3 and Appendix]. For other nonlinearities $f(x,u)$ the reader is referred to [8]. ∎

Acknowledgement. This paper was written while I enjoyed the hospitality and financial support of the School of Mathematics, Georgia Institute of Technology. I am gratefully indebted to P. Waltman for providing me with copies of of [3,4,5] and [11] and for suggesting the problem to me.

References:

[1] Brezis,H. *Operateurs maximaux monotones et semigroupes non-lineaires*, Paris 1971, North Holland

[2] Diaz,J.I., J.M.Morel & L.Oswald, An elliptic equation with singular nonlinearity. Commun. in Partial Differ. Equations **12** (1987) 1333-1344.

[3] Fink,A.M., Gatica,J.A., Hernandez,G.E. & P.Waltman, Approximation of solutions of singular second order boundary value problems. SIAM J. Math. Anal. **22** (1991) 440-462.

[4] Gatica,J.A., Hernandez,G.E. & P.Waltman, Radially symmetric solutions of a class of singular elliptic equations, Proc. Edinburgh Math. Soc. **33** (1990) 169-180.

[5] Gatica,J.A., Oliker,V. & P.Waltman, Iterative procedures for nonlinear second order boundary value problems. Annali di Mat. Pura Applic. IV **157** (1990) 1-25.

[6] Gatica,J.A., Oliker,V. & P.Waltman, Singular nonlinear boundary value problems for second order differential equations. J. Differ. Equations **79** (1989) 62-79.

[7] B.Kawohl, A remark on Korevaar's concavity maximum principle and on the asymptotic uniqueness of solutions to the plasma problem. Math. Methods Appl. Sci. **8** (1986) 93-101.

[8] B.Kawohl, When are solutions to nonlinear elliptic boundary value problems convex? Comm. Partial Differ. Equations **10** (1985) 1213-1225.

[9] P.L.Lions, On the existence of positive solutions of semilinear elliptic equations. SIAM Review **24** (1982) 441-467.

[10] M.Schatzman, Problemes aux limites nonlineaires non coercifs. Ann. Sc. Norm. Sup. Pisa **27** (1973) 641-686.

[11] A.C.Lazer & P.J.McKenna, On a singular nonlinear elliptic boundary value problem. Proc. Amer. Math. Soc. **111** (1991) 451-453.

[12] M.G.Crandall, P.H.Rabinowitz & L.Tartar, On a Dirichlet problem with a singular nonlinearity. Commun. in Partial Differ. Equations **2** (1977) 193-222.

Bernhard Kawohl, SBF 123, Universität Heidelberg, Im Neuenheimer Feld 294, D 6900 Heidelberg, Germany

N KENMOCHI, T KOYAMA AND A VISINTIN

On a class of variational inequalities with memory terms

§0. Introduction.

This paper is concerned with a parabolic variational inequality of the form:

$$\begin{cases} u(\cdot,t) \in K(t), \quad 0 < t < T, \\ u(\cdot,0) = u_0, \\ \int_Q (u_t + Au + w - f)(u-v)dxdt + \int_0^T \hat{B}(u)dt \le \int_0^T \hat{B}(v)dt \\ \qquad \text{for all } v \in L^p(0,T;W^{1,p}(\Omega)) \text{ with } v(t) \in K(t), \ 0 < t < T. \end{cases} \tag{0.1}$$

Here $0 < T < \infty$, $2 \le p < \infty$, Ω is a bounded domain in $\mathbf{R}^N$ ($N \ge 1$), A is a (nonlinear) elliptic operator from $W^{1,p}(\Omega)$ into its dual space $W^{1,p}(\Omega)'$, $\hat{B}$ is a proper lower semi-continuous (l.s.c.) convex function on $L^2(\Omega)$, $K(t)$ ($0 \le t < \infty$) is a nonempty closed convex subset of $W^{1,p}(\Omega)$, and $u_0 \in L^2(\Omega)$ and $f \in L^2(Q)$ ($Q = (0,T) \times \Omega$) are data; u is the unknown and w is the source term, involving nonlinear memory effects, which is described in terms of a memory operator $\mathcal{F}$ (see section 1 for the definition) as

$$w(x,t) = [\mathcal{F}(u(x,\cdot); w_0(x))](t) \ \text{ for all } t \in [0,T] \text{ and a.e. } x \in \Omega. \tag{0.2}$$

In this case, $w(x,t)$ depends not only on the evolution of $u(x,\cdot)$ on $[0,t]$ but also on the initial value $w_0(x)$ for a.e. $x \in \Omega$.

As important examples of such a sort of memory operators, there are several classes such as (i) integral operators, (ii) nonlinear operators with time delay, and (iii) hysteresis operators. So far as nonlinear operators of class (i) or (ii) are concerned, we know many interesting results about nonlinear functional differential equations of Volterra type including them in the source terms (for instance, see Crandall-Nohel [4], Attouch-Damlamian [2], Aizicovici [1], Mitidieri-Vrabie [11]). In this paper we are especially interested in the last class (iii). Hysteresis models arising in physics and mechanics were systematically studied by Krasnosel'skii-Pokrovskii [10], and recently some types of evolution equations with hysteresis have been discussed by Visintin [13-16], Friedman-Hoffmann [5], Friedman-Jiang [6], Hilpert [7], Hoffmann-Kenmochi [8] and Kenmochi-Koyama [9].

If the system (0.1)-(0.2) is regarded as a nonlinear process of the heat conduction with time dependent constraint $K(t)$, then (0.2) can represent, for instance, as a distribution of thermostats characterized by continuous hysteresis loops. In the papers [9, 15, 16] quoted above, for evolution problems with hysteresis in source terms, the existence of solutions is mainly discussed as well as the uniqueness in some special cases. In this paper, we shall show that the existence of a solution of problem (0.1)-(0.2) is a direct consequence of the abstract result [9; Theorem 1.2] and its uniqueness is obtained in a more general setting than those of [15; Theorem 3] and [9; section 7].

NOTATIONS. In general, for a (real) Banach space V we denote by $|\cdot|_V$ the norm in V. By a (possibly multivalued) operator $A : V \to W$ for two spaces V, W we mean that A assigns to each $v \in V$ a subset Av of W; the set $D(A) := \{v \in V; Av \neq \phi\}$ is its domain and $R(A) := \cup_{v \in V} Av$ is its range. For a proper l.s.c. convex function $\varphi : V \to \mathbf{R} \cup \{+\infty\}$, V being a Banach space, $D(\varphi)$ denotes the effective domain, i.e. $D(\varphi) = \{v \in V; \varphi(v) < \infty\}$ and $\partial\varphi$ denotes the subdifferential.

§1. Formulation of the problem.

We begin with the definition of memory operators. Let $\mathcal{L}$ be a non-empty closed subset of $\mathbf{R}^2$ whose projection onto the first axis of $\mathbf{R}^2$ coincides with the whole line $\mathbf{R}$. Then $\mathcal{F}$ is called a (continuous) memory operator if the following conditions $(m.1) \sim (m.4)$ are fulfilled:

$(m.1)$ $\mathcal{F} \; : \; D(\mathcal{F}) := \{(v, \xi) \in C([0,T]) \times \mathbf{R} : (v(0), \xi) \in \mathcal{L}\} \to C([0,T])$.

$(m.2)$ (Causality) If $(v_i, \xi) \in D(\mathcal{F})$, $i = 1, 2$, and $v_1 = v_2$ on $[0, t)$ for $0 < t \leq T$, then $[\mathcal{F}(v_1, \xi)](t) = [\mathcal{F}(v_2, \xi)](t)$.

$(m.3)$ If $(v, \xi) \in D(\mathcal{F})$, then $[\mathcal{F}(v, \xi)](0) = \xi$ and $(v(t), [\mathcal{F}(v, \xi)](t)) \in \mathcal{L}$ for any $t \in [0, T]$.

$(m.4)$ (Continuity) If $\{(v_n, \xi_n)\} \subset D(\mathcal{F})$, $v_n \to v$ uniformly on $[0, T]$ and $\xi_n \to \xi$, then $\mathcal{F}(v_n, \xi_n) \to \mathcal{F}(v, \xi)$ uniformly on $[0, T]$ as $n \to \infty$.

Let Ω be a bounded domain in $\mathbf{R}^N$ $(N \geq 1)$ with boundary Γ of Lipschitz class, and let

$$0 < T < \infty, \quad 2 \leq p < \infty, \quad Q := \Omega \times (0, T).$$

For simplicity, set

$$H := L^2(\Omega), \quad V := W^{1,p}(\Omega), \quad V' := \text{the dual space of } W^{1,p}(\Omega),$$

and denote by $(\cdot, \cdot)_H$ the inner product in H, and by $\langle \cdot, \cdot \rangle$ the duality pairing between V and V'. Then by identifying H with its dual, we see that

$$V \subset H \subset V',$$

and the injections are compact.

Next, in order to introduce a nonlinear elliptic operator $A : D(A) = V \to V'$, let $\alpha_i : \mathbf{R} \to \mathbf{R}$, $i = 1, 2, \cdots, N$, be increasing and continuous functions such that

$$a_1|\xi|^p \leq \alpha_i(\xi)\xi \leq a_2|\xi|^p \quad \text{for all } \xi \in \mathbf{R},$$

where a_1 and a_2 are positive constants. Clearly, $\alpha_i(v) \in L^{p'}(\Omega)$ (where $p^{-1} + p'^{-1} = 1$) for all $v \in L^p(\Omega)$. As a typical example of α_i, we have $\alpha_i(\xi) = |\xi|^{p-2}\xi$.

With the family of functions $\{\alpha_i\}$, we define a maximal monotone operator $A : V \to V'$ by

$$\langle Au, v\rangle := \sum_{i=1}^{N} \int_\Omega \alpha_i(u_{x_i})v_{x_i}dx \quad \text{for} \quad u, v \in V.$$

Associated with A, we consider a function $\hat{A} : V \to \mathbf{R}^+$ given by

$$\hat{A}(z) := \sum_{i=1}^{N} \int_\Omega \hat{\alpha}_i(z_{x_i})dx \text{ for } z \in V, \tag{1.1}$$

where $\hat{\alpha}_i : \mathbf{R} \to \mathbf{R}^+$ is the primitive of α_i with $\hat{\alpha}_i(0) = 0$, i.e.

$$\hat{\alpha}_i(\xi) := \int_0^\xi \alpha_i(\eta)d\eta, \text{ for } \xi \in \mathbf{R}.$$

Clearly $\hat{A}$ is convex and continuous on V.

Moreover, let β be a maximal monotone graph in $\mathbf{R}^2$ such that

$$0 \in R(\beta), \tag{1.2}$$

and define a function $\hat{B} : H \to [0, \infty]$ by

$$\hat{B}(z) := \int_\Omega \hat{\beta}(z)dx \text{ for } z \in H, \tag{1.3}$$

where $\hat{\beta}$ is a primitive of β. By (1.2), without loss of generality, we may suppose $\hat{\beta} \geq 0$ on $\mathbf{R}$ and hence $\hat{B} \geq 0$ on H.

Finally, let $K(t)$ be a nonempty closed and convex subset of V for every $t \in [0, T]$, and let $u_0, w_0 \in H$ and $f \in L^2(Q)$.

Then our problem is formulated as follows:

Problem (P). Find a function $u \in W^{1,2}(0, T; H) \cap L^\infty(0, T; V)$ such that

$$u(\cdot, t) \in K(t) \text{ for all } t \in [0, T],$$

$$u(\cdot, 0) = u_0 \text{ in } H,$$

$$w(x, t) = [\mathcal{F}(u(x, \cdot), w_0(x))](t) \text{ for all } t \in [0, T] \text{ and a.e. } x \in \Omega, \tag{1.4}$$

and

$$(u'(t) + w(t) - f(t), u(t) - v)_H + \langle Au(t), u(t) - v\rangle + \hat{B}(u(t)) \le \hat{B}(v)$$
$$\text{for all } v \in K(t) \text{ and a.e. } t \in [0,T],$$

where $u' = (d/dt)u$.

REMARK 1. Under compatibility condition $(u_0(x), w_0(x)) \in \mathcal{L}$ (a.e. $x \in \Omega$) for initial data u_0 and w_0, condition (1.4) makes sense. In fact, we have

$$[\mathcal{F}(u(x,\cdot), w_0(x))] \in C([0,T]) \text{ for a.e. } x \in \Omega,$$

since $u \in W^{1,2}(0,T;H) \cap L^\infty(0,T;V)$ implies $u(x,\cdot) \in C([0,T])$ for a.e. $x \in \Omega$; this fact is due to the following Lemma 1.

LEMMA 1. *$Y := W^{1,2}(0,T;H) \cap L^\infty(0,T;V) \subset L^2(\Omega; C([0,T])) := X$ and the injection from Y into X is compact.*

PROOF. Fix any number $1/2 < \kappa < 1$. Then

$$Y \subset W^{1,2}(Q) \subset W^{\kappa,2}(Q) \subset W^{\kappa,2}(0,T;H) = L^2(\Omega; W^{\kappa,2}(0,T)) \subset X$$

and all injections are continuous. In particular, the injection from $W^{1,2}(Q)$ to $W^{\kappa,2}(Q)$ is compact so that $Y \subset X$ with compact embedding. □

§2. Existence result.

The solvability of problem (P) is discussed under the following smoothness condition $(\star)$ on the mapping $t \mapsto K(t)$.

$(\star)$ There are functions $c_1 \in W^{1,2}(0,T)$ and $c_2 \in W^{1,1}(0,T)$ such that for each $s,t \in [0,T]$ with $s \le t$ and $z \in K(s)$, there is $\tilde{z} \in K(t)$ satisfying

$$|\tilde{z} - z|_H \le |c_1(t) - c_1(s)|(1 + (\hat{A}(z) + \hat{B}(z))^{1/2})$$

and

$$(\hat{A}(\tilde{z}) + \hat{B}(\tilde{z})) - (\hat{A}(z) + \hat{B}(z)) \le |c_2(t) - c_2(s)|(1 + \hat{A}(z) + \hat{B}(z)).$$

THEOREM 1. *Let $\mathcal{F}$ be a memory operator characterized by $(m.1) \sim (m.4)$, and A, $\hat{B}$ be as defined in the previous section. Also, let $\{K(t)\}$ be a family of non-empty closed and convex subsets of V for which $(\star)$ holds. Furthermore, assume that*

$$\begin{cases} \text{There exists a constant } C > 0 \text{ such that} \\ |[\mathcal{F}(v,\xi)](t)| \le C \text{ for all } (v,\xi) \in D(\mathcal{F}) \text{ and } t \in [0,T], \end{cases} \tag{2.1}$$

$$u_0 \in K(0),\ \hat{B}(u_0) < \infty,\ w_0 \in H,\ (u_0(x), w_0(x)) \in \mathcal{L} \text{ for a.e. } x \in \Omega,$$

and

$$f \in L^2(Q).$$

Then problem (P) has at least one solution u with $w \in X$.

Under the same assumptions as in Theorem 1, we prove:

LEMMA 2. *Put*

$$X(T) := \cup_{0<t\le T} L^2(\Omega; C([0,t])),$$
$$X_0(T) := \{v \in X(T); v(\cdot,0) = u_0\}$$

and consider the operator $\tilde{\mathcal{F}} : X_0(T) \to X(T)$ which is given by

$$\tilde{\mathcal{F}}(v)(x,\cdot) := \mathcal{F}(v(x,\cdot), w_0(x))(\cdot) \text{ on } [0,t] \text{ for a.e.} x \in \Omega,$$
$$\text{all } v \in X_0(T) \text{ and } t \in (0,T].$$

Then, for each $t \in (0,T]$, $\tilde{\mathcal{F}}$ maps $L^2(\Omega; C([0,t])) \cap X_0(T)$ into $L^2(\Omega; C([0,t]))$ continuously with respect to the topology of $L^2(\Omega; C([0,t]))$.

PROOF. Let $0 < t \le T$, and let $\{v_n\}$ be a sequence in $L^2(\Omega; C([0,t])) \cap X_0(T)$ such that $v_n \to v$ in $L^2(\Omega; C([0,t]))$. Now let $\{v_{n_k}\}$ be any subsequence of $\{v_n\}$ such that

$$v_{n_k}(x,\cdot) \to v(x,\cdot) \text{ in } C([0,t]) \text{ as } k \to \infty$$

for a.e. $x \in \Omega$. Then, by conditon $(m.4)$,

$$\mathcal{F}(v_{n_k}(x,\cdot); w_0(x)) \to \mathcal{F}(v(x,\cdot); w_0(x)) \text{ in } C([0,t]) \text{ as } k \to \infty$$

for a.e. $x \in \Omega$. Besides, on account of (2.1) and Lebesgue's dominated covergence theorem, we have

$$\tilde{\mathcal{F}}(v_{n_k}) \to \tilde{\mathcal{F}}(v) \text{ strongly in } L^2(\Omega; C([0,t])).$$

Consequently, without extracting any subsequence of $\{v_n\}$, we see that $\tilde{\mathcal{F}}(v_n) \to \tilde{\mathcal{F}}(v)$ in $L^2(\Omega; C([0,t]))$ as $n \to \infty$. □

PROOF OF THEOREM 1. First, define a family $\{\varphi^t\}_{0\le t\le T}$ of proper l.s.c. convex functions φ^t on H by

$$\varphi^t(z) := \begin{cases} \hat{A}(z) + \hat{B}(z), & \text{if } z \in K(t), \\ \infty, & \text{otherwise.} \end{cases}$$

Then we see from formula (1.1) and condition $(\star)$ that

(φ.1) For each $s, t \in [0, T]$ with $s \le t$ and each $z \in D(\varphi^s)$, there is $\tilde{z} \in D(\varphi^t)$ such that

$$|\tilde{z} - z|_H \le |c_1(t) - c_1(s)|(1 + \varphi^s(z)^{1/2})$$

and

$$\varphi^t(\tilde{z}) - \varphi^s(z) \le |c_2(t) - c_2(s)|(1 + \varphi^s(z));$$

(φ.2) For each $R \ge 0$, the set

$$\cup_{0 \le t \le T}\{z \in H; \varphi^t(z) \le R, |z|_H \le R\}$$

is relatively compact in H.

lso, from condition (2.1) and Lemma 2, we see that $\tilde{\mathcal{F}}$ maps $L^2(\Omega; C([0,t]))$ into $L^\infty(\Omega; C([0,t]))$ d is continuous in $L^2(\Omega; C([0,t]))$ for every $t \in (0, T]$. Therefore, the mapping $G(v) :=$ $(v) - f$ satisfies conditions $(G1) \sim (G3)$ in Kenmochi-Koyama [9] and hence [9; Theorem 2] implies that problem

$$\begin{cases} u'(t) + \partial\varphi^t(u(t)) + G(u)(t) \ni 0 \text{ in } H, \ 0 < t < T, \\ u(0) = u_0, \end{cases} \tag{2.2}$$

as at least one solution $u \in W^{1,2}(0, T; H)$ such that $\varphi^{(\cdot)}(u) \in L^\infty(0, T)$. Here note that $\in \partial\varphi^t(z)$ if and only if $z \in K(t)$, $z^* \in H$ and

$$(z^*, v - z)_H + \langle Az, z - v\rangle + \hat{B}(z) \le \hat{B}(v) \text{ for all } v \in K(t).$$

ccordingly, evolution problem (2.2) is nothing but (P) and the above existence result ves a complete proof of Theorem 1. □

REMARK 2. As far as the existence of a solution of (P) is concerned, it is enough to ppose the following condition (2.3) instead of (2.1):

$$\begin{cases} \text{There is a constant } C' > 0 \text{ such that} \\ \quad |[\mathcal{F}(v, \xi)](t)| \le C'(|v|_{C([0,t])} + 1) \\ \text{for any } (v, \xi) \in D(\mathcal{F}) \text{ and } t \in [0, T]. \end{cases} \tag{2.3}$$

this case, the existence of a solution of (P) is not obtained as a direct application of [9; heorem 1.2], but the proof is quite similar.

§3. Uniqueness result and L^∞-estimate.

The uniqueness result is stated as follows:

THEOREM 2. *In addition to the assumptions of Theorem 1, suppose that*

$$\begin{cases} \text{for any } t \in [0,T], z_1, z_2 \in K(t) \text{ and } z \in V, \\ \text{if } \min\{z_1, z_2\} \le z \le \max\{z_1, z_2\}, \text{ then } z \in K(t), \end{cases} \tag{3.1}$$

and

$$\begin{cases} \text{there is a constant } L > 0 \text{ such that} \\ |\mathcal{F}(v_1,\xi) - \mathcal{F}(v_2,\xi)|_{C([0,t])} \le L|v_1 - v_2|_{C([0,t])}, \\ \text{for any } (v_i,\xi) \in D(\mathcal{F}),\ i = 1,2, \text{ and } t \in [0,T]. \end{cases} \tag{3.2}$$

Then problem (P) has at most one solution.

This theorem is obtained easily from the following Lemma on the auxiliary problem $(P)_\omega$, without memory term, for each $\omega \in L^2(Q)$:

$$\begin{cases} u(t) \in K(t) \text{ for all } t \in [0,T], \\ (u'(t) + \omega(t) - f(t), u(t) - v)_H + \langle Au(t), u(t) - v\rangle + \hat{B}(u(t)) \le \hat{B}(v) \\ \qquad \text{for all } v \in K(t) \text{ and a.e. } t \in [0,T], \\ u(0) = u_0. \end{cases} \tag{$(P)_\omega$}$$

As is mentioned in the proof of Theorem 1, this problem $(P)_\omega$ has one and only one solution $u \in W^{1,2}(0,T;H) \cap L^\infty(0,T;V)$ for each $\omega \in L^2(Q)$, and it is denoted by $k(\omega)$, i.e. $u = k(\omega)$.

LEMMA 3. *In addition to the assumptions of Theorem 1, suppose* (3.1) *holds. Then, for any* ω_1, $\omega_2 \in L^\infty(Q)$, *we have*

$$|(k(\omega_1) - k(\omega_2))^+|_{L^\infty(Q_t)} \le \int_0^t |(\omega_1 - \omega_2)^-|_{L^\infty(Q_\tau)} d\tau, \tag{3.3}$$

for all $t \in [0,T]$, *where* $Q_t := \Omega \times (0,t)$.

PROOF. For any $M > 0$ and $q > 2$ we define functions μ_M and $\mu_{M,q}$ on $\mathbf{R}$ by

$$\mu_M(\xi) := \min(|\xi|, M)\mathrm{sgn}(\xi),$$
$$\mu_{M,q}(\xi) := \int_0^\xi |\mu_M(\eta)|^{q-1}\mathrm{sgn}(\eta)d\eta = \begin{cases} q^{-1}|\xi|^q, & \text{if } |\xi| \le M, \\ q^{-1}M^q + M^{q-1}(|\xi| - M), & \text{otherwise.} \end{cases}$$

Clearly

$$-M^{q-2}\xi^- \le \mu'_{M,q}(\xi) = |\mu_M(\xi)|^{q-1}\mathrm{sgn}(\xi) \le M^{q-2}\xi^+.$$

Now, put $u_i = k(\omega_i)$, $(i = 1,2)$. Then

$$u_1 \ge u_1 - M^{-q+2}\mu'_{M,q}((u_1 - u_2)^+) \ge \min\{u_1, u_2\}.$$

Hence, (3.1) implies that

$$v_1 := u_1(t) - M^{-q+2}\mu'_{M,q}((u_1 - u_2)^+(t)) \in K(t).$$

Similarly $v_2 := u_2 + M^{-q+2}\mu'_{M,q}((u_1 - u_2)^+) \in K(t)$. Here, substituting v_i as the test function $v \in K(t)$ in the variational inequality of $(P)_{\omega_i}$, $i = 1, 2$, we have

$$\begin{aligned}&(u_1'(t) + \omega_1(t) - f(t), M^{-q+2}\mu'_{M,q}((u_1 - u_2)^+)(t))_H \\ &\quad + \langle Au_1(t), M^{-q+2}\mu'_{M,q}((u_1 - u_2)^+)(t)\rangle + \hat{B}(u_1(t)) \\ &\quad \le \hat{B}(u_1(t) - M^{-q+2}\mu'_{M,q}((u_1 - u_2)^+)(t)).\end{aligned}$$

and

$$\begin{aligned}&(u_2'(t) + \omega_2(t) - f(t), -M^{-q+2}\mu'_{M,q}((u_1 - u_2)^+)(t))_H \\ &\quad + \langle Au_2(t), -M^{-q+2}\mu'_{M,q}((u_1 - u_2)^+)(t)\rangle + \hat{B}(u_2(t)) \\ &\quad \le \hat{B}(u_2(t) + M^{-q+2}\mu'_{M,q}((u_1 - u_2)^+)(t)).\end{aligned}$$

Summing both sides of these inequalities yields

$$\begin{aligned}&(u_1'(t) - u_2'(t) + \omega_1(t) - \omega_2(t), M^{-q+2}\mu'_{M,q}((u_1 - u_2)^+)(t))_H \\ &\quad + \langle Au_1(t) - Au_2(t), M^{-q+2}\mu'_{M,q}((u_1 - u_2)^+)(t)\rangle + \hat{B}(u_1(t)) + \hat{B}(u_2(t)) \\ &\le \hat{B}(u_1(t) - M^{-q+2}\mu'_{M,q}((u_1 - u_2)^+)(t)) \\ &\quad + \hat{B}(u_2(t) + M^{-q+2}\mu'_{M,q}((u_1 - u_2)^+)(t)).\end{aligned} \tag{3.4}$$

The monotonicity of α_i and convexity of $\hat{\beta}$ imply

$$\langle Au_1 - Au_2, \mu'_{M,q}((u_1 - u_2)^+)\rangle \ge 0,$$

and

$$\hat{\beta}(u_2 + M^{-q+2}\mu'_{M,q}((u_1 - u_2)^+)) - \hat{\beta}(u_2) \le \hat{\beta}(u_1) - \hat{\beta}(u_1 - M^{-q+2}\mu'_{M,q}((u_1 - u_2)^+)),$$

respectively. From these inequalities and (3.4) it follows that

$$(u_1'(t) - u_2'(t), \mu'_{M,q}((u_1 - u_2)^+)(t))_H \le -(\omega_1(t) - \omega_2(t), \mu'_{M,q}((u_1 - u_2)^+)(t))_H. \tag{3.5}$$

Noting that $|\mu'_{M,q}(\xi)|^{q/(q-1)} = |\mu_M(\xi)|^q \le q\mu_{M,q}(\xi)$, we see from (3.5) that

$$\begin{aligned}&\left(\int_\Omega \mu_{M,q}((u_1 - u_2)^+(t))dx\right)^{(q-1)/q}\frac{d}{dt}\left(\int_\Omega \mu_{M,q}((u_1 - u_2)^+(t))dx\right)^{1/q} \\ &= \frac{d}{dt}\frac{1}{q}\int_\Omega \mu_{M,q}((u_1 - u_2)^+(t))dx \\ &= \frac{1}{q}(\mu'_{M,q}((u_1 - u_2)^+(t)), u_1'(t) - u_2'(t))_H \\ &\le q^{-1/q}|(\omega_1 - \omega_2)^-(t)|_{L^q(\Omega)}\left(\int_\Omega \mu_{M,q}((u_1 - u_2)^+(t))dx\right)^{(q-1)/q},\end{aligned}$$

i.e.

$$\frac{d}{dt}\left(\int_\Omega \mu_{M,q}((u_1 - u_2)^+(t))dx\right)^{1/q} \le q^{-1/q}|(\omega_1 - \omega_2)^-(t)|_{L^q(\Omega)},$$

and thus

$$|\mu_M((u_1 - u_2)^+(t))|_{L^q(\Omega)} \le \int_0^t |(\omega_1 - \omega_2)^-(\tau)|_{L^q(\Omega)}d\tau.$$

Letting $q \to \infty$ and $M \to \infty$ give

$$|(u_1 - u_2)^+(t)|_{L^\infty(\Omega)} \le \int_0^t |(\omega_1 - \omega_2)^-(\tau)|_{L^\infty(\Omega)} d\tau.$$

Accordingly, (3.3) holds. □

PROOF OF THEOREM 2. Let u_1 and u_2 be two solutions of (P). Then, by Lemma 3 and condition (3.2),

$$|u_1 - u_2|_{L^\infty(Q_t)} \le L \int_0^t |(u_1 - u_2)|_{L^\infty(Q_\tau)} d\tau,$$

for all $t \in [0, T]$. Hence Gronwall's Lemma implies $|u_1 - u_2|_{L^\infty(Q_t)} = 0$, i.e. $u_1 = u_2$. □

As to the L^∞-estimate for solutions we have the following theorem.

THEOREM 3. *In addition to the assumptions of Theorem* 1, *suppose that* $u_0 \in L^\infty(\Omega)$, $f \in L^\infty(Q)$, (3.1) *holds and*

$$\begin{cases} \text{there are constants } m_* \text{ and } m^* \text{ such that} \\ \min\{z, m_*\} \in K(t), \max\{z, m^*\} \in K(t) \text{ for any } t \in [0,T] \text{ and } z \in K(t). \end{cases} \tag{3.6}$$

Then any solution u *of* (P) *satisfies*

$$|u|_{L^\infty(Q)} \le T(C|\Omega|^{1/2} + |f|_{L^\infty(Q)}) + |u_0|_{L^\infty(\Omega)} + k_0, \tag{3.7}$$

where k_0 *is a constant depending only on* m_*, m^* *and* β.

PROOF. Let u be a solution of (P) and r_β be a number so that $0 \in \beta(r_\beta)$. Putting $m'_* := \max\{m_*, r_\beta\}$, with the same notations as before we see that

$$u \ge u - M^{-q+2}\mu'_{M,q}((u - m'_*)^+) \ge \min\{u, m'_*\} \ge \min\{u, m_*\},$$

for any $M > 0$ and $q > 2$. This shows by (3.1) and (3.6) that

$$u(t) - M^{-q+2}\mu'_{M,q}((u - m'_*)^+(t)) \in K(t).$$

Therefore from the variational inequality of (P) we derive

$$\begin{aligned} &(u'(t) + \tilde{\mathcal{F}}(u)(t) - f(t), M^{-q+2}\mu'_{M,q}((u - m'_*)^+)(t))_H \\ &+\langle Au(t), M^{-q+2}\mu'_{M,q}((u - m'_*)^+(t))\rangle + \hat{B}(u(t)) \\ &\le \hat{B}(u(t) - M^{-q+2}\mu'_{M,q}((u - m_*)^+(t))). \end{aligned} \tag{3.8}$$

Here, note that

$$\langle Au, \mu'_{M,q}((u - m'_*)^+)\rangle \ge 0,$$

and

$$\hat{\beta}(u - M^{-q+2}\mu'_{M,q}((u - m'_*)^+) \le \hat{\beta}(u),$$

since $m'_* \ge r_\beta$ and $0 \in \beta(r_\beta)$. Now, in a calculation similar to that in the proof of Lemma 3, from (3.8) with the above two inequalities it follows that

$$|(u - m'_*)^+(t))|_{L^\infty(\Omega)} \le |(u_0 - m'_*)^+|_{L^\infty(\Omega)} + t|\tilde{\mathcal{F}}(u) - f|_{L^\infty(Q_t)}$$

for all $t \in [0, T]$, which shows

$$|u^+(t)|_{L^\infty(\Omega)} \le \max\{|u_0|_{L^\infty(\Omega)}, m_*, r_\beta\} + t(C|\Omega|^{1/2} + |f|_{L^\infty(Q_t)}).$$

for all $t \in [0, T]$. A similar estimate for $|u^-(t)|_{L^\infty(\Omega)}$ is obtained. Thus (3.7) holds for $k_0 = \max\{m_*, -m^*, |r_\beta|\}$. □

§4. Examples.

In this section we give some examples of A, $\hat{B}$, $\mathcal{F}$ and $\{K(t)\}$ in typical nonlinear variational inequalities.

The simplest example of A is given by (1.1) with $a_i(\xi) = |\xi|^{p-2}\xi$, that is

$$\langle Au, v\rangle := \sum_{i=1}^{N} \int_\Omega |u_{x_i}|^{p-2} u_{x_i} v_{x_i} dx, \quad u, v \in V.$$

Also as a convex function $\hat{B} : H \to [0, \infty]$, consider the function given by (1.3) with $\hat{\beta} = |\xi|^l$ $(l \ge 1)$.

Next, as a memory operator $\mathcal{F}$ consider the Preisach model for hysteresis (cf. [12]); for its mathematical representation and investigation of fundamental properties including (3.2), we refer to Visintin [15] and Brokate-Visintin [3]. In the remainder of this paper, let A, $\hat{B}$ and $\mathcal{F}$ be as specified above, and consider nonlinear parabolic variational problems with time-dependent constraints $K(t)$.

EXAMPLE 1. Let $g \in W^{1,2}(0, T; H) \cap L^\infty(0, T; V) \cap L^\infty(Q)$ and define $K(t)$, $0 \le t \le T$, by

$$K(t) = \{z \in V; z = g(\cdot, t) \text{ on } \Gamma \text{ in the sense of traces }\}.$$

In this case, conditions $(\star)$, (3.1) and (3.6) are satisfied. To verify condition $(\star)$, it is enough to take $z - g(s) + g(t)$ as $\tilde{z}$. Therefore, Theorems 1,2 and 3 are applicable to (P) associated with the above $K(t)$. It is easy to see that (P) can be written in the following form:

$$\begin{cases} u_t - \sum_{i=1}^{N} (|u_{x_i}|^{p-2} u_{x_i})_{x_i} + w + \beta(u) \ni f \text{ in } Q, \\ w(x,t) = [\mathcal{F}(u(x,\cdot), w_0(x))](t) \text{ in } Q, \\ u = g \text{ on } \Sigma := \Gamma \times (0,T), \\ u(\cdot, 0) = u_0 \text{ in } \Omega. \end{cases} \tag{4.1}$$

EXAMPLE 2. Let g be as in Example 1, and consider for each $t \in [0,T]$

$$K(t) = \{z \in V; z \geq g(\cdot,t) \text{ on } \Gamma\}.$$

Then, conditions $(\star)$, (3.1) and (3.6) are satisfied in this case too. And Theorems 1, 2 and 3 are applicable to the corresponding variational inequality (P) which is written in the form (4.1) with the boundary condition replaced by

$$\begin{cases} u \geq g \text{ on } \Sigma, \\ \sum_{i=1}^{N} |u_{x_i}|^{p-2} u_{x_i} n_i \geq 0 \text{ on } \Sigma, \\ \sum_{i=1}^{N} |u_{x_i}|^{p-2} u_{x_i} n_i (u-g) = 0 \text{ on } \Sigma, \end{cases}$$

where $n(x) = (n_1(x), n_2(x), \cdots, n_N(x))$ is the outward unit normal vector to Γ at $x \in \Gamma$.

EXAMPLE 3. Let g_i, $i = 1, 2$, be Lipschitz continuous functions in $\bar{Q}$ such that $g_2 - g_1 \geq \delta_0$ on $\bar{Q}$ for a positive constant δ_0 and $\nabla g_i \in W^{1,2}(0,T; L^p(\Omega))$, $i = 1, 2$. Put

$$K(t) = \{z \in V; g_1(\cdot,t) \leq z \leq g_2(\cdot,t) \text{ a.e. on } \Omega\}, 0 \leq t \leq T.$$

Then conditions $(\star)$, (3.1) and (3.6) hold; to verify $(\star)$ it is enough to take

$$\frac{g_2(t) - g_1(t)}{g_2(s) - g_1(s)}(z - g_1(s)) + g_1(t)$$

as $\tilde{z} \in K(t)$ for any $z \in K(s)$. Therefore, applying Theorems 1 and 2 to (P) associated with the above $K(t)$, we see that the problem has one and only one solution u. As is easily seen, this problem is formally equivalent to the following system:

$$g_1 \leq u \leq g_2 \text{ in } Q,$$

$$u_t - \sum_{i=1}^{N} (|u_{x_i}|^{p-2} u_{x_i})_{x_i} + w + \tilde{\beta} \begin{cases} = f & \text{in } \{g_1 < u < g_2\} \cap Q, \\ \geq f & \text{in } \{u = g_1\} \cap Q, \\ \leq f & \text{in } \{u = g_2\} \cap Q, \end{cases}$$

$$\sum_{i=1}^{N} |u_{x_i}|^{p-2} u_{x_i} n_i \begin{cases} = 0 & \text{on } \{g_1 < u < g_2\} \cap \Sigma, \\ \leq 0 & \text{on } \{u = g_1\} \cap \Sigma, \\ \geq 0 & \text{on } \{u = g_2\} \cap \Sigma, \end{cases}$$

$$w(x,t) = [\mathcal{F}(u(x,\cdot), w_0(x))](t) \text{ in } Q,$$

$$u(\cdot, 0) = u_0 \text{ in } \Omega.$$

References

[1] S. Aizicovici, Time-dependent Volterra integrodifferential equations, J. Integral Equations **10**, 45-60 (1985).

[2] H.Attouch and A. Damlamian, A nonlinear Volterra equations in variable domain, MRC Tech. Summary Rep. # 2028, Madison, Wisconsin, 1979.

[3] M. Brokate and A. Visintin, Properties of the Preisach model of hysteresis, J. Reine Angew. Math. **402**, 1-40 (1989).

[4] M. G. Crandall and J. Nohel, An abstract functional differential equation and a related nonlinear Volterra equation, Israel J. Math. **29**, 313-328 (1978).

[5] A. Friedman and K.-H. Hoffmann, Control of free boundary problems with hysteresis, SIAM J. Control Optim. **26**, 42-55 (1988).

[6] A. Friedman and Li-Shang Jiang, Periodic solutions for a thermostat control problem, Comm. P. D. E. **13**, 515-550 (1988).

[7] M. Hilpert, On uniqueness for evolution problems with hysteresis, Mathematical Models for Phase Change Problems, ed. J. F. Rodrigues, Intern. Ser. Numer. Math. Vol. 88 ,377-388, Birkhäuser, Basel-Boston-Berlin, 1989.

[8] K.-H. Hoffmann and N. Kemmochi, Two-phase Stefan problems with feedback controls, Mathematical Models for Phase Change Problems, ed. J. F. Rodrigues, Intern. Ser. Numer. Math. Vol. 88, 239-260, Birkhäuser, Basel-Boston-Berlin, 1989.

[9] N. Kenmochi and T. Koyama, Nonlinear functional variational inequalities governed by time-dependent subdifferentials, to appear in Nonlinear Anal.

[10] M. A. Krasnoselskiǐ and A. V. Pokrovskiǐ, Systems with hysteresis (Russian), Nauka, Moskow, 1983. English translation: Springer, Berlin, 1989.

[11] E. Mitidieri and I. Vrabie, A class of strongly nonlinear functional differential equations, Ann. Mat. pura appl. 151, 125-147 (1988).

[12] F. Preisach, Über die magnetische Nachwirkung, Z. Phys. **94**, 277-302 (1935).

[13] A. Visintin, A model for hysteresis of distributed systems, Ann. Mat, pura appl, **131**, 203-231 (1982).

[14] A. Visintin, A phase transition problem with delay, Control Cyb. **11**, 5-18 (1982).

[15] A. Visintin, On the Preisach model for hysteresis, Nonlinear Anal. **8**, 977-996 (1984).

[16] A. Visintin, Evolution problems with hysteresis in source term, SIAM J. Math. Anal. **17**, 1113-1138 (1986).

Nobuyuki KENMOCHI : Department of Mathematics, Faculty of Education, Chiba University, Yayoi-cho, Chiba, 260 Japan.

Tetsuya KOYAMA : Department of Mathematics, Hiroshima Institute of Technology, Miyake, Saiki-ku, Hiroshima, 731-51 Japan.

Augusto VISINTIN : Dipartimento di Matematica, Universita degli Studi di Trento, 38050 Povo (Trento), Italy.

N KUTEV

Global solvability and boundary gradient blow up for one dimensional parabolic equations

The aim of this paper is to investigate the global solvability and the blow up of the solutions of the Dirichlet problem for nonlinear parabolic equations

$$u_t-u_{xx}=f(t,x,u,u_x) \quad \text{in } Q=(a,b)x(0,\infty),$$
$$u(a,t)=A(t),u(b,t)=B(t) \text{ for } t\geq 0,u(x,0)=\psi(x) \text{ for } x\in[a,b] \tag{1}$$

where $\psi(x)\in C^{2,\alpha}[a,b]$ and $A(t),B(t)\in C^{2,\alpha}[0,\infty)$ are bounded functions, $\alpha\in(0,1)$. The motivation to consider this problem is the natural interest to describe the necessary and sufficient conditions for the classical solvability of nonlinear parabolic equation (1) in terms of global solvability or blow ups of the solutions of (1). In this way a precise explanation of the classical nonexistence results for elliptic and parabolic equations /see for example [1],[8],[10]/ will be given by means of the boundary gradient blow up phenomenon.

Let us note that the global solvability and the blow up of the amplitude of the solutions of parabolic equations with nonlinear terms depending only on the unknown function u, was during the last ten years object of investigation for many mathematicians /see for example the references in [2]/. As for the general parabolic equations depending additionaly on the gradient of the solution, only in [2],[4],[7] the role of the damping term $|Du|^q$, $q\leq 2$, was studied but it was again in connection with the blow up of the amplitude of the solutions. Firstly, the gradient blow up phenomenon was indirectly pointed out in a counterexample in [5] in the one dimensional case and later this example was generalized in the multidimensional case in [3] where existence of a special domain and data was shown for which boundary gradient blow up of the solution appears.

Unlike the above papers, in the present one we will give precise conditions for global existence of a classical solution of

(1) and if these conditions fail then it will be proved that either the amplitude of the solution at infinity or the gradient of the solution on the boundary for a finite time /when the difference of the data $B(t)-A(t)$ is sufficiently large/, blows up. In fact, (1) is a uniformly parabolic equation and the critical growth of $f(t,x,u,u_x)$ with respect to u_x in order to obtain boundary gradient blow ups is slightly greater than two /see (5), (8)/. As for fully nonlinear nonuniformly parabolic equations

$$\begin{aligned} &u_t - F(t,x,u,Du,D^2u) = 0 \quad \text{in } Q=\Omega\times(0,\infty) \\ &u=\psi \quad \text{on the parabolic boundary} \end{aligned} \tag{2}$$

where Ω is a bounded domain in R^n, it will be shown in a forthcoming paper that the boundary gradient blow up phenomenon depends on the behaviour of the principal symbol at infinity, the geometry of the domain Ω and the higher norms of the data. More precisely, the sign of the following Bernstein-Serrin function

$$B(\psi,\phi) = F(t,x,\psi,D\psi+s\nu,D^2\psi+sD\nu-s^2\phi(s^2)\nu\otimes\nu.\text{sign } s)-\psi_t$$

for sufficiently large values of the parameter s defines the critical growth of (2). Here x is in a sufficiently small neighbourhood of $\partial\Omega$, $t\geq 0$, $\nu(x)$ is the inner normal to $\partial\Omega$ at the point $y(x)\in\partial\Omega$ nearest to x, the positive nondecreasing function $\phi\in C[0,\infty)$ satisfies $(3)_i$ and $\nu\otimes\nu$ is the matrix $\{\nu_i\nu_j\}$, $i,j=1,2,\ldots n$. In order to formulate the main results in the present paper let us denote with ϕ and Φ, resp., continuous, positive nondecreasing functions which satisfy the growth conditions

$$\int_1^\infty dt/t\phi(t)=\infty \qquad \text{or} \qquad \int_1^\infty dt/t\Phi(t)<\infty , \tag{3}$$

resp. For example, for $\phi(t)=\ln(2+t)$ and $\Phi(t)=\ln^{1+\varepsilon}(2+t)$, $\varepsilon>0$, condition (3) holds. Suppose that $f(t,x,u,u_x)$ satisfies the condition

$$f(t,x,z,p)\text{sign}(z)\leq g(x)/h(p) \quad \text{where } \int_a^b g(x)dx<\int_{-\infty}^\infty h(p)dp \tag{4}$$

for $t\geq 0$, $x\in[a,b]$, $|z|\geq m_0$, $p\in R$, for some positive functions $g\in C[a,b]$, $h\in C(R)$ and for some positive constant m_0. Moreover, in a sufficient small neighbourhood N_a and N_b of the endpoints a,b we need the additional assumptions

$$|f(t,x,z,p)|\leq\phi^{1+\beta}(p^2)(x-a)^\beta(1+p^2)^{1+\beta/2} \quad \text{for } x\in N_a \tag{5$_a$}$$

$$|f(t,x,z,p)| \leq \phi^{1+\beta}(p^2)(b-x)^{\beta}(1+p^2)^{1+\beta/2} \quad \text{for } x \in N_b \qquad (5)_b$$

and for $t \geq 0$, $\inf u \leq z \leq \sup u$, $p \in R$. Here β is some nonnegative constant and the function ϕ satisfies $(3)_i$.

Theorem 1: Suppose $f \in C^1$ satisfies (4),(5), the compatibility conditions $\psi(a)=A(0)$, $\psi(b)=B(0)$, $A'(0)-\psi''(a)-f(0,a,\psi(a),\psi'(a))=0$, $B'(0)-\psi''(b)-f(0,b,\psi(b),\psi'(b))=0$, as well as at least one of the following conditions

$$|f(t,x,z,p)| \leq Cp^2 \; ; \qquad (6)_i$$

$$f_z(t,x,z,p)+f_x(t,x,z,p)/p \leq 0 \; ; \qquad (6)_{ii}$$

$f_z+f_x/p \leq |pf_p-f|/2(C+1)\operatorname{osc} u$ and pf_p-f has a constant $(6)_{iii}$ sign for $x \in [a,b]$, $t \geq 0$, $\inf u \leq z \leq \sup u$, $|p| \geq L_0$, where L_0, C are some positive constants.

Then (1) has a unique bounded global $C^{2+\alpha,1+\alpha/2}([0,\infty)\times[a,b])$ solution.

The idea of the proof of theorem 1 is to find out global with respect to t a priori estimates for the amplitude and the gradient of the solutions of (1) and to apply the method of continuity or some fixed point theorems. For this purpose by means of (4) global sup- and subsolutions of (1) are given using the solutions of the ordinary differential equations $v''=\pm g(x)/h(u_x)$ in (a,b). As for the boundary gradient estimates, for example at the end point a, the local barriers in a neighbourhood of a are of the form $w(x)=A(t)+(B(t)-A(t))(x-a)/(b-a)+v(x-a)$, where $v(s) \in C^2[0,\varepsilon]$ for some $\varepsilon>0$ has the following properties: $v(0)=0$, $v' \geq M$, $v(\varepsilon) \geq K$, $v''(s)=-\phi^{1+\beta}(v'^2)s^{\beta}(1+v'^2)^{1+\beta/2}$ for some sufficiently large nonnegative constants M,K. Differentiating (1) with respect to x and multiplying with u_x we obtain second order differential inequality for u_x^2 which gives us by means of (6) the desired global gradient estimates.

Condition (4) guarantees global a priori estimates for the amplitude of the solutions of (1) and is the best possible one. Indeed, if (4) fails, but (5),(6) still hold, then the solution of (1) blows up at infinity. More precisely, we have the following result.

Theorem 2: Suppose $f \in C^1$ satisfies (5), **at** least one of the conditions (6) as well as the condition

$$|f(t,x,z,p)|=g(x)/h(p) \quad , \int_a^b g(x)dx=\int_{-\infty}^{\infty} h(p)dp \tag{7}$$

for some positive functions $g\in C^1[a,b], h\in C^1(R)$.

Then (1) has a unique global $C^{2+\alpha,1+\alpha/2}([0,\infty)x[a,b])$ solution which blow up at infinity i.e. $\lim u(t,x)=\infty$ for $t\to\infty$.

Sketch of the proof:By means of theorem 1 global sup- and subsolutions of (7) can be foun out as solutions of the equations $v_t-v_{xx}=\pm(g(x)-\varepsilon)/h(u_x)$,for some $\varepsilon>0$ and for some suitable initial and boundary data.Then the proof of theorem 2 follows from the fact that $v_\varepsilon(t,x)\to\infty$ when $t\to\infty,\varepsilon\to 0$.

Let us note that the blow up of the solutions in theorem 2 is not due to the source term f(u) as in the well-investigated cases of semilinear parabolic equations.In fact,(4) which is a sufficient condition for a priori estimates of the amplitude of the solutions of (1) is necessary for gradient a priori estimates and if (4) fails i.e. (7) holds then gradient blow ups of the solutions appear.More precisely,the gradient of the solution in theorem 2 blows up on the boundary at infinity and becouse of condition (5) this necessarily leads to blow up of the amplitude of the solution at infinity,too.However,if (5) does not hold,as in theorem 3,then only boundary gradient blow up of the solutions of (1) holds i.e. the amplitude of the solutions remains bounded.The same effect was investigated by Giusti [6] and Urbas [11] for nonuniformly elliptic equations of curvature type without boundary conditions.They proved that under conditions similar to those in (7) the solutions are vertical on the boundary.**Unlike the elliptic** case,for parabolic equations the above phenomenon appear only at infinity.

As for (5),it guarantees global a priori estimates for the gradient of the solution of (1) at the endpoints a,b,provided that the amplitude of the solutions has been already estimated. When one of conditions $(5)_a$ or $(5)_b$ fails,then a boundary gradient phenomenon appears.In order to formulate the results let us assume that either

$$|f(t,x,z,p)|\geq\Phi^{1+\beta}(p^2)(x-a)^\beta(1+p^2)^{1+\beta/2} \quad \text{for } x\in N_a \text{ ,or} \tag*{$(8)_a$}$$

$$|f(t,x,z,p)|\geq\Phi^{1+\beta}(p^2)(b-x)^\beta(1+p^2)^{1+\beta/2} \quad \text{for } x\in N_b \tag*{$(8)_b$}$$

holds for $t>0, z\in R, p\in R$, for some constant $\beta>0$ and for some function Φ satisfying $(3)_{ii}$. Moreover, in order to localize the super-growth condition (8) in a neighbourhood N_a or N_b of the points a and b, we need the following assumption for the sign of f: either

$$f(t,x,z,p)>0 \quad \text{for } t>0, x\in[a,b], z>m_1, p<-L_1, \text{ or} \tag*{$(9)_i$}$$

$$f(t,x,z,p)<0 \quad \text{for } t>0, x\in[a,b], z<-m_1, p>L_1 \tag*{$(9)_{ii}$}$$

for some nonnegative constants m_1, L_1.

Theorem 3: Suppose $f\in C^1$ satisfies (4),(8),(9) as well as one of conditions (6). Then for all data A(t),B(t), whose difference is sufficiently large the gradient of the solution of (1) blows up on the boundary for a finite time.

Remark: An estimate from below and from above for the critical difference $|B-A|$ and for the time of the blow up of the gradient on the boundary can be given in theorem 3, too.

The idea of the proof of theorem 3 is, using (9), to localize the problem in a neighbourhood of the end point a /or b/. Then by means of the barrier function of the form $w(t,x)=v(x)+\psi(x)+C(t-T)$, where C is a suitably chosen positive constant and $v(a)=0$, $v'(a)=\infty$, to prove by means of (8) that w(t,x) is a subsolution of (1). Thus the solution of (1) blows up at the end point a at the latest at the time T.

Acknowledgement: This work was accomplished during the author's research period as A.v.Humboldt fellow at the University of Karlsruhe.

REFERENCES

1. S.Bernstein, Conditions necessaires et suffisantes pour la possibilite du probleme de Dirichlet, C.R.Acad.Sci.Paris, 150 (1910), 514-515.

2. M.Chipot, F.Weissler, Some blowup results for a nonlinear parabolic equations with a gradient term, SIAM J.Math.Anal. 20 (1989), 886-907.

3. T.Dlotko, Examples of parabolic problems with blowing-up derivatives, J.Math.Anal.Appl., 154 (1991), 226-237.

4. M.Fila, Remarks on blow up for a nonlinear parabolic equation with a gradient term, Proc.Amer.Math.Soc. 111 (1991), 795-801.

5.A.F.Filipov,Conditions for the existence of a solution to a quasilinear parabolic equation,DAN SSSR,141 (1961),568-570.

6.E.Giusti,On the equation of surface of prescribed mean curvature,Invent.Math.,46 (1978),111-137.

7.B.Kawohl,L.A.Peletier,Observations on blow up and dead cores for nonlinear parabolic equations,Math.Z.,202 (1989),207-217.

8.G.Lieberman,The first initial-boundary value problem for quasilinear second order parabolic equations,Ann.Scuola Norm. Sup.Pisa,13 (1986),347-387.

9.R.Redheffer,W.Walter,Counterexamples for parabolic differential equations,Math.Z.,153 (1977),229-236.

10.J.Serrin,The problem of Dirichlet for quasilinear elliptic differential equations with many independent variables,Philos. Trans.Roy.Soc.London Ser.A 264 (1969),413-496.

11.J.Urbas,The equation of prescribed Gauss curvature without boundary conditions,J.Differential Geometry,20 (1984),311-327.

Nickolai Kutev
Institute of Mathematics
Bulgarian Academy of Sciences
P.O.Box 373,Sofia,Bulgaria

J LIANG

Weakly-coercive quasilinear elliptic equations with inhomogeneous measure data

Abstract. We extend a result of existence of weak solutions to a Dirichlet problem for a class of quasilinear elliptic operators which are weakly coercive with the lower order terms and depend on u and Du nonlinearly. The free term of the equation only belongs to the set of bounded measures.

§0 Introduction

In this paper, we consider the following problem:

$$(0.1) \qquad \begin{cases} -div(A(x,u,Du)) + A_0(x,u,Du) = f & \text{in } \Omega \\ u = 0 & \text{on } \partial\Omega \end{cases}$$

where $f \in M(\Omega)$, and Ω is a bounded open set of $\mathbf{R}^N$. $M(\Omega)$ denotes the set of bounded measures on Ω.

The existence of the solution for this kind of problem is well know when A and A_0 are linear (see [S]). In [BG], the authors have studied the nonlinear case when A is coercive and is independent of u and the lower order term A_0 is independent of Du, with A satisfying some condition which is stronger than strictly monotonicity. The existence of the weak solution for the problem has been obtained. In the remarks, the paper also mentions that A_0 can depends on Du. Recently, in [R], the author studied the case when A is coercive, depends on u nonlinearly and only satisfies a strictly monotone condition without lower order term involved.

We will study the more general case in which A is weakly coercive and is nonlinear both to u and Du, the lower order term A_0 depends on u and Du. We also discuss the sign conditions.

The following hypotheses on $A = (A_1, \cdots, A_N)$ and A_0 will be assumed in this paper:

(H1) Growth condition on the main term:

$$|A_i(x,u,\xi)| \le g_i(x) + e_i(x)|u|^{\rho_i} + b_i(x)|\xi|^{p-1}$$

where $g_i \in L_{p'}$, $e_i(x) \in L_{r_i}$, $0 \le \rho_i < \frac{N(p-1)(r_ip-r_i-p)}{(N-p)r_ip} \wedge (p-1)$ and $r_i > \frac{p}{p-1}$, $i = 1, 2, \cdots, N$. If $r_i = \infty$, and $p \ge \sqrt{N}$, then ρ_i can reach $p-1$. $a \wedge b = \min\{a,b\}$, and correspondingly, $a \vee b = \max\{a,b\}$.

(H2) Weakly coercive condition:

This work is supported by Fundação Oriente in Portugal.

$$A_i(x,u,\xi)\xi_i \geq \alpha|\xi|^p - g_i(x)|\xi| - e_i(x)|u|^{\rho_i}|\xi|$$

where $\alpha > 0$ is a constant. g_i, e_i, and ρ_i, $i = 1, 2, \cdots, N$ are as above.

(H3) Monotonous condition:

$$(A_i(x,u,\xi) - A_i(x,u,\eta))(\xi_i - \eta_i) > \mu|\xi - \eta|$$

where $\mu > 0$.

(H4) Growth condition on the lower order term:

$$|A_0(x,u,\xi)| \leq e_0(x)|u|^{\rho_0} + b_0(x)|\xi|^{\omega}$$

where $e_0 \in L_{r_0}$, $b_0 \in L_\lambda$, and $r_0 \geq \frac{N}{p}$, $0 \leq \rho_0 < p-1$, $\lambda > N$, $0 \leq \omega < p-1$.

(H5) Continuity condition:

$A_i(x,u,\xi)$ and $A(x,u,\xi)$ are continuous with respect to all their arguments in the sense given by [LU] Chap.4, Theorem 8.8.

An example of A_i and A_0 satisfying our hypotheses is:

$$\begin{cases} A_i(x,u,\xi) = \alpha_i(x)|\xi|^{p-2}\xi_i + \beta_i(x)|u||\xi|^{p-2} + \gamma_i(x,u) \\ A_0(x,u,\xi) = \sigma(x,u)f(\xi) \end{cases}$$

When $p > N$, we know that $M(\Omega) \subset W^{-1,p'}$, and in this case, the existence of a solution is already known. (See [BG]). So we are just interested in the case of $p \leq N$. We will discuss this problem in four sections. In §1, we give some auxiliary lemmas; in §2, we obtain the *a priori* estimate of the solution in terms of $\|f\|_{L_1}$, and in §3, the limit process is discussed. Some remarks are included in §4.

§1 Some Auxiliary Lemma

LEMMA 1. *If f is smooth enough, there exists a solution to problem (0.1) under the hypotheses (H1)-(H5).*

This result can be found in [LU], Chap. 4 Theorem 8.8. The only thing we need is to verify the conditions of the Theorem, and that is not difficult to do.

LEMMA 2. *Let u_n be a sequence of $W_0^{1,p}(\Omega)$ having the following properties:*

a) There exists q, $1 < q \leq p$ such that u_n remains in a bounded set of $W_0^{1,q}(\Omega)$, and u_n converges to u (weakly pointwise).

b) For all $k > k_0$, z_n^k remains in a bounded set of $W_0^{1,p}(\Omega)$ as n goes to infinity, where $z_n^k = [k - (k - |u_n|)^+]sign(u_n)$ (respectively $z^k = [k - (k - |u|)^+]sign(u)$).

c) For all $k > k_0$, there exists a real function c_k such that for all $0 < \epsilon < \epsilon_0$,

$$\limsup_n \int_\Omega h_\epsilon(|u_n - z_k|)[A(x,u_n,Du_n)][D(u_n - z^k)]dx \leq c_k(\epsilon),$$

with $c_k(\epsilon)$ going to zero when ϵ goes to zero. Here h_ϵ is the characteristic function of $[0,\epsilon]$, and A satisfies the hypotheses (H1)-(H3). Then, the sequence Du_n converges to Du almost everywhere.

This Lemma was given by [R] when $A(x,u,Du)$ is coercive. In our case when A is weakly coercive, the proof is the same.

§2 The *A Priori* Estimate Of The Solution

In this section, we suppose $f \in L_1(\Omega)$.

Let us now get an *a priori* estimate and verify the condition of Lemma 2. Here we use the truncation method, used by Boccardo-Gallouët to obtain the estimate of $\|u\|_{W^{1,q}}$ for some $q < p$ in terms of $\|f\|_{L_1}$. (See [BG].)

We also consider the method used by Rakotoson to obtain the estimate. (See [R].) The results obtained by these two methods have slightly different limits on the growth of u in A_i (ρ_i).

Let

$$(2.1)\qquad \begin{cases} u &= \sum_{m=1}^{\infty} u_m \\ u_m &= \{[(u-(m-1))\vee 0]\wedge 1\} \\ &\quad + \{[(u+(m-1))\wedge 0]\vee(-1)\} \\ \Omega_m &= \{x\in\Omega, m-1<|u(x)|\le m, |Du(x)|>0\} \\ R_m &= \{x\in\Omega, |u(x)|>m-1\} \end{cases}$$

Choose q such that:

$$\begin{aligned} &1\vee(p-1)\vee\frac{\lambda(p-1)}{\lambda-1}\vee\frac{\rho_0 N}{N-p-\rho_0}\vee\max_i\{\frac{\sigma_i N}{N+\sigma_i}\} \\ &<q<\frac{N(p-1)}{N-1}\le p \end{aligned}$$

where $q^* = \frac{Nq}{N-q}$, $\sigma_i = \rho_i\frac{r_i p}{r_i p - r_i - p}$.

This kind of q can be found because from the hypotheses, we have $2-\frac{1}{N}<p\le N$, $\lambda > N$, $\rho_0 < p-1$, and $\sigma_i = \rho_i\frac{r_i p}{r_i p-r_i-p} < \frac{N(p-1)}{N-p}$, then $\frac{N(p-1)}{N-1} > 1\vee(p-1)\vee\frac{\lambda(p-1)}{\lambda-1}\vee\frac{\rho_0 N}{N-p-\rho_0}\vee\max_i\{\frac{\sigma_i N}{N+\sigma_i}\}$.

Now from (0.1) and the hypotheses (H1)-(H5), we take u_m as a test function to multiply the first equation of (0.1), then integrate it by parts to obtain:

$$
\begin{aligned}
&\alpha\|Du_m\|_p^p \\
&\le \int_{\Omega_m} A_i(x,u,Du)D_iu_m dx \\
&+ \int_{\Omega_m} g_i|D_iu_m|dx + \int_{\Omega_m} |e_i||u|^{\rho_i}|Du_m|dx \\
&\le -\int_{R_m} A_0(x,u,Du)u_m dx + \int_{R_m} fu_m dx \\
&+ \int_{\Omega_m} g_i|D_iu_m|dx + \int_{\Omega_m} |e_i||u|^{\rho_i}|Du_m|dx \\
&\le \int_{R_m} |e_0||u|^{\rho_0}|u_m|dx + \int_{R_m} |b_0||Du|^{\omega}|u_m|dx + \|f\|_1 \\
&+ \|g_i\|_{p'}\|D_iu_m\|_p + \|e_i\chi(\Omega_m)\|_{r_i}\|u\chi(\Omega_m)\|_{\sigma_i}^{\rho_i}\|Du_m\|_p \\
&\le \int_{R_m} |e_0||u|^{\rho_0}dx + \left(\int_{R_m} |b_0|^{\frac{q}{q-\omega}}dx\right)^{\frac{q-\omega}{q}}\left(\int_{R_m} |Du|^q dx\right)^{\frac{\omega}{q}} \\
&+ \|f\|_1 + \frac{\alpha}{2}\|Du_m\|_p^p \\
&+ C_\alpha\sum_i \|g_i\|_{p'}^{p'} + C_\alpha\|e_i\chi(\Omega_m)\|_{r_i}^{\frac{p}{p-1}}\|u\chi(\Omega_m)\|_{\sigma_i}^{\rho_i\frac{p}{p-1}}
\end{aligned}
$$

$\chi(Q)$ is the character function of the set Q, for $Q \subset \mathbf{R}^N$, that is,

$$
\chi(Q) = \begin{cases} 1 & x \in Q \\ 0 & x \notin Q \end{cases}
$$

And $\|b_0\|_{\frac{q}{q-\omega}} \le C\|b_0\|_\lambda$, since $q > \frac{\lambda(p-1)}{\lambda-1}$ and $\omega < p-1$.

Then, let us denote by $\theta_i = \frac{p}{p-1}\frac{\rho_i}{\sigma_i} = \frac{r_ip-r_i-p}{r_i(p-1)}$. From the above formula, we have the following estimate:

$$
\begin{aligned}
\frac{\alpha}{2}\|Du_m\|_p^p &\le C\left[1 + \sum_{n=m}^{\infty}\int_{\Omega_n} |e_0||u|^{\rho_0}dx\right] \\
&+ C\left[\left(\sum_{n=m}^{\infty}\|Du_n\|_q^q\right)^{\frac{\omega}{q}} + \|e_i\chi(\Omega_m)\|_{r_i}^{\frac{p}{p-1}}(\|u\chi(\Omega_m)\|_{\sigma_i}^{\sigma_i})^{\theta_i}\right]
\end{aligned} \tag{2.2}
$$

where C depends only on the given data and $\|f\|_1$.

Since $q < p$, where q is as before, we can use Hölder's Inequality to obtain:

$$
\begin{aligned}
&\left(\frac{\alpha}{2}\right)^{\frac{q}{p}} \|Du_m\|_q^q \\
\le &\left(\frac{\alpha}{2}\right)^{\frac{q}{p}} \left(\|Du_m\|_p^p\right)^{\frac{q}{p}} \left(mes(\Omega_m)\right)^{1-\frac{q}{p}} \\
\le &C^{\frac{q}{p}} \left(mes(\Omega_m)\right)^{1-\frac{q}{p}} + C^{\frac{q}{p}} \left(\sum_{n=m}^{\infty} \int_{\Omega_n} |e_0||u|^{\rho_0} dx \right)^{\frac{q}{p}} \left(mes(\Omega_m)\right)^{1-\frac{q}{p}} \\
&+ C^{\frac{q}{p}} \left(\sum_{n=m}^{\infty} \|Du_n\|_q^q \right)^{\frac{\omega}{p}} \left(mes(\Omega_m)\right)^{1-\frac{q}{p}} \\
&+ C^{\frac{q}{p}} \left[\|e_i \chi(\Omega_m)\|_{r_i}^{\frac{p}{p-1}} \left(\|u\chi(\Omega_m)\|_{\sigma_i}^{\sigma_i}\right)^{\theta_i} \right]^{\frac{q}{p}} \left(mes(\Omega_m)\right)^{(1-\frac{q}{p})} \\
\le &I_1 + I_2 + I_3 + I_4
\end{aligned}
$$

Because in Ω_m, $u > m - 1$. It follows that:

$$
mes(\Omega_m) \le \frac{C}{m^\alpha} \int_\Omega |u|^\alpha dx
$$

for some $\alpha > 0$.

Then we have:

$$
\begin{aligned}
I_1 \le &C \left(\frac{1}{m^{q^*}} \int_{\Omega_m} |u|^{q^*} dx \right)^{1-\frac{q}{p}} \\
\le &C \left(\frac{1}{m^\beta} \right)^{\frac{q}{p}} \left(\int_{\Omega_m} |u|^{q^*} dx \right)^{1-\frac{q}{p}}
\end{aligned}
$$

where

$$
\beta = \frac{N(p-q)}{N-q} \begin{cases} = N > 1 & p = N \\ > N \dfrac{p - \frac{N}{N-1}(p-1)}{N - \frac{N}{N-1}(p-1)} = 1 & p < N \end{cases}
$$

And,

$$
\begin{aligned}
I_2 \le &C \left(\sum_{n=m}^{\infty} \int_{\Omega_n} |e_0||u|^{\rho_0} dx \right)^{\frac{q}{p}} \frac{1}{m^{\frac{\gamma q}{p}}} \left(\int_{\Omega_m} |u|^{\frac{\gamma q}{p-q}} dx \right)^{1-\frac{q}{p}} \\
\le + &C \left(\frac{1}{m^\gamma} \sum_{n=m}^{\infty} \int_{\Omega_n} |e_0||u|^{\rho_0} dx \right)^{\frac{q}{p}} \left(\int_{\Omega_m} |u|^{\frac{\gamma q}{p-q}} dx \right)^{1-\frac{q}{p}}
\end{aligned}
$$

where $1 < \gamma \le \beta \wedge \{p - \omega\} \wedge \{p - \rho_0\}$, for $\beta > 1$, $p - \omega > 1$, and $\rho_0 < p - 1$, γ can be chosen.

Also,

$$
\begin{aligned}
I_3 \leq & C\left(\sum_{n=m}^{\infty} \|Du_n\|_q^q \right)^{\frac{\omega}{p}} \frac{1}{m^{\frac{\gamma q}{p}}} \left(\int_{\Omega_m} |u|^{\frac{\gamma q}{p-q}} dx \right)^{1-\frac{q}{p}} \\
\leq & C\left(\frac{1}{m^{\gamma}} \sum_{n=m}^{\infty} \|Du_n\|_q^q \right)^{\frac{\omega}{p}} \left(\int_{\Omega_m} |u|^{\frac{\gamma q}{p-q}} dx \right)^{1-\frac{q}{p}}
\end{aligned}
$$

where γ is chosen as above.

As well as,

$$
\begin{aligned}
I_4 \leq & C\left[\left(\int_{\Omega_m} |e_i|^{r_i} dx \right)^{\frac{p}{r_i(p-1)}} \left(\int_{\Omega_m} |u|^{\sigma_i} dx \right)^{\theta_i} \right]^{\frac{q}{p}} (mes(\Omega_m))^{(1-\frac{q}{p})} \\
\leq & C\left[\left(\int_{\Omega_m} |e_i|^{r_i} dx \right)^{(1-\theta_i)} \left(\int_{\Omega_m} |u|^{\sigma_i} dx \right)^{\theta_i} \right]^{\frac{q}{p}} (mes(\Omega_m))^{(1-\frac{q}{p})}
\end{aligned}
$$

noting that, $1-\theta_i = 1 - \frac{r_i p - r_i - p}{r_i(p-1)} = \frac{p}{r_i(p-1)}$.

So, by Hölder's Inequality, summing up on m from n_0 to infinity, we have:

$$
\begin{aligned}
\sum_{m=n_0}^{\infty} I_1 \leq & C\left(\sum_{m=n_0}^{\infty} \frac{1}{m^{\beta}} \right)^{\frac{q}{p}} \left(\sum_{m=n_0}^{\infty} \int_{\Omega_m} |u|^{q^*} dx \right)^{1-\frac{q}{p}} \\
\leq & C_{n_0} \left(\int_{\Omega} |u|^{q^*} dx \right)^{1-\frac{q}{p}} = C_{n_0} \|u\|_{q^*}^q + C
\end{aligned}
$$

where $C_{n_0} \to 0$ as $n_0 \to \infty$, because of the convergence of the series ($\gamma, \beta > 1$). And $q^*(1-\frac{q}{p}) \leq \frac{qp}{p-q}\frac{p-q}{p} = q$.

$$
\begin{aligned}
\sum_{m=n_0}^{\infty} I_2 \leq & C\left(\sum_{m=n_0}^{\infty} \frac{1}{m^{\gamma}} \sum_{n=m}^{\infty} \int_{\Omega_n} |e_0||u|^{\rho_0} dx \right)^{\frac{q}{p}} \left(\sum_{m=n_0}^{\infty} \int_{\Omega_m} |u|^{\frac{\gamma q}{p-q}} dx \right)^{1-\frac{q}{p}} \\
\leq & C\left(\sum_{n=n_0}^{\infty} \int_{\Omega_n} |e_0||u|^{\rho_0} dx \sum_{m=n_0}^{n} \frac{1}{m^{\gamma}} \right)^{\frac{q}{p}} \left(\int_{\Omega} |u|^{\frac{\gamma q}{p-q}} dx \right)^{1-\frac{q}{p}} \\
\leq & C_{n_0} \left(\int_{\Omega} |e_0||u|^{\rho_0} dx \right)^{\frac{q}{p}} \left(\int_{\Omega} |u|^{\frac{\gamma q}{p-q}} dx \right)^{1-\frac{q}{p}} \\
\leq & C_{n_0} \|e_0\|_{r_0}^{\frac{q}{p}} \|u\|_{r_0'\rho_0}^{\frac{\rho_0 q}{p}} \|u\|_{\frac{\gamma q}{p-q}}^{\frac{\gamma q}{p}} \leq C_{n_0} \|u\|_{q^*}^q + C
\end{aligned}
$$

where C_{n_0} is as above. Here, we have changed the order of the sums and used the fact that $r_0 \geq \frac{N}{p}$, $q > \frac{\rho_0 N}{N-p-\rho_0}$, thus $r_0'\rho_0 \leq q^*$, and $\frac{\gamma q}{p-q} \leq q^*$, $q\frac{\rho_0+\gamma}{p} \leq q$.

$$\sum_{m=n_0}^{\infty} I_3 \leq C\left(\sum_{m=n_0}^{\infty}\frac{1}{m^\gamma}\left(\sum_{n=m}^{\infty}\|Du_n\|_q^q\right)^{\frac{\omega}{q}}\right)^{\frac{q}{p}}\left(\sum_{m=n_0}^{\infty}\int_{\Omega_m}|u|^{\frac{\gamma q}{p-q}}dx\right)^{1-\frac{q}{p}}$$

$$\leq C\left[\left(\sum_{m=n_0}^{\infty}\frac{1}{m^\gamma}\right)^{1-\frac{\omega}{q}}\left(\sum_{n=n_0}^{\infty}\|Du_n\|_q^q\sum_{m=n_0}^{n}\frac{1}{m^\gamma}\right)^{\frac{\omega}{q}}\right]^{\frac{q}{p}}$$

$$\cdot\left(\sum_{m=n_0}^{\infty}\int_{\Omega_m}|u|^{\frac{\gamma q}{p-q}}dx\right)^{1-\frac{q}{p}}$$

$$\leq C_{n_0}\left(\sum_{n=n_0}^{\infty}\|Du_n\|_q^q\right)^{\frac{\omega}{p}}\|u\|_{\frac{\gamma q}{p-q}}^{\frac{\gamma q}{p}}$$

$$\leq \frac{1}{2}\left(\frac{\alpha}{2}\right)^{\frac{q}{p}}\|Du\|_q^q + C_{\alpha n_0}\|u\|_{q^*}^{\frac{\gamma q}{p-\omega}} + C$$

$$\leq \frac{1}{2}\left(\frac{\alpha}{2}\right)^{\frac{q}{p}}\|Du\|_q^q + C_{\alpha n_0}\|u\|_{q^*}^{q} + C$$

where $\frac{\gamma q}{p-q} \leq q^*$, $\frac{\gamma q}{p-\omega} \leq q$.

$$\sum_{m=n_0}^{\infty} I_4 \leq C\left[\left(\sum_{m=n_0}^{\infty}\int_{\Omega_m}|e_i|^{r_i}dx\right)^{1-\theta_i}\left(\sum_{m=n_0}^{\infty}\int_{\Omega_m}|u|^{\sigma_i}dx\right)^{\theta_i}\right]^{\frac{q}{p}}$$

$$\cdot\left[\sum_{m=n_0}^{\infty}mes(\Omega_m)\right]^{1-\frac{q}{p}}$$

$$\leq C\left[\left(\sum_{m=n_0}^{\infty}\int_{\Omega_m}|e_i|^{r_i}dx\right)^{1-\theta_i}\left(\int_{\Omega}|u|^{\sigma_i}dx\right)^{\theta_i}\right]^{\frac{q}{p}}(mes(\Omega))^{1-\frac{q}{p}}$$

$$\leq C_{n_0}\|u\|_{\sigma_i}^{\frac{q\rho_i}{p-1}} \leq C_{n_0}\|u\|_{q^*}^q + C$$

where $\sigma_i \leq q^*$, $\rho_i \leq p-1$.

So,

$$\sum_{m=n_0}^{\infty}\left(\frac{\alpha}{2}\right)^{\frac{q}{p}}\|Du_m\|_q^q = \sum_{m=n_0}^{\infty}\sum_i I_i$$

$$\leq C_{n_0}\|u\|_{q^*}^q + \frac{1}{2}\left(\frac{\alpha}{2}\right)^{\frac{q}{p}}\|Du\|_q^q + C$$

When $m \leq n_0$, we can obtain the following estimate from (2.2) without difficulty:

$$\|Du_m\|_q^q \leq C(n_0,\epsilon) + \frac{\epsilon}{n_0}(\|u\|_{q^*}^q + \|Du\|_q^q)$$

where n_0 will be determined below and ϵ can be chosen as small enough.

That is, for all together, we have:

$$\|Du\|_q^q = \sum_{m=1}^{\infty} \|Du_m\|_q^q \leq (C_{n_0} + \epsilon)\|u\|_{q^*}^q + C$$

By Sobolev's Inequality [A], we have:

$$\|u\|_{q^*}^q \leq C_{q^*}^q \|Du\|_q^q \leq C_{q^*}^q [C + (C_{n_0} + \epsilon)\|u\|_{q^*}^q]$$

Taking n_0 big enough, then taking ϵ small enough, we have:

$$\|u\|_{q^*}^q \leq C$$

and

$$\|Du\|_q^q \leq C$$

where C depends only on the given data and $\|f\|_1$.

These are the estimates we need.

Moreover, from the calculation, we can see that if $r_i = \infty$, $\rho_i = p - 1$, $\sigma_i = p$. So the condition $\sigma_i \leq q^*$ becomes $p \leq q^*$. Thus, if we have the condition $p \geq \sqrt{N}$, and choose $\frac{N}{1+\sqrt{N}} \leq q \leq \frac{N}{N-1}(p-1)$, the condition $p \leq q^*$ can be satisfied. That is, if $p \geq \sqrt{N}$, and $r_i = \infty$, ρ_i can reach $p - 1$. But in this method, when $p < \sqrt{N}$, even if $r_i = \infty$, ρ_i is only allowed to be smaller than $\frac{N(p-1)^2}{(N-p)p}$, which is strictly smaller than $p - 1$.

Now let us prove that if $r_i = \infty$, ρ_i also can close to $p - 1$ arbitrarily.

Let us use Rakotoson method ([R]) to obtain the estimate of $\|u\|_{W^{1,q}}$ for the same $q < p$ in terms of $\|f\|_{L_1}$.

For $\delta > 0$, let us define

$$\phi(t) = \frac{1}{\delta}\left(1 - \frac{1}{(1+|t|)^\delta}\right)\mathrm{sign}(t)$$

Taking $v = \phi(u)$ as a test function, we have:

$$\begin{aligned}
&\alpha \int_\Omega \frac{|Du|^p}{(1+|u|)^{1+\delta}} dx \\
&\leq - \int_\Omega A_0(x, u, Du)\phi(u) dx + \int_\Omega f\phi(u) dx \\
&\quad + \int_\Omega \frac{g_i(x)|Du|}{(1+|u|)^{1+\delta}} dx + \int_\Omega \frac{e_i(x)|u|^{\rho_i}|Du|}{(1+|u|)^{1+\delta}} dx \\
&\leq \int_\Omega \frac{|u|^{\sigma_i}}{(1+|u|)^{1+\delta}} dx + \frac{\alpha}{2}\int_\Omega \frac{|Du|^p}{(1+|u|)^{1+\delta}} dx \\
&\quad + \frac{C}{\delta}\int_\Omega |u|^{\rho_0} dx + \frac{C}{\delta}\int_\Omega |Du|^\omega dx + C\Big(\|f\|_1 + \sum_i \|g_i\|_{p'}^{p'}\Big)
\end{aligned}$$

Notice that here,

$$
\begin{aligned}
&\int_\Omega |u|^\nu dx \le \int_\Omega |Du|^\nu dx \\
\le &\epsilon \int_\Omega \frac{|Du|^p}{(1+|u|)^{1+\delta}} dx + C_\epsilon \int_\Omega (1+|u|)^{(1+\delta)\frac{\nu}{p-\nu}} dx \\
\le &C_\epsilon + \epsilon \int_\Omega \frac{|Du|^p}{(1+|u|)^{1+\delta}} dx + \frac{1}{2} \int_\Omega |u|^\nu dx
\end{aligned}
$$

as long as $\nu < p - 1 - \delta$.

Since we can choose δ small enough, the growth rate of u and Du in A_0 has to be strictly smaller than $p - 1$. And in A_i we find that if $r_i = \infty$, the condition on the growth of u should be strictly smaller than $p - 1$, but it can approach $p - 1$ as close as possible. Then we have:

$$
\int_\Omega \frac{|Du|^p}{(1+|u|)^{1+\delta}} dx \le C
$$

As the proof in [R] Lemma 1, if $q < \frac{N}{N-1}(p-1)$, $\delta < \frac{N(p-q-1)}{N-q}$, we can obtain:

$$
\|Du\|_q \le C \qquad \text{and} \qquad \|u\|_{q^*} \le C \tag{2.3}
$$

where C depends only on the given data and $\|f\|_1$.

So the two methods get different conditions on the growth of u in A_i.

This means that we obtained the estimate we wanted.

§3 The Limit Process

In this section, we suppose that $f \in M(\Omega)$. We consider a sequence of $\{f_n\}$ which converges to f in the distributional sense and whose members belong to L_1 with uniformly bounded norms. It is further assumed that:

$$
\|f_h\|_{L_1} \le B = \|f\|_{M(\Omega)}
$$

Let u_h be the solution of (0.1) responding to f_h. Then, for every integer h, we have:

$$
\|u_h\|_{W^{1,q}} \le C
$$

where C depends on B, and independent of n. Then there exists u in $W_0^{1,q}(\Omega)$ and some subsequence (still denote by $\{u_h\}$) such that:

$$
\begin{aligned}
u_h &\rightharpoonup u \qquad w\text{-} \text{ in } W_0^{1,q}(\Omega) \\
u_h &\to u \qquad \text{in } L_q(\Omega) \\
u_h &\to u \qquad a.e. \text{ in } \Omega
\end{aligned}
$$

Then using Lemma 2, we can obtain:

$$Du_h \to Du \qquad L_q(\Omega)$$

as long as we verify the assumptions a)-c). In §2, we have obtained a). If we sum up m from 1 to k, in the formula (2.2) and use the estimate (2.3), we can obtain b). To obtain c), let us consider the function:

$$S_\epsilon(t) = \begin{cases} t & \text{if } |t| \le \epsilon \\ \epsilon \,\text{sign}(t) & \text{otherwise} \end{cases}$$

Let us take $S_\epsilon(u_h - z_k)$ as a test function, we obtain:

$$\begin{aligned} &\int_\Omega h_\epsilon(|u_h - z_k|)A(x, u_h, Du_h)D(u_h - z_k)dx \\ \le &\int_\Omega A_0(x, u_h, Du_h)S_\epsilon(u_h - z_k)dx + \int_\Omega f_h S_\epsilon(u_h - z_k)dx \\ \le &\epsilon\bar{C} \end{aligned}$$

Here we have used the growth condition of the operator and the estimate (2.3). So $\bar{C}$ is independent of h. This is, assumption c) of Lemma 2 is satisfied. Then passing to the limit and letting h go to infinity we obtain the following result:

THEOREM 1. *Under the hypotheses of (H1)-(H6), there exists at least one weak solution for the Dirichlet problem (0.1) of nonlinear elliptic equation with the free term belonging to* $M(\Omega)$.

§4 Some Remarks

REMARK 1. *A similar result can be posed for the parabolic equation by using basically the same method.*

REMARK 2. *For the lower order term* A_0*, we can see from our proof that if the order of* u *is smaller than* $p-1$*, we do not need to give the sign condition. But if it is not smaller than* $p-1$*, the sign condition is necessary. That means that if we have an additional hypothesis on* A_0*, for example:*

1)

$$A_0(x, u, \xi)u \ge 0 \qquad \text{for a.e. } x \in \Omega$$

2)

$$(A_0(x, u, \xi) - A_0(x, v, \xi))(u - v) \ge 0 \qquad \text{for a.e. } x \in \Omega$$

We can reduce the growth order of u *to:*

$$\rho_0 \le \frac{(p-1)N}{N-p} \tag{4.1}$$

Let us show this Remark in short words. With condition 1) above, when we estimate the solution by the method used by Rakotoson, we can get rid of the term of $A_0(\cdot,u,\cdot)\phi(u)$ because $A_0(\cdot,u,\cdot)\text{sign}(u)\left(1-\frac{1}{(1+|u|)^\delta}\right)\geq 0$. And with the condition 2) above, we use the truncation method. This time, in stead of estimating $A_0(x,u,Du)u_m$, we can use the condition to estimate $A_0(x,0,Du)u_m$. It will be much simpler. In the both cases, condition (4.1) guarantee that $A_0(x,u_h,Du_h)$ converges to $A_0(x,u,Du)$ in $L_1(\Omega)$. And the limit process is still valid.

ACKNOWLEDGE

This work has pofited from the helpful suggestions by Prof. J.F. Rodrigues. The authoress wishes to thank him here.

∎

REFERENCES

[A] Adams, R.A., Sobolev Spaces, *New York Academic Press* (1975).

[BG] Boccardo, L.& Gallouët, T., Non-linear Elliptic and Parabolic Equations involving Measure Data, *J. Funct. Anal.* **87** (1989), 149-169.

[LU] Ladyzhenskaja, O.A. & Ural'tseva, N.N., "Linear and Quasilinear Elliptic Equations", Nauka Press, Moscow,(1964); English Translation: Academic Press New York and London (1968).

[R] Rakotoson, J. M., Quasilinear Elliptic Problems with Measures as Data, *Diff. and Int. Equ.* **4** No. **3** (May, 1991), 449-457.

[S] Stampacchia, G., Equations Elliptiques du Second Ordre à Coefficients Discontinus, in *"Séminaire de Mathématiques Supérieures, Université de Montréal,* (1965)".

Av. Prof. Gama Pinto 2, 1699 Lisboa Codex; PORTUGAL.

G M LIEBERMAN

Existence of solutions to the first initial-boundary value problem for parabolic equations via elliptic regularization

There are several ways to show that the first initial-boundary value problem for a parabolic equation is solvable. We refer to [5, Sections III.4, III.16, IV.5-8] and [3, Chapter 4] for a more complete description of these. Each has certain advantages and disadvantages, but they all have one undesirable feature in common, which we want to avoid here. In each case, the first initial-boundary value problem is solved in some class of domains for the boundary values lying in some finite- dimensional function space, and then specific approximation arguments are used to show that the boundary values can be taken from a more usual function space, such as the space of continuous functions.

For elliptic equations, the Hilbert space approach is probably the easiest way to show the solvability of the Dirichlet problem, but there seems to be no corresponding direct way of showing that some simple parabolic problem is solvable. Here we consider the heat equation as a special degenerate elliptic equation to get a simple proof of the solvability of the first initial-boundary value problem.

We start in Section 1 with the Hilbert space solution for elliptic equations to show that the Dirichlet problem for the heat equation is solvable in sufficiently small space-time balls with sufficiently smooth boundary values. After proving some elementary estimates in Section 2, we modify Perron's process in Section 3 (using only continuous subsolutions, unlike prior workers [2, 6, etc.] who needed semicontinuous ones) to consider more general domains. Along the way, we use basic ideas about weak solutions of elliptic and parabolic equations to prove the solvability.

It should be noted that the existence of solutions to the Dirichlet problem in space-time balls with arbitrary radius is already known, but the usual proof (see [1] or [2]) requires the full machinery for solvability of the Dirichlet problem for the heat equation. We prove this result directly for sufficiently small balls as the basis for solvability in more general domains.

1. LOCAL SOLVABILITY.

Our first step is to solve a family of approximating problems in a ball. To state our results, we write x for points in $\mathbb{R}^n, n \geq 1, t$ for real numbers, and (x,t) for points in $\mathbb{R}^{n+1}$. We write Du for the spatial gradient of a function u, but we use ∇u or the phrase "gradient of u" to mean its space-time gradient. Similarly D^2u denotes the matrix of second spatial derivaties of u, and $\nabla^2 u$ is the matrix of all second space-time derivatives.

We denote by B the ball $\{|x|^2 + t^2 < R^2\}$ for some fixed positive R and for $\varepsilon \in \mathbb{R}$, we define the operator L_ε by

$$L_\varepsilon u = \Delta u + \varepsilon u_{tt} - u_t,$$

where Δ denotes the spatial Laplacian. For brevity, we write H for L_0, which is the heat operator. A function u in $W^{1,2}_{\text{loc}}(\Omega)$, the space of functions with locally square integrable weak gradient, is a weak solution of the equation $L_\varepsilon u = f$ in Ω if u satisfies the integral identity

$$\int_\Omega Du \cdot Dv + \varepsilon u_t v_t + u_t v + fv \; dxdt = 0$$

for all $W^{1,2}(\Omega)$ functions v which vanish on $\partial\Omega$. Note that it suffices to check this integral identity only for C^2 functions.

We begin with a maximum principle, which is a special case of standard results (see, for example, [5, Theorem III.7.2] and [4, Theorem 8.1] when $b + c = 0$). We give here a simple proof, which combines the elliptic and parabolic elements of the equation.

Lemma 1.1. *Let $\varepsilon \geq 0$, and let Ω be a domain in $\mathbb{R}^{n+1}$. If u is a continuous $W^{1,2}$ function such that $L_\varepsilon u \geq 0$ in Ω and $u \leq 0$ on $\partial\Omega$, then $u \leq 0$ in Ω.*

Proof. Set $u_+ = \max\{u, 0\}$ and use $(T-t)u_+$ as test function in the weak form of the inequality $L_\varepsilon u \geq 0$ to see that

$$\int_\Omega (T-t)Du \cdot D(u_+) + \varepsilon(T-t)u_t(u_+)_t - \varepsilon u_t u_+ + (T-t)u_t u_+ \; dxdt$$

is nonpositive. Recalling that the gradient of u_+ equals the gradient of u where $u > 0$ and is zero elsewhere, we infer that

$$\begin{aligned} 0 &\geq \int_\Omega (T-t)|D(u_+)|^2 + \varepsilon(T-t)(u_+)_t{}^2 \; dxdt \\ &\quad + \int_\Omega -\frac{1}{2}\varepsilon(u_+)^2{}_t + (T-t)\frac{1}{2}(u_+{}^2)_t \; dxdt \\ &\geq \int_\Omega -\frac{1}{2}\varepsilon(u_+)^2{}_t + (T-t)\frac{1}{2}(u_+{}^2)_t \; dxdt \\ &= \frac{1}{2}\int_\Omega u_+{}^2 \; dxdt; \end{aligned}$$

the equality is just integration by parts. Thus u_+ is everywhere zero. □

We now deduce a global L^∞ gradient bound and an existence result.

Lemma 1.2. *Let φ be a C^2 function and let f be a C^1 function with*

$$|\varphi| + |\nabla\varphi| + |\nabla^2\varphi| \leq \Phi, \;\; |f| + |\nabla f| \leq F \text{ in } B \tag{1.1}$$

for some nonnegative constants Φ and F. Then, for every positive ε, there is a unique weak solution u_ε of the boundary value problem

$$L_\varepsilon u_\varepsilon = f \text{ in } B, u_\varepsilon = \varphi \text{ on } \partial B \tag{1.2}$$

If $R \le 1/2$, then u_ε is globally Lipschitz and

$$|Du_\varepsilon| \le C(n)(F + \Phi). \tag{1.3}$$

Proof. The existence follows from the standard theory of weak solutions [4, Chapter 8]. It suffices to solve $L_\varepsilon v_\varepsilon = f - L_\varepsilon \varphi$ in Ω, $v_\varepsilon = 0$ on $\partial\Omega$, and this problem has a unique weak solution by virtue of the Riesz representation theorem. Note that uniqueness of the solution is a simple consequence of the maximum principle.

To prove (1.3), we use a simple variant of the maximum principle. Setting $w = R^2 - |x|^2 - t^2$, we see that $L_\varepsilon w = -2n - 2\varepsilon + 2t \le -1$ in B (because $R \le 1/2$) and $w = 0$ on ∂B. It follows that, for $K = (n+1)\Phi + F$,

$$L_\varepsilon (u_\varepsilon - \varphi - Kw) \ge 0 \text{ in } B, \; u_\varepsilon - \varphi - Kw = 0 \text{ on } \partial B,$$

and then the maximum principle implies that $u_\varepsilon - \varphi \le Kw$ in B, and similarly $u_\varepsilon - \varphi \ge -Kw$ in B. These two inequalities give both a bound on u_ε itself inside B and a boundary Lipschitz estimate. The global Lipschitz bound now follows from another easy maximum principle argument (see, e.g., [9]). For any vector τ in $\mathbb{R}^{n+1}$, define $B_\tau = \{(x,\, t) \in B : (x,\, t) + \tau \in B\}$, and for $(x,\, t) \in B_\tau$, $v_\tau(x,\, t) = u_\varepsilon((x,\, t) + \tau) - u_\varepsilon(x,\, t)$. From the boundary Lipschitz estimate, we have $|v_\tau| \le K|\tau|$ on ∂B_τ and $|L_\varepsilon v_\tau| \le F|\tau|$. Therefore, $h = v_\tau - K|\tau| - F|\tau|w$ satisfies $L_\varepsilon h \ge 0$ in B_τ, $h \le 0$ on ∂B_τ, and hence $h \le 0$ in B_τ. Similarly, applying the maximum principle to $h^* = -v_\tau - K|\tau| - F|\tau|w$ shows that $|v_\tau| \le [K + Fw]|\tau| \le C(n)[F + \varphi]|\tau|$. This inequality easily implies the desired gradient bound. $\square$.

From this uniform global gradient bound, we easily infer the existence of a unique solution to the Dirichlet problem for the heat equation in B for smooth enough f and φ. In the next section, we show how to relax the hypotheses on φ.

Lemma 1.3. *Under the hypotheses of Lemma 1.2, there is a unique solution of*

$$Hu = f \text{ in } B, \; u = \varphi \text{ on } \partial\, B, \tag{1.4}$$

and (1.3) *is valid for the solution.*

Proof. From Lemma 1.2, the solutions of (1.2) are uniformly bounded and equicontinuous, so there is a sequence $(\varepsilon(k))$ decreasing to 0 such that $(u_{\varepsilon(k)})$ converges uniformly to some function u, and u satisfies (1.3). Now for any function $v \in C^2(\bar{B})$ with compact support, we can write the weak form of the equation $L_{\varepsilon(k)}u_{\varepsilon(k)} = f$ as

$$\int\limits_B u_{\varepsilon(k)}\Delta v \; dxdt - \varepsilon(k) \int\limits_B (u_{\varepsilon(k)})_t v_t \; dxdt$$
$$+ \int\limits_B u_{\varepsilon(k)} v_t \; dxdt = \int\limits_B fv \; dxdt$$

As $\varepsilon(k)$ goes to zero, $u_{\varepsilon(k)}\Delta v + u_{\varepsilon(k)}v_t$ converges uniformly to $u\Delta v + uv_t$, while the second integral is uniformly bounded, so its product with $\varepsilon(k)$ converges to zero as $\varepsilon(k)$ goes to zero. Hence

$$\int\limits_B u\Delta v + uv_t \; dxdt = \int\limits_B fv \; dxdt.$$

Next we integrate by parts in the first integral to see that u is a weak solution of $Hu = f$ in B. The uniform Lipschitz estimate guarantees that u also satisfies the boundary condition. The maximum principle implies the uniqueness of the solution. □

2. APRIORI ESTIMATES.

In this section we prove some local estimates for the gradients of weak solutions to (1.2).

Lemma 2.1. *Under the hypotheses of Lemma 1.3, even if φ is merely continuous,*

$$\int\limits_{B(R/2)} |\nabla u|^2 \; dxdt \le C(R, n) \int\limits_B u^2 + f^2 \; dxdt. \tag{2.1}$$

Furthermore

$$|\nabla u| \le C(R, n) \max(|u| + |f| + |\nabla f|) \;\; in \; B(R/2). \tag{2.2}$$

Proof. We first estimate the integral of $|Du|^2$ in the "usual way". For w from the proof of Lemma 1.2, we take uw^2 as test function to see that

$$\int_B |Du|^2 w^2 \, dxdt + \int_B u_t u w^2 \, dxdt + 2\int_B wDu \cdot Dwu \, dxdt = \int_B w^2 fu \, dxdt.$$

We now estimate the third and fourth integrals in this equation by Cauchy's inequality and transform the second integral as follows:

$$\int_B u_t u w^2 \, dxdt = \frac{1}{2}\int_B (u^2)_t w^2 \, dxdt = -\int_B u^2 w w_t \, dxdt.$$

Since w and its derivatives are bounded by constants depending only on R, we see that

$$\int_B w^2 |Du|^2 \, dxdt \le C(n, R)\int_B u^2 + f^2 \, dxdt. \tag{2.3}$$

Noting that w is bounded away from zero on $B(R/2)$ gives the desired spatial derivative estimate, but we use the full force of this estimate for our time derivative estimate.

To estimate the time derivative, for $\mu \in (0, 1/4)$, we define $h(x, t) = (((1-\mu)R)^2 - |x|^2 - t^2)_+^2$ and for each $\sigma \in (0, \mu/2)$, we use as test function $h(x, t)^2\{u(x, t+\sigma) - u(x, t)\}$ in the equation for $u(x, t)$ and in the equation for $u(x, t+\sigma)$. (The idea here is really to use $h^2 u_t$, but that may not be a legitimate test function.) If we add together the resulting integral identities and suppress the x dependence from the notation, our equation then becomes

$$\begin{aligned} &\int_B Du(t) \cdot \{Du(t+\sigma) - Du(t)\}h(t)^2 \, dxdt \\ &\quad + \int_B Du(t+\sigma) \cdot \{Du(t+\sigma) - Du(t)\}h(t)^2 \, dxdt \\ &\quad + 2\int_B Du(t) \cdot Dh(t)\{u(t+\sigma) - u(t)\}h(t) \, dxdt \end{aligned}$$

$$+2\int_B Du(t+\sigma)\cdot Dh(t)\{u(t+\sigma)-u(t)\}h(t)\,dxdt$$
$$+\int_B u_t(t)\{u(t+\sigma)-u(t)\}h(t)^2\,dxdt$$
$$+\int_B u_t(t+\sigma)\{u(t+\sigma)-u(t)\}h(t)^2\,dxdt$$
$$=\int_B f(t)\{u(t+\sigma)-u(t)\}h(t)^2\,dxdt$$
$$+\int_B f(t+\sigma)\{u(t+\sigma)-u(t)\}h(t)^2\,dxdt.$$

Now we label these integrals $I_1,\ldots,I_8$ and see what happens when we divide by σ and let σ tend to zero. First

$$\begin{aligned}I_1+I_2&=\int_B\{|Du(t+\sigma)|^2-|Du(t)|^2\}h(t)^2\,dxdt\\&=\int_B|Du(t+\sigma)|^2h(t)^2\,dxdt-\int_B|Du(t)|^2h(t)^2\,dxdt\\&=\int_B|Du(t)|^2h(t-\sigma)^2\,dxdt-\int_B|Du(t)|^2h(t)^2\,dxdt\\&=\int_B|Du(t)|^2\{h(t-\sigma)^2-h(t)^2\}\,dxdt,\end{aligned}$$

so as σ goes to zero, $(I_1+I_2)/\sigma$ goes to $-2\int_B|Du|^2h_th\,dxdt$. Next we note that $\{u(x,\,t+\sigma)-u(x,\,t)\}/\sigma$ is uniformly bounded and converges almost everywhere to $u_t(x,t)$, so $(I_3+I_4)/\sigma$ goes to $2\int_B Du\cdot Dhu_th\,dxdt$. Arguing as for I_3+I_4 also shows that $(I_5+I_6)/\sigma$ goes to $2\int_B u_t{}^2h^2\,dxdt$, and $(I_7+I_8)/\sigma$ goes to $2\int_B u_th^2f\,dxdt$. It follows that

$$\begin{aligned}\int_B u_t{}^2h^2\,dxdt&=\int_B|Du|^2h_th\,dxdt-2\int_B Du\cdot Dhhu_t\,dxdt+\int_B u_th^2f\,dxdt\\&\le\frac{1}{2}\int_B u_t{}^2h^2\,dxdt+C(R)\int_B|Du|^2h\,dxdt+2\int_b h^2f^2\,dxdt.\end{aligned}$$

Now send μ to zero, and note that h converges uniformly to w^2. The resulting inequality, along with our previous estimate on Du, leads easily to (2.1).

For the pointwise gradient bound, we first define u' to be the solution of $Lu' = f$ in $B, u' = 0$ on ∂B, and we set $U = u - u'$. Then, from the gradient bound on u' given in Lemma 1.3, we see that it suffices to estimate the gradient of U. For this estimate, we first note that each component of the gradient of U is in $W^{1,2}_{\text{loc}}(\Omega)$ because (2.1) gives uniform $W^{1,2}_{\text{loc}}(\Omega)$ estimates for the corresponding difference quotients. Hence, for $k = 1, \dots, n, H(D_k U) = 0$. It follows that $v = w^2|DU|^2$ is a weak solution of

$$Hv = w^2|D^2U|^2 + H(w^2)|DU|^2 + D(w^2) \cdot D|DU|^2$$

and therefore $Hv \geq -C(n)|DU|^2$. Now we observe that $H(U^2) = 2|DU|^2$, and we set $C' = C(n) \max |U|^2$. Then $H(v + C(n)U^2 - C') \geq 0$ in B, and $v + C(n)U^2 - C' \leq 0$ on ∂B, so the maximum principle implies that $v \leq C' - C(n)U^2 \leq C'$ in B. Hence we have

$$|DU| \leq C' \text{ in } B(7R/8). \tag{2.4}$$

To see that this estimate is valid if φ is merely continuous, note that if (φ_m) is a sequence of C^2 functions converging uniformly to φ, then the corresponding solutions of (1.2) converge uniformly to a limiting function u, which clearly satisfies the boundary condition. In addition, if we integrate by parts, we can write the integral identity for u_m as

$$\int_B u_m \Delta v + u_m v_t \, dxdt = \int_B fv \, dxdt.$$

Sending m to infinity gives

$$\int_B u\Delta v + uv_t \, dxdt = \int_B fv \, dxdt,$$

and then integrating by parts (which is permissible because ∇u is L^∞) in the first integral shows that u is a solution of $Hu = f$. Since (2.4) holds uniformly in m, it also holds for u.

We now write e_k for the n-dimensional vector with $k-th$ component 1 and all others zero. If $\sigma \in (0,\, 1/8)$, then $\pm\{U(x+\sigma e_k,\, t) - U(x,\, t)\}/\sigma$ are weak solutions of $HU' = 0$ in $B(3R/4)$, and these functions are bounded by $C(R)\max|U|$. The argument leading to (2.4) now implies that $|DU'|^2 \leq C \max|U'|^2$ in $B(R/2)$. Sending σ to zero gives an estimate on $|D^2U|$ in $B(R/2)$. Integration by parts shows us that $\Delta U - U_t = f$ *a. e.* and we use this equation to estimate $|U_t|$ there. □

By applying the Arzela-Ascoli theorem, we also obtain a parabolic version of Harnack's first convergence theorem.

Lemma 2.2. *If* (u_n) *is a bounded sequence of weak solutions to* $Hu_n = f$ *in* B, *with* $f \in C^1(B)$, *then there is a subsequence which converges uniformly in* $B(R/2)$ *to a solution* u *of* $Hu = f$ *in* $B(R/2)$. □

3. GLOBAL EXISTENCE.

Now we are ready to apply the Perron process to infer the existence of solutions to the heat equation with Dirichlet data on suitable portions of the boundary.

We now fix a domain Ω in $\mathbb{R}^{n+1}$, and functions $\varphi \in C(\bar{\Omega}),\ f \in C^1(\Omega)$. Our goal is to solve the problem

$$Hu = f \text{ in } \Omega,\ u = \varphi \text{ on } \partial\Omega. \tag{3.1}$$

We say that a continuous function v is a <u>subsolution</u> <u>(supersolution)</u> if $v \leq \varphi (v \geq \varphi)$ on $\partial\Omega$ and if for any ball $B = B(X,\ R)$ with center X and radius $R \leq \frac{1}{2}$ which lies entirely in Ω, the solution of $Hw = f$ in $B, w = v$ on ∂B given by Lemma 1.3 satisfies $w \geq v (w \leq v)$ in B.

Our starting point is the well-known parabolic maximum principle.

Lemma 3.1. *Suppose* $0 < t < T$ *in* Ω, *and let* $\tau \in (0, T)$. *If* $Hu \geq 0$ *in* $\Omega(\tau) = \Omega \cap (t < \tau)$ *and* $u \leq 0$ *on* $\partial\Omega(\tau) = \partial\Omega \cap \{t < \tau\}$, *then* $u \leq 0$ *in* $\Omega(\tau)$.

Proof. We proceed as in Lemma 1.1 with $T - t$ replaced by $(\tau - t)_+$. □

From this maximum principle, we deduce the strong maximum principle for the heat equation, in which we write $\Omega'(\tau)$ for the set $\{x\colon (x,\ \tau) \in \Omega\}$.

Lemma 3.2. *Under the hypotheses of Lemma 3.1, if $Du(\cdot,\tau)$ is continuous, then either $u(\cdot,\tau) < 0$ in $\Omega'(\tau)$ or u is identically zero in $\Omega'(\tau)$.*

Proof. We imitate the proof of the strong maximum principle for elliptic equations given in [4, Theorem 3.5]. First we prove a boundary point lemma just like [4, Lemma 3.4].

For fixed $X \in \mathbb{R}^{n+1}$ and $\rho, r > 0$, define $Q(X,r,\rho) = \{Y = (y,s)\colon |x-y| < r, t-\rho < s < t\}$ and $Q'(X,r,\rho) = \{Y = (y,s)\colon r/2 < |x-y| < r, t-\rho < s < t\}$ and $w'(Y) = \exp(-2n|x-y|^2/r^2) - \exp(-2n)$. Then $Hw' \geq 0$ and $w' \geq 0$ in $Q'(X,r)$. If $u < 0$ in $Q(X,r)$ with $u = 0$ somewhere on $\{|x-y| = r, t-\rho < s < t\}$, say at X_0, set $\rho' = (t_0+\rho-t)/2$. Then there is a positive σ so that $u+\sigma w' \leq 0$ on $\partial\{Q'(X,r,\rho')\}(t_0)$ (that is, on $\partial\Omega(\tau)$ for $\Omega = Q'$ and $\tau = t_0$) and $H(u+\sigma w') \geq 0$ in $\{Q'(X,r,\rho')\}(t_0)$ so $u + \sigma w' \leq 0$ in $\{Q'(X,r,\rho)\}(t_0)$. It follows that, for $\nu = (x - x_0)/r$, the inner normal derivative $\partial u/\partial\nu$ is strictly positive.

Now suppose there is a point x_1 on the boundary of the closed set $\Sigma = \{y\colon u(y,\tau) = 0\}$ with $x_1 \in \Omega'(\tau)$. Then there is x such that $u(x,\tau) < 0$, so if x_0 is the nearest point in Σ to x, we infer from the boundary point lemma that $Du(x_0,\tau)$ is nonzero, but $u(\cdot,\tau)$ attains its maximum at x_0 so $Du(x_0,\tau)$ must be zero. From this contradiction we deduce the result. □

The strong maximum principle gives a comparison principle for subsolutions and supersolutions.

Lemma 3.3. *If w is a subsolution and v is a supersolution, then $w \leq v$ in Ω.*

Proof. If not, then $w - v$ has an interior positive maximum with value ε. Let $\Sigma = \{X\colon (w-v)(X) = \varepsilon\}$, and choose τ so that there is a point $X \in \Sigma$ such that $X = (x,\tau)$. If W and V are solutions of $HW = f$ in $B(X,R)$, $W = w$ on $\partial B(X,R)$ and $HV = f$ in $B(X,R)$, $V = v$ on $\partial B(X,R)$, then $H(W-V-\varepsilon) = 0$ in $B(X,R)$ and $W-V-\varepsilon \leq 0$ on $\partial B(X,R)$. From the gradient bounds in Lemma 2.1, $D^2(W-V-\varepsilon)$ is bounded, so $D(W-V-\varepsilon)(\cdot,\tau)$ is continuous. Therefore the strong maximum principle implies that $W-V-\varepsilon < 0$ at X or $W-V-\varepsilon$ is identically zero in $B(X,R)'(\tau)$. Since

$w - v - \varepsilon \le W - V - \varepsilon$, the first case can't occur, so $W - V - \varepsilon$ is identically zero and therefore $w - v - \varepsilon = 0$ on $\partial B(X, R)'(\tau)$. Because R is arbitrary, it follows that x is an interior point of Σ and therefore $\Sigma = \{x: (x, \tau) \in \Omega\}$. In this case $w - v = \varepsilon$ at some point of $\partial\Omega$ by continuity of w and v, contradicting the assumption $w \le v$ on $\partial\Omega$. Hence $w \le v$ in Ω. □

To produce our solution, we define S to be the set of all subsolutions and we define the function U by $U(x,t) = \sup_{v \in S} v(x,t)$.

Theorem 3.4. *$HU = f$ in Ω.*

Proof. Note first that $|f|_0 t + |\varphi|_0$ is a supersolution and $-(|f|_0 t + |\varphi|_0)$ is a subsolution so U is bounded above and below, by Lemma 3.3.

Now fix $X = (x,t)$ in Ω and $R \le 1/4$ so that the ball $B(X,R)$ is a subset of Ω, and choose a countable dense subset (X_m) of $B(X,R)$. For each X_m, take a sequence (v_{mk}) in S so that $v_{mk}(X_m)$ converges to $U(X_m)$, and note that $v_k = \max\{v_{1k}, \ldots, v_{kk}\}$ is also a subsolution. If we define V_k to be the solution of $HV_k = f$ in $B, V_k = v_k$ on ∂B, and extend V_k to be v_k on $\Omega \setminus B$, it follows that $V_k \in S$ also. Since $V_k \ge v_k$, it follows that $V_k(X_m)$ converges to $U(X_m)$ for all m. In addition, the V_n's are uniformly bounded in $B(X, R/2)$, so some subsequence converges uniformly to a weak solution V of $HV = f$ in $B(X,R/2)$ by Lemma 2.2, and $V = U$ on the dense subset $\{X_m\}$. To see that $V = U$ everywhere, let X_0 be any other point in $B(X, R/2)$ and let V' be the function obtained by applying the preceeding construction to the sequence (Y_m) defined by $Y_m = X_{m-1}$. Then $U(X_0) = V'(X_0)$ by construction and $V'(X_0) = V(X_0)$ since these two functions are continuous. □

To show that u attains the prescribed boundary values, we introduce the concept of barriers. We say that a sequence (w_i^+) of supersolutions is an <u>upper</u> <u>barrier</u> at $X_0 \in \partial\Omega$ if $w_i^+(X_0)$ converges to $\varphi(X_0)$ as i goes to infinity and a sequence (w_i^-) of subsolutions is a <u>lower</u> <u>barrier</u> if $w_i^-(X_0)$ converges to $\varphi(X_0)$ as i goes to infinity. When both upper and lower barriers exist at X_0, we just say that there is a barrier at X_0.

Lemma 3.5. *If there is a barrier at $X_0 \in \partial\Omega$, then U is continuous at X_0.*

Proof. Fix $\varepsilon > 0$, and choose k so large that $|w_i^{\pm}(X_0) - \varphi(X_0)| < \varepsilon/2$. From the maximum principle Lemma 3.3, we have $w_k^+ \geq U \geq w_k^-$ in Ω. Hence if $\delta > 0$ is so small that $|w_k^{\pm}(X) - w_k^{\pm}(X_0)| < \varepsilon/2$ for $|X - X_0| < \delta$, it follows that $|U(X) - \varphi(X_0)| < \varepsilon$ for $|X - X_0| < \delta$. $\square$

Now we can give a sufficient geometric condition for solvability of the Dirichlet problem for the heat equation. We say that a domain Ω satisfies a tusk condition at $X_0 \in \partial\Omega$ if there are positive constants ρ, τ and a vector v such that $\{X\colon |x - x_0 - v(t_0 - t)^{\frac{1}{2}}| \leq \rho(t_0 - t)^{\frac{1}{2}}, t_0 - \tau \leq t \leq t_0\}$ meets Ω only at X_0.

Theorem 3.6. *If Ω satisfies a tusk condition at each point of its boundary, then there is a unique solution u of the Dirichlet problem*

$$Hu = f \text{ in } \Omega, u = \varphi \text{ on } \partial\Omega \tag{3.3}$$

for any continuous f and φ.

Proof. Note that the maximum principle automatically gives the uniqueness of solutions, so we only show their existence.

Suppose first that $f \in C^1$. In this case, the existence of u follows from the Perron process once we show the existence of appropriate barriers at any $X_0 \in \partial\Omega$. Now according to [7], if we introduce coordinates (r, θ) by first defining $v' = (|x - x_0|^2 + t_0 - t)^{\frac{1}{2}}/2^{\frac{1}{2}}|v|$, and then $r = (|x - x_0 - v'v|^2 + t_0 - t)^{\frac{1}{2}}, \theta = \arcsin(t_0 - t)^{\frac{1}{2}}/r$, then there are positive constants α and $\beta > 1$, and a positive function g such that $w' = r^\alpha g(\theta)$ satisfies $Lw' \leq -1$ in $\Omega' = \{X \in \Omega\colon t_0 - \tau < t < t_0\}, w'(X) \leq \beta$ if $t = t_0$, and $w'(X) \geq 1$ if $X \in \Omega$ and $t = t_0 - \tau$. Now we take $W(t)$ to be any positive, continuously differentiable function on $[0, t_0]$ such that $W(t_0) > \beta, W(t_0 - \tau) < 1$, and $W' > 0$ on the closed interval, and define w by $w(x) = (|x - x_0|^2 + (t - t_0)^2)^{\alpha/2} + (t - t_0)^\alpha$ if $t > t_0, w(X) = \min\{w'(X), W(t)\}$ in Ω' and $w(X) = W(t)$ if $t < t_0 - \tau$. It is easy to check that $w_k^{\pm} = \pm kKw \pm 1/k + \varphi(X_0)$ provides a barrier as long as the positive constant K is sufficiently large.

If f is merely continuous, we approximately f uniformly by C^1 functions f_m and let u_m be the corresponding solutions of the Dirichlet problem. By the maximum principle, the $u_m's$ converge uniformly to some function u, and the interior L^2 derivative bounds show that u also satisfies the differential equation. □

Existence of solutions to the first boundary value problem in various domains is now an easy consequence of this result. To describe the admissible class of domains easily, we assume that Ω lies in the strip $\{0 < t < T\}$ and we use $P\Omega$ to denote the closure of the set of points $X \in \partial\Omega$ such that $t < T$.

Corollary. *Suppose Ω satisfies a tusk condition at each point of $P\Omega$ and that $\partial\Omega \setminus P\Omega$ is a relatively open subset of $\{t = T\}$. If φ is continuous on $P\Omega$ and f is continuous on Ω, then there is a unique solution u of*

$$Hu = f \text{ in } \Omega, u = \varphi \text{ on } P\Omega. \tag{3.4}$$

Proof. Choose δ so large that Ω is a subset of $\{|x| < \delta\}$, let $\alpha(t) = 2-(t/T)$, set $\Omega' = \{X \colon |x| < \alpha(t)\delta, T \leq t \leq 2T\}$, and extend f and φ as continuous functions to $\Omega'' = \Omega \cup \Omega'$. Theorem 3.6 gives a solution to the Dirichlet problem $Hu = f$ in $\Omega'', u = \varphi$ on $\partial\Omega''$, which therefore is a solution to (3.4). □

In particular the corollary applies with Ω is a cylinder with cross section satisfying an $n-$ dimensional exterior sphere (or exterior cone) condition.

By using the deeper estimates of Schauder type and those introduced by Ladyzhenskaya and Ural'tseva for studying weak solutions of parabolic equations, the full existence theory for equations with variable coefficients can be deduced from the results given here. We omit the details.

REFERENCES

1. E.G. Effros and J.L. Kazdan, *On the Dirichlet problem for the heat equation*, Indiana Univ. Math. J. **20** (1971), 683–693.
2. L.C. Evans and R.F. Gariepy, *Wiener's criterion for the heat equation*, Arch. Rational Mech. Anal **78** (1982), 293–314.
3. A. Friedman, *Partial Differential Equations of Parabolic Type*, Krieger, Malabar, FL, 1983.
4. D. Gilbarg and N.S. Trudinger, *Elliptic Partial Differential Equations of Second Order, 2nd Ed.*, Springer-Verlag, New York, 1983.
5. O.A. Ladyzhenskaya, V.A. Solonnikov, and N.N. Ural'ceva, *Linear and Quasilinear Equations of Parabolic Type*, Amer. Math. Soc., 1968.
6. G.M. Lieberman, *Intermediate Schauder estimates for second order parabolic equations. II. Existence, uniqueness, and regularity*, J. Differential Equations **63** (1986), 32–57.
7. G.M. Lieberman, *Intermediate Schauder theory for second order parabolic equations. III. The tusk condition*, Appl. Anal **33** (1989), 25–43.
8. J. Lukeš, *A new type of generalized solution of the Dirichlet problem for the heat equation*, Nonlinear Evolution Equations and Potential Theory, Plenum Press, London and New York, 1975.
9. N.S. Trudinger, *Lipschitz continuous solutions of elliptic equations of the form* $A(Du)D^2u = 0$, Math. Z. **109** (1969), 211–216.

Department of Mathematics
Iowa State University
Ames, Iowa 50011

J-P LOHÉAC

Artificial boundary conditions for advection-diffusion equations

0. Introduction.

When computing the solution of a partial differential equation in an unbounded domain, one often introduces an artificial boundary in order to limit the computational cost.
We deal here with the advection-diffusion problem:

$$(1)\qquad \begin{cases} u_t + au_{\mathbf{x}_1} - \nu\Delta u = 0\,, \text{ in } \mathbf{R}^n \times \mathbf{R}^+; \\ u(0) = u^0\,, \text{ in } \mathbf{R}^n; \end{cases}$$

This equation arises when modelling the concentration of a coloured product in natural flows. On the other hand, this work is a first stage to extend results obtained in [2] about Oseen problem.
We shall denote $\mathbf{x}_1$ by x and $(\mathbf{x}_2, \mathbf{x}_3, \ldots, \mathbf{x}_n)$ by X. We define the left half-space and the right half-space of $\mathbf{R}^n$:

$$\mathbf{R}^n_- = \{\mathbf{x} \in \mathbf{R}^n \,:\, x < 0\}\,; \quad \mathbf{R}^n_+ = \{\mathbf{x} \in \mathbf{R}^n \,:\, x > 0\}\,.$$

Let us suppose: the initial data u^0 vanish in the right half-space $\mathbf{R}^n_+$, the advection coefficient a is a positive function of $\mathbf{x}$ and the viscosity ν is small.
We introduce the artificial boundary:

$$\Sigma = \left(\{0\} \times \mathbf{R}^{n-1}\right) \times \mathbf{R}^+;$$

and we replace (1) by an artificial problem in the left half-space:

$$(2)\qquad \begin{cases} u_t + au_x - \nu\Delta u = 0\,, \text{ in } \mathbf{R}^n_- \times \mathbf{R}^+; \\ u(0) = u^0\,, \text{ in } \mathbf{R}^n_-; \\ \text{Artificial Boundary Condition on } \Sigma. \end{cases}$$

By using a Fourier method, L. Halpern has built a family of artificial boundary conditions in the case of a constant coefficient a (see [1]). The p-th condition ($p \geq 1$) can be written:

$$\left(\frac{\partial}{\partial t} + a\frac{\partial}{\partial x}\right)^p u = 0\,, \text{ on } \Sigma.$$

This condition is such that, when replacing (1) by (2), we introduce an error in $\mathbf{R}^n_- \times \mathbf{R}^+$ which is an $O(\nu^{2p})$.
Let us suppose now that a is a non-constant function. In order to extend the previous estimates, we shall study the artificial boundary conditions by means of transmission problems.

1. The first artificial boundary condition.

This condition can be written:

$$u_t + au_x = 0\,, \text{ on } \Sigma\,.$$

Hence we replace (1) by the following problem:

$$(3)\qquad \begin{cases} u_t + au_x - \nu\Delta u = 0\,, \text{ in } \mathbf{R}^n_- \times \mathbf{R}^+; \\ u(0) = u^0\,, \text{ in } \mathbf{R}^n_-; \\ u_t + au_x = 0\,, \text{ on } \Sigma\,. \end{cases}$$

This artificial problem is associated to the following transmission problem:

$$(4)\qquad \begin{cases} u_t + au_x - \nu\Delta u = 0\,, \text{ in } \mathbf{R}^n_- \times \mathbf{R}^+; \\ u_t + au_x = 0\,, \text{ in } \mathbf{R}^n_+ \times \mathbf{R}^+; \\ u(0) = u^0\,, \text{ in } \mathbf{R}^n; \\ u,\ u_t + au_x \text{ are continuous through } \Sigma. \end{cases}$$

Assume that (1) and (4) have solutions which are denoted by v and w, respectively. Then $w|_{\mathbf{R}^n_- \times \mathbf{R}^+}$ satisfies (3) and we may define the error by: $e = v - w$. We can see that e satisfies the following advection-diffusion problem:

$$(5)\qquad \begin{cases} e_t + ae_x - \nu\Delta e = 0\,, \text{ in } \mathbf{R}^n_- \times \mathbf{R}^+; \\ e_t + ae_x - \nu\Delta e = \nu\Delta w\,, \text{ in } \mathbf{R}^n_+ \times \mathbf{R}^+; \\ e(0) = 0\,, \text{ in } \mathbf{R}^n. \end{cases}$$

Now we have to estimate e in the left half-space. The proof of error estimates consists of two stages:
1) "Advective estimates": we estimate the solution of an advection-diffusion equation with null initial data and a right hand-side which vanishes in the left half-space (subsection 1.2).
2) Error estimates (theorem 8 and corollary 9): we estimate Δw in the right half-space and we use advective estimates (subsection 1.3).
We start by proving the well-posedness of (3) and (4). Firstly, we assume that a and u^0 are regular enough:

$$(6)\qquad \exists \alpha > 0 : \quad \forall \mathbf{x} \in \mathbf{R}^n\,, \quad \alpha \le a(\mathbf{x})\,;$$

$$(7)\qquad \exists \beta > 0 : \quad \sup\left\{ \sup_{\mathbf{x}} |a(\mathbf{x})|\,, \sup_{i,\mathbf{x}} \left|\frac{\partial a}{\partial \mathbf{x}_i}(\mathbf{x})\right|\,, \sup_{i,j,\mathbf{x}} \left|\frac{\partial^2 a}{\partial \mathbf{x}_i \partial \mathbf{x}_j}(\mathbf{x})\right| \right\} \le \beta\,.$$

$$(8)\qquad u^0 \in \mathrm{H}^4(\mathbf{R}^n)\,; \quad \mathrm{supp}(u^0) \subset \mathbf{R}^n_-\,.$$

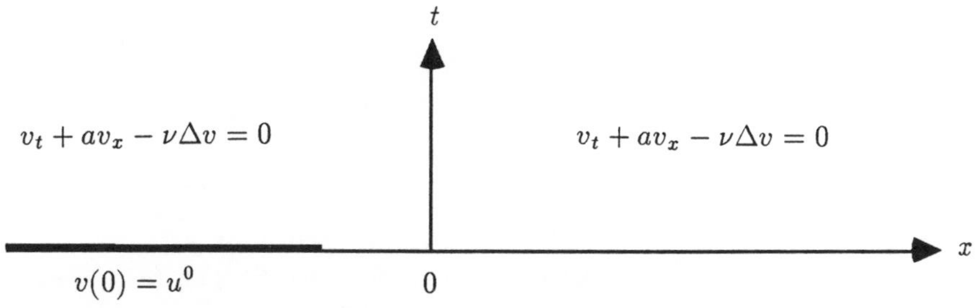

Fig. 1: the full-space problem.

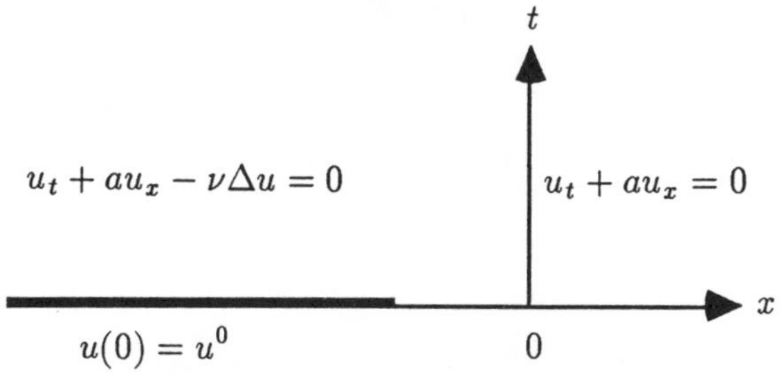

Fig. 2: the first artificial problem.

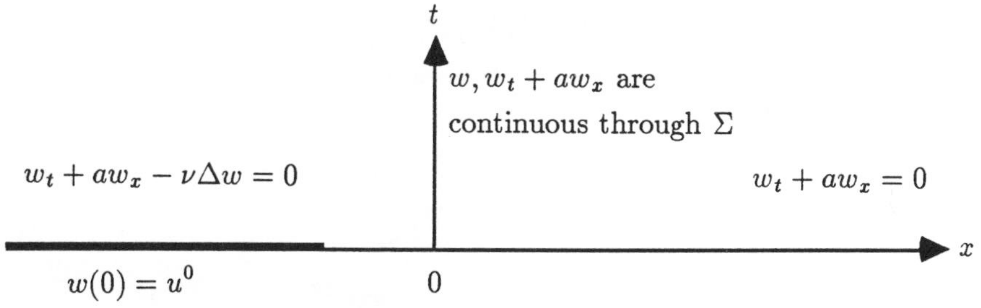

Fig. 3: the first transmission problem.

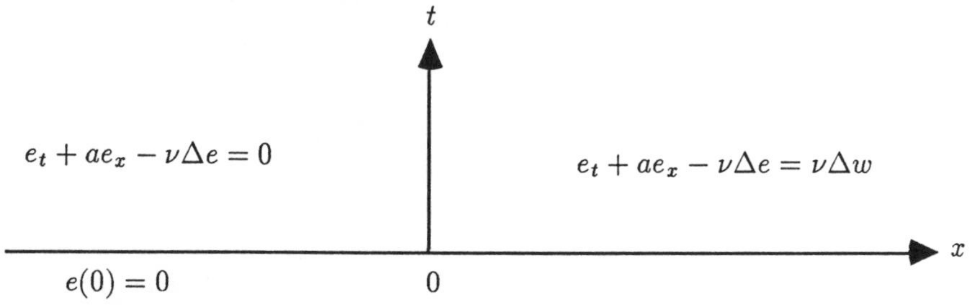

Fig. 4: the error problem ($e = v - w$).

Under assumptions (6), (7), (8), problems (1), (3) and (4) are well-posed: for (1), it is a standard result (see [3] or [6], for instance), the solution v of (1) is such that:

$$\forall T > 0, \quad v \in \mathrm{L}^\infty(0,T;\mathrm{L}^2(\mathbf{R}^n)).$$

The following subsection 1.1 is concerned with the well-posedness of (3) and (4).

1.1. The artificial problem and the transmission problem.
We are going to show that (3) is well posed if assumptions (6), (7), (8) are satisfied. In order to prove error estimates, we will study a more general inhomogeneous problem. We consider two functions p and q from $\mathbf{R}^n_-$ to $\mathbf{R}$ such that:

(9) $$\exists q_0 > 0 : \quad \forall \mathbf{x} \in \mathbf{R}^n_-, \quad q_0 \leq q(\mathbf{x});$$

(10) $$\exists \gamma > 0 : \quad \sup\left\{\sup_{\mathbf{x}} |p(\mathbf{x})|, \sup_{\mathbf{x}} |q(\mathbf{x})|, \sup_{\mathbf{x}} \left|\frac{\partial q}{\partial x}(\mathbf{x})\right|\right\} \leq \gamma.$$

Let u^0, f, g be such that:

(11) $$u^0 \in \mathrm{L}^2(\mathbf{R}^n_-);$$

(12) $$\forall T > 0, \quad \begin{cases} f \in \mathrm{L}^2(0,T;\mathrm{L}^2(\mathbf{R}^n_-)); \\ g \in \mathrm{L}^2(0,T;\mathrm{L}^2(\mathbf{R}^{n-1})). \end{cases}$$

We consider the following problem:

(13) $$\begin{cases} u_t + pu + qu_x - \nu\Delta u = f, \text{ in } \mathbf{R}^n_- \times \mathbf{R}^+; \\ u_t + pu + qu_x = g, \text{ on } \Sigma; \\ u(0) = u^0, \text{ in } \mathbf{R}^n_-. \end{cases}$$

Notations. *Assume that for $T > 0$, u is such that:*

$$u \in \mathrm{L}^\infty(0,T;\mathrm{L}^2(\mathbf{R}^n_-)); \quad u|_\Sigma \in \mathrm{L}^\infty(0,T;\mathrm{L}^2(\mathbf{R}^{n-1}));$$

then we define:

$$N_{\ell,\nu,T}(u) = \|u\|_{\mathrm{L}^\infty(0,T;\mathrm{L}^2(\mathbf{R}^n_-))} + \nu^{1/2}\|u|_\Sigma\|_{\mathrm{L}^\infty(0,T;\mathrm{L}^2(\mathbf{R}^{n-1}))}.$$

If u^0, f, g satisfy (11), (12), then we define:

$$K_{\ell,\nu,T}(u^0,f,g) = \|u^0\|_{\mathrm{L}^2(\mathbf{R}^n_-)} + \|f\|_{\mathrm{L}^2(0,T;\mathrm{L}^2(\mathbf{R}^n_-))} + \nu^{1/2}\|g\|_{\mathrm{L}^2(0,T;\mathrm{L}^2(\mathbf{R}^{n-1}))}.$$

Let us give estimates of the solution of (13) with respect to u^0, f and g.

Proposition 1. *Under assumptions* (9), (10), (11), (12), *problem* (13) *has one and only one solution* u *such that for all* $T > 0$:

$$u \in \mathrm{L}^\infty(0,T;\mathrm{L}^2(\mathbf{R}^n_-)) \cap \mathrm{L}^2(0,T;\mathrm{H}^1(\mathbf{R}^n_-))\,; \quad u|_\Sigma \in \mathrm{L}^\infty(0,T;\mathrm{L}^2(\mathbf{R}^{n-1}))\,;$$

there exists $C_1 > 0$ *(independent of* ν*) such that :*

$$N_{\ell,\nu,T}(u) \leq C_1\, K_{\ell,\nu,T}(u^0, f, g)\,.$$

The proof of this proposition is standard (see [5]):
1) we build a variationnal formulation and we give a priori estimates;
2) we build a solution by using Galerkin method;
3) the uniqueness is a consequence of the linearity of (13).
When applying this proposition with $p = 0$, $q = a$, $f = 0$ and $g = 0$, we obtain the well-posedness of (3):

Proposition 2. *Under assumptions* (6), (7), (8), *the artificial problem* (3) *has one and only one solution* u *such that for all* $T > 0$:

$$u \in \mathrm{L}^\infty(0,T;\mathrm{L}^2(\mathbf{R}^n_-)) \cap \mathrm{L}^2(0,T;\mathrm{H}^1(\mathbf{R}^n_-))\,; \quad u|_\Sigma \in \mathrm{L}^\infty(0,T;\mathrm{L}^2(\mathbf{R}^{n-1}))\,;$$

there exists $C_2 > 0$ *(independent of* ν*) such that :*

$$N_{\ell,\nu,T}(u) \leq C_2\, \|u^0\|_{\mathrm{L}^2(\mathbf{R}^n)}\,.$$

Now we are going to prove the well-posedness of (4) under assumptions (6), (7), (8). As above, we consider a more general inhomogeneous problem.
Let r be a function from $\mathbf{R}^n$ to $\mathbf{R}$ such that:

(14) $$\exists r_0 > 0 : \quad \forall \mathbf{x} \in \mathbf{R}^n\,, \quad r_0 \leq r(\mathbf{x})\,.$$

(15) $$\exists \delta > 0 : \quad \sup\left\{\sup_{\mathbf{x}} |r(\mathbf{x})|\,,\ \sup_{\mathbf{x}} \left|\frac{\partial r}{\partial x}(\mathbf{x})\right|\right\} \leq \delta\,.$$

Let u^0, f be such that:

(16) $$u^0 \in \mathrm{L}^2(\mathbf{R}^n)\,; \quad \mathrm{supp}(u^0) \subset \mathbf{R}^n_-\,;$$

(17) $$\forall T > 0\,, \quad \begin{cases} f \in \mathrm{L}^2(0,T;\mathrm{L}^2(\mathbf{R}^n))\,; \\ f|_\Sigma \in \mathrm{L}^2(0,T;\mathrm{L}^2(\mathbf{R}^{n-1}))\,. \end{cases}$$

We consider the following problem:

$$\text{(18)} \qquad \begin{cases} u_t + r u_x - \nu \Delta u = f \,, \text{ in } \mathbf{R}^n_- \times \mathbf{R}^+; \\ u_t + r u_x = f \,, \text{ in } \mathbf{R}^n_+ \times \mathbf{R}^+; \\ u(0) = u^0 \,, \text{ in } \mathbf{R}^n; \\ u,\ u_t + r u_x - f \text{ are continuous through } \Sigma \,. \end{cases}$$

Notations. *Assume that for* $T > 0$, u *is such that:*

$$u \in \mathrm{L}^\infty(0,T;\mathrm{L}^2(\mathbf{R}^n))\,; \quad u|_\Sigma \in \mathrm{L}^\infty(0,T;\mathrm{L}^2(\mathbf{R}^{n-1}))\,;$$

then we define:

$$N_{\nu,T}(u) = \|u\|_{\mathrm{L}^\infty(0,T;\mathrm{L}^2(\mathbf{R}^n))} + \nu^{1/2}\|u|_\Sigma\|_{\mathrm{L}^\infty(0,T;\mathrm{L}^2(\mathbf{R}^{n-1}))}\,.$$

If u^0 *and* f *satisfy* (16) *and* (17), *then we define:*

$$K_{\nu,T}(u^0,f) = \|u^0\|_{\mathrm{L}^2(\mathbf{R}^n)} + \|f\|_{\mathrm{L}^2(0,T;\mathrm{L}^2(\mathbf{R}^n))} + \nu^{1/2}\|f|_\Sigma\|_{\mathrm{L}^2(0,T;\mathrm{L}^2(\mathbf{R}^{n-1}))}\,.$$

Proposition 3. *Under assumptions* (14), (15), (16), (17), *problem* (18) *has one and only one solution* u *such that for all* $T > 0$:

$$u \in \mathrm{L}^\infty(0,T;\mathrm{L}^2(\mathbf{R}^n))\,; \quad u|_{\mathbf{R}^n_- \times \mathbf{R}^+} \in \mathrm{L}^2(0,T;\mathrm{H}^1(\mathbf{R}^n_-))\,; \quad u|_\Sigma \in \mathrm{L}^\infty(0,T;\mathrm{L}^2(\mathbf{R}^{n-1}))\,.$$

We build a solution of (18) by using proposition 2 and by integration along the characteristic curves of $\partial/\partial t + r\partial/\partial x$. We deduce the uniqueness of this solution from the linearity of (18) (see [5]).
We can build an energy estimate of u. This will be useful to get error estimates. Let us assume futhermore:

$$\text{(19)} \qquad f(0) \in \mathrm{L}^2(\mathbf{R}^n)\,; \quad \operatorname{supp}(f(0)) \subset \mathbf{R}^n_-\,.$$

The energy estimate is given in the following proposition (see [5] for proof).

Proposition 4. *Assume that* (14), (15), (16), (17), (19) *are satisfied and let* u *be the solution of* (18); *then for all* $T > 0$, *there exists* $C_4 > 0$ *(independent of* ν*) such that:*

$$N_{\nu,T}(u) \le C_4\, K_{\nu,T}(u^0,f)\,.$$

We may now say that (4) is well-posed. We apply propositions 3 and 4 with $r = a$, $f = 0$. Let us denote its solution by w.

Proposition 5. *Under assumptions* (6), (7), (8), *the transmission problem* (4) *has one and only one solution* w *such that* $w|_{\mathbf{R}^n_- \times \mathbf{R}^+}$ *is the solution of* (3) *and such that for all* $T > 0$:

$$w \in \mathrm{L}^\infty(0,T;\mathrm{L}^2(\mathbf{R}^n))\,; \quad w|_{\mathbf{R}^n_- \times \mathbf{R}^+} \in \mathrm{L}^2(0,T;\mathrm{H}^1(\mathbf{R}^n_-))\,; \quad w|_\Sigma \in \mathrm{L}^\infty(0,T;\mathrm{L}^2(\mathbf{R}^{n-1}))\,;$$

there exists $C_5 > 0$ *(independent of* ν*) such that:*

$$N_{\nu,T}(w) \leq C_5 \, \|u^0\|_{\mathrm{L}^2(\mathbf{R}^n)}\,.$$

1.2. Advective estimates.

Let T be a positive real number and consider two functions $\mathbf{a} = (\mathbf{a}_1, \ldots, \mathbf{a}_n)$ from $\mathbf{R}^n \times [0,T]$ to $\mathbf{R}^n$ and b from $\mathbf{R}^n \times [0,T]$ to $\mathbf{R}$ such that:

$$\text{(20)} \qquad \exists \alpha_1 > 0 : \quad \forall (\mathbf{x},t) \in \mathbf{R}^n \times [0,T]\,, \quad \alpha_1 \leq \mathbf{a}_1(\mathbf{x},t)\,;$$

$$\text{(21)} \qquad \exists \beta_1 > 0 : \quad \sup \left\{ \sup_{(\mathbf{x},t)} |b(\mathbf{x},t)|\,, \ \sup_{i,(\mathbf{x},t)} |\mathbf{a}_i(\mathbf{x},t)|\,, \ \sup_{i,j,(\mathbf{x},t)} \left| \frac{\partial \mathbf{a}_i}{\partial \mathbf{x}_j}(\mathbf{x},t) \right| \right\} \leq \beta_1\,.$$

Let g be such that:

$$\text{(22)} \qquad g \in \mathrm{L}^\infty(0,T;\mathrm{L}^2(\mathbf{R}^n))\,; \quad \mathrm{supp}(g) \subset \mathbf{R}^n_+ \times [0,T]\,.$$

Let us consider the problem:

$$\text{(23)} \qquad \begin{cases} u_t + bu + \mathbf{a}.\nabla u - \nu \Delta u = g\,, \text{ in } \mathbf{R}^n \times [0,T]\,; \\ u(0) = 0\,, \text{ in } \mathbf{R}^n. \end{cases}$$

This problem is well-posed (see [3] or [6] for instance): its solution, denoted by u, belongs to $\mathrm{L}^\infty(0,T;\mathrm{L}^2(\mathbf{R}^n))$.
We denote by $u(x,t)$ the function: $X \mapsto u(x,X,t)$.

Theorem 6. *Under assumptions* (20), (21), (22), *there exist* $h > 0$, $M > 0$ *(independent of* ν*) such that for almost every* (x,t) *in* $(-\infty,0) \times [0,T]$, *the solution* u *of* (23) *satisfies:*

$$\|u(x,t)\|_{\mathrm{L}^2(\mathbf{R}^{n-1})} \leq M \, \nu^{1/2} \, \|g\|_{\mathrm{L}^\infty(0,T;\mathrm{L}^2(\mathbf{R}^n))} \inf \left\{ 1, \left(\frac{\nu}{|x|} \right)^{1/2} \right\} \exp\!\left(\frac{\alpha_1 x}{2h\nu} \right).$$

The proof consists of two stages (see[4]):
1) the result is proved for the advection-diffusion operator with constant coefficients $\partial/\partial t + \alpha \partial/\partial x - \nu\Delta$; we use its elementary solution. Futhermore we have here: $h = 1$, $\alpha_1 = \alpha$.

2) in the general case, we construct and estimate a fundamental solution of equation (23); the form of this estimate is similar to the above elementary solution and we may apply the previous result.

1.3. Error estimates.

In order to apply theorem 6 to (5), we estimate $\Delta w|_{\mathbf{R}^n_+ \times \mathbf{R}^+}$. To this end, we have to estimate some derivatives of w. We proceed as follows:
1) if w is differentiated only with respect to $\mathbf{x}_i$ ($i \geq 2$) or t, we use propositions 3 and 4 (transmission problem);
2) if w is differentiated once or twice with respect to x, we estimate the considered derivative of w firstly in the left half-space, secondly in the right half-space: we use proposition 1 (artificial problem) to build estimates in $\mathbf{R}^n_- \times \mathbf{R}^+$; estimates in $\mathbf{R}^n_+ \times \mathbf{R}^+$ will be derived from estimates of w_t, w_{tt} or $w_{\mathbf{x}_i t}$ ($i \geq 2$) in the full space.
After intermediate lemmas (see [5]), we obtain:

Proposition 7. *Under assumptions* (6), (7), (8), *let w be the solution of* (4), *then Δw satisfies for all $T > 0$, for all $\nu_0 > 0$:*

$$\Delta w|_{\mathbf{R}^n_+ \times \mathbf{R}^+} \in \mathrm{L}^\infty(0, T; \mathrm{L}^2(\mathbf{R}^n_+))\,;$$

there exists $C_7 > 0$ (independent of ν in $(0, \nu_0)$) such that:

$$\|\Delta w|_{\mathbf{R}^n_+ \times \mathbf{R}^+}\|_{\mathrm{L}^\infty(0,T;\mathrm{L}^2(\mathbf{R}^n_+))} \leq C_7 \, \|u^0\|_{\mathrm{H}^4(\mathbf{R}^n)}\,.$$

Now we apply theorem 6 and proposition 7 to problem (5) and we can give error estimates (for notations, see theorem 6).

Theorem 8. *Under assumptions* (6), (7), (8), *for all $T > 0$, for all $\nu_0 > 0$, there exist $h > 0$, $m > 0$ (independent of ν such that $0 < \nu < \nu_0$) such that for almost every (x, t) in $(-\infty, 0) \times [0, T]$:*

$$\|e(x,t)\|_{\mathrm{L}^2(\mathbf{R}^{n-1})} \leq m\, \nu^{3/2}\, \|u^0\|_{\mathrm{H}^4(\mathbf{R}^n)} \inf\left\{1, \left(\frac{\nu}{|x|}\right)^{1/2}\right\} \exp\left(\frac{\alpha x}{2h\nu}\right).$$

By integration with respect to x in $(-\infty, 0)$, we get that the error is an $O(\nu^2)$ in the left half-space:

Corollary 9. *Under assumptions* (6), (7), (8), *for all $T > 0$, for all $\nu_0 > 0$, there exists $k > 0$ (independent of ν such that $0 < \nu < \nu_0$) such that:*

$$\left\| e|_{\mathbf{R}^n_- \times \mathbf{R}^+} \right\|_{\mathrm{L}^\infty(0,T;\mathrm{L}^2(\mathbf{R}^n_-))} \leq k\, \nu^2\, \|u^0\|_{\mathrm{H}^4(\mathbf{R}^n)}\,.$$

2. The second artificial boundary condition.

This study is still in progress and will be reported in a forthcoming paper. Here we will only explain the method when a is constant:

$$\forall \mathbf{x} \in \mathbf{R}^n, \quad a(\mathbf{x}) = \alpha, \text{ with : } \alpha \in (0, +\infty). \tag{24}$$

The second artificial boundary condition can be written:

$$N^2 u = 0, \text{ on } \Sigma;$$

with:

$$N = \frac{\partial}{\partial t} + \alpha \frac{\partial}{\partial x}.$$

Let us recall the full-space problem:

$$\begin{cases} Nu - \nu \Delta u = 0, \text{ in } \mathbf{R}^n \times \mathbf{R}^+; \\ u(0) = u^0, \text{ in } \mathbf{R}^n; \end{cases} \tag{1}$$

2.1. The artificial problem and the transmission problem.

The second artificial problem is:

$$\begin{cases} Nu - \nu \Delta u = 0, \text{ in } \mathbf{R}^n_- \times \mathbf{R}^+; \\ u(0) = u^0, \text{ in } \mathbf{R}^n_-; \\ N^2 u = 0, \text{ on } \Sigma. \end{cases} \tag{25}$$

It will be associated to the following transmission problem:

$$\begin{cases} Nu - \nu \Delta u = 0, \text{ in } \mathbf{R}^n_- \times \mathbf{R}^+; \\ N^2 u = 0, \text{ in } \mathbf{R}^n_+ \times \mathbf{R}^+; \\ u(0) = u^0, \text{ in } \mathbf{R}^n_-; \\ u,\ Nu,\ N^2 u \text{ are continuous through } \Sigma. \end{cases} \tag{26}$$

We suppose now:

$$u^0 \in \mathrm{H}^6(\mathbf{R}^n); \quad \mathrm{supp}(u^0) \subset \mathbf{R}^n_-; \tag{27}$$

As above problem (1) is well-posed, let us denote its solution by v ($v \in \mathrm{L}^\infty(0, T; \mathrm{L}^2(\mathbf{R}^n))$). Futhermore we have:

Proposition 10. *Under assumptions* (24) *and* (27), *problems* (25) *and* (26) *are well-posed. The solution* w *of* (26) *belongs to* $\mathrm{L}^\infty(0, T; \mathrm{L}^2(\mathbf{R}^n))$ *and* $w|_{\mathbf{R}^n_- \times \mathbf{R}^+}$ *satisfies* (25).

If u is a solution of (25) (resp. (26)), then Nu is a solution of a problem similar to (3) (resp. (4)) with initial data: $\nu\Delta u^0$. We can apply proposition 2 (resp. 5). We get a solution of (25) (resp. (26)) by integration along characteristic curves of N. The uniqueness of w is derived from the linearity of (25) (resp. (26)).

2.1. Error estimates.
We denote by v and w the solutions of (1) and (26), respectively. As above we define the error: $e = v - w$.

Theorem 11. *Under assumptions (24) and (27), for all $T > 0$, for all $\nu_0 > 0$, there exists $M > 0$ (independent of ν such that $0 < \nu < \nu_0$) such that for almost every (x,t) in $(-\infty,0) \times [0,T]$:*

$$\|e(x,t)\|_{L^2(\mathbf{R}^{n-1})} \leq M\,\nu^{7/2}\,\|u^0\|_{H^6(\mathbf{R}^n)}\,\inf\left\{1,\left(\frac{\nu}{|x|}\right)^{1/2}\right\}\exp\left(\frac{\alpha x}{2\nu}\right).$$

Consider $\tilde{v} = Nv$ and $\tilde{w} = Nw$. We can see that $\tilde{v}$ satisfies:

$$(28)\qquad \begin{cases} Nu - \nu\Delta u = 0\,, \text{ in } \mathbf{R}^n \times \mathbf{R}^+;\\ u(0) = \nu\Delta u^0\,, \text{ in } \mathbf{R}^n; \end{cases}$$

and that $\tilde{w}$ satisfies:

$$(29)\qquad \begin{cases} Nu - \nu\Delta u = 0\,, \text{ in } \mathbf{R}^n_- \times \mathbf{R}^+;\\ Nu = 0\,, \text{ in } \mathbf{R}^n_+ \times \mathbf{R}^+;\\ u(0) = \nu\Delta u^0\,, \text{ in } \mathbf{R}^n_-;\\ u,\ Nu \text{ are continuous through } \Sigma. \end{cases}$$

We can see that (28) and (29) are similar to (1) and (4),respectively. We apply theorem 8 to $\tilde{e} = \tilde{v} - \tilde{w} = Ne$. We get:
For all $T > 0$, for all $\nu_0 > 0$, there exists $m > 0$ (independent of ν such that $0 < \nu < \nu_0$) such that for almost every (x,t) in $(-\infty,0) \times [0,T]$:

$$\|\tilde{e}(x,t)\|_{L^2(\mathbf{R}^{n-1})} \leq m\,\nu^{5/2}\,\|u^0\|_{H^6(\mathbf{R}^n)}\,\inf\left\{1,\left(\frac{\nu}{|x|}\right)^{1/2}\right\}\exp\left(\frac{\alpha x}{2\nu}\right).$$

We have:

$$\begin{cases} Ne = \tilde{e}\,, \text{ in } \mathbf{R}^n \times \mathbf{R}^+\,;\\ e(0) = 0\,, \text{ in } \mathbf{R}^n\,. \end{cases}$$

Then we can get e by integration along characteristic curves of N:

$$e(x,t) = \int_0^t \tilde{e}(x - \alpha s, t - s)\,ds\,.$$

Then:

$$\|e(x,t)\|_{L^2(\mathbf{R}^{n-1})} \leq m\,\nu^{5/2}\,\|u^0\|_{H^6(\mathbf{R}^n)} \int_0^t \inf\left\{1, \left(\frac{\nu}{|x-\alpha s|}\right)^{1/2}\right\} \exp\left(\frac{\alpha(x-\alpha s)}{2\nu}\right) ds$$

$$\leq m\,\nu^{5/2}\,\|u^0\|_{H^6(\mathbf{R}^n)} \inf\left\{1, \left(\frac{\nu}{|x|}\right)^{1/2}\right\} \int_0^{+\infty} \exp\left(-\frac{\alpha^2 s}{2\nu}\right) ds\,.$$

We get the result with: $M = 2m/\alpha^2$.
As above, by integration with respect to x in $(-\infty, 0)$, we get that the error is an $O(\nu^4)$ in the left half-space:

Corollary 12. *Under assumptions (24) and (27), for all $T > 0$, for all $\nu_0 > 0$, there exists $k > 0$ (independent of ν such that $0 < \nu < \nu_0$) such that:*

$$\left\| e|_{\mathbf{R}^n_- \times \mathbf{R}^+} \right\|_{L^\infty(0,T;L^2(\mathbf{R}^n_-))} \leq k\,\nu^4\,\|u^0\|_{H^6(\mathbf{R}^n)}\,.$$

3. Numerical experiments.

We deal here with a one-dimensional problem as in [1]:

$$\begin{cases} u_t + a u_x - \nu \Delta u = 0\,, \text{ in } (-1,1)\times[0,T]\,; \\ u = g\,, \text{ on } \{-1\}\times[0,T]\,; \\ u_x = 0\,, \text{ on } \{1\}\times[0,T]\,; \\ u(0) = 0\,, \text{ in } (-1,1)\,. \end{cases}$$

We approximate this problem in the slab $(-1,0)\times[0,T]$ by an artificial problem:

$$\begin{cases} u_t + a u_x - \nu \Delta u = 0\,, \text{ in } (-1,0)\times[0,T]\,; \\ u = g\,, \text{ on } \{-1\}\times[0,T]\,; \\ \text{Artificial Boundary Condition on } \{0\}\times[0,T]\,. \\ u(0) = 0\,, \text{ in } (-1,0)\,. \end{cases}$$

As artificial boundary condition, we propose:

(A.B.C. 0)	$u_x = 0\,;$
(A.B.C. 1)	$u_t + a u_t = 0\,;$
(A.B.C. 2)	$u_{tt} + 2a u_{tx} + a^2 u_{xx} = 0\,;$
(A.B.C. 3)	$u_{tt} + 2a u_{tx} + a^2 u_{xx} + a a_x u_x = 0\,.$

Futhermore, we choose: $g(t) = \dfrac{\sin t}{\sqrt{t^2+1}}$.
These problems are approximated by finite differences in space and we use a Crank-Nicolson scheme in time (see [1] and [5]).

We use : $T = 5$, $\Delta t = \Delta x = 0.004$ (time and space meshes).
We compute firstly the "exact" solution in the "large" domain, secondly the "artificial" solution in $(-1,0) \times [0,5]$ and finally the error e; we define:

$$\text{Nmax }(e) = \sup_{t\in[0,5]} \left(\int_{-1}^{0} |e(x,t)|^2 \, dx \right)^{1/2} .$$

We have performed experiments for several functions a; numerical results are very similar: the most significant results are shown in figures 5 and 6 where we plot $-\log(\text{Nmax }(e))$ as a function of $-\log \nu$. We can see the quadratic dependence on ν of the error as stated in corollary 9 (A.B.C.1). Futhermore, we can see that (A.B.C. 3) is a better condition than (A.B.C. 2) (Fig. 6).

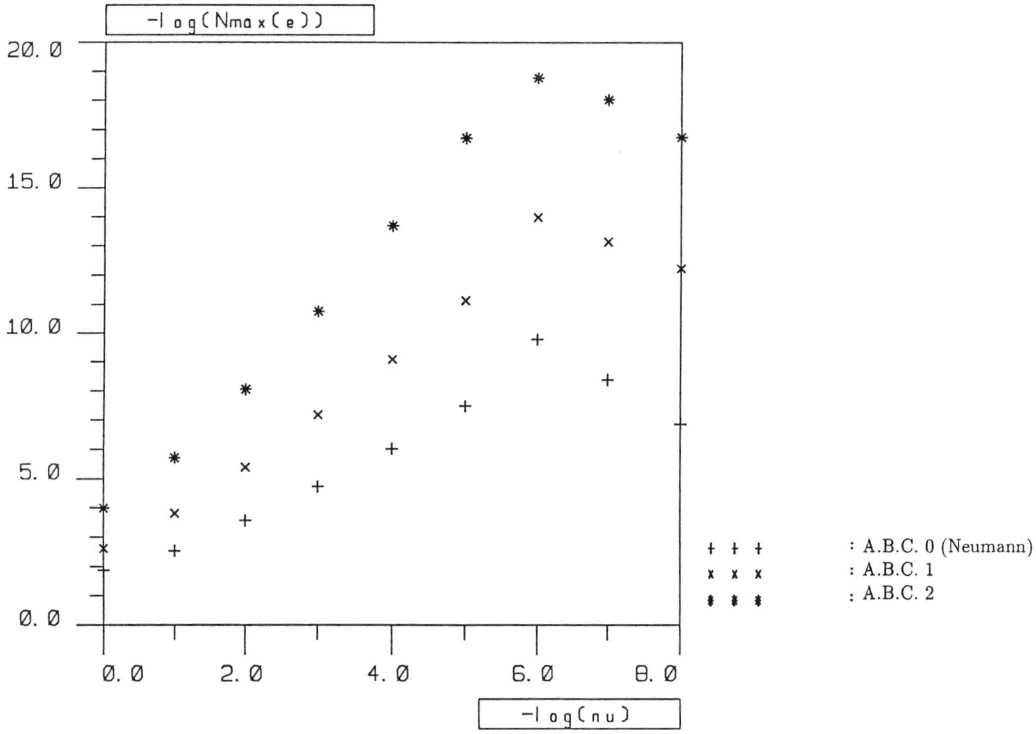

Fig. 5: $a(x) = 1$.

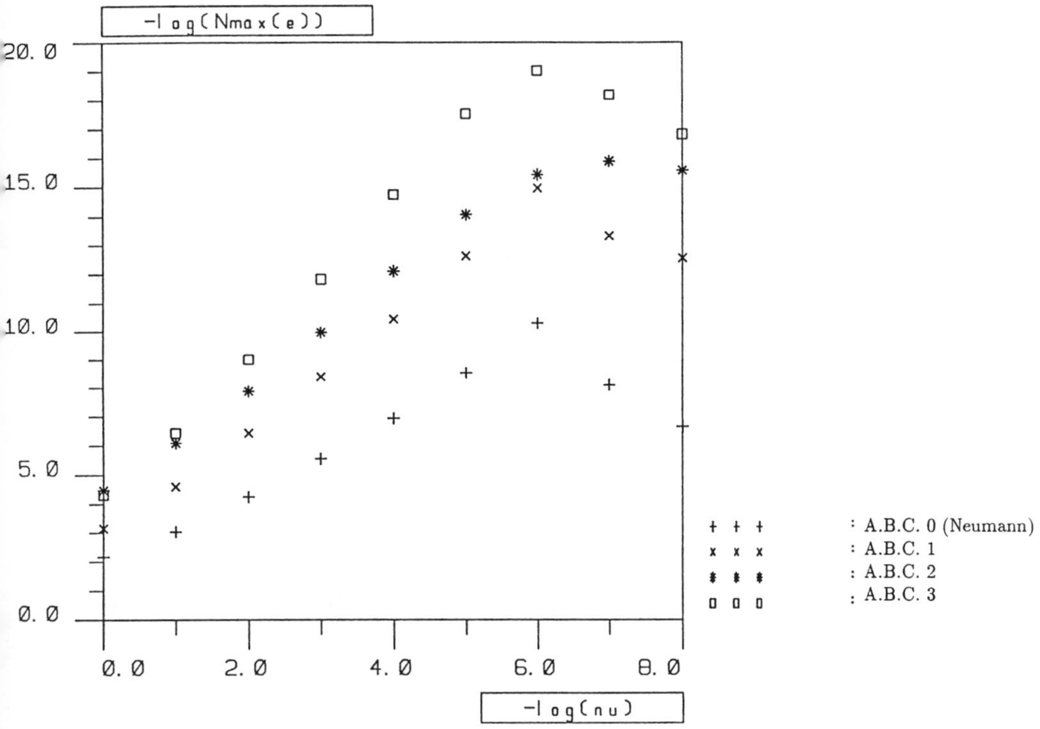

Fig. 6: $a(x) = (3 + \sin x)/2$.

References.

1. Halpern, L., Artificial boundary conditions for the linear advection-diffusion equation, *Math. Comp.* , Vol. 46 (1986).
2. Halpern, L. and Schatzman, M., Artificial boundary conditions for incompressible viscous flows, *SIAM J. Math. Analysis*, Vol. 20 (1989).
3. Lions, J.-L. and Magenes, E., *Problèmes aux limites non homogènes et applications*, Dunod, Paris (1968).
4. Lohéac, J.-P., Advective estimates, *Math. Meth. in the Appl. Sci.*, Vol. 13 (1990).
5. Lohéac, J.-P., An artificial boundary condition for an advection-diffusion equation, *Math. Meth. in the Appl. Sci.*, Vol. 14 (1991).
6. Raviart, P.-A. and Thomas, J.-M., *Introduction à l'analyse numérique des équations aux dérivées partielles*, Masson, Paris (1983).

École Centrale de Lyon, Département M.I.S.
B.P. 163, 69131 ÉCULLY Cedex, FRANCE

R PÜTTER

Bounds for Neumann eigenvalues of n-dimensional balls and second eigenfunctions on ellipsoids

1 Introduction

Let $\Omega = \Omega_a = \{x \in \mathbb{R}^n : \frac{x_1^2}{a_1^2} + \frac{x_2^2}{a_2^2} + \cdots + \frac{x_n^2}{a_n^2} < 1\}$ be an n-dimensional ellipsoid of volume $\omega_n = \mathrm{vol}(B^n)$, where $a = (a_1, a_2, \ldots, a_n) \in A_n := \{(a_1, \ldots, a_n) : a_i \in \mathbb{R}^+, 1 \leq i \leq n, \prod_{i=1}^n a_i = 1\}$. Consider the eigenvalue problem for the Laplacian on Ω with Neumann boundary conditions:

$$\begin{cases} \text{Find } \mu \in \mathbb{R},\ u \in C^2(\Omega) \cap C^1(\overline{\Omega}) \text{ such that} \\ \Delta u + \mu u = 0 \text{ in } \Omega, \\ \partial_\nu u = 0 \text{ on } \partial\Omega. \end{cases} \tag{1.1}$$

It is well-known that the eigenvalues may be arranged in a nondecreasing sequence $0 = \mu_1 < \mu_2 \leq \mu_3 \leq \ldots$ tending to infinity with corresponding eigenfunctions $\{u_i\}_{i\in\mathbb{N}} \subset C^\infty(\overline{\Omega})$. By Courant's max-min principle (see [2]), the second eigenvalue may be characterized as

$$\mu_2 = \inf\left\{ \frac{\int_\Omega |\nabla v|^2 dx}{\int_\Omega v^2 dx} : v \in H^1(\Omega) \setminus \{0\},\ \int_\Omega v dx = 0 \right\}. \tag{1.2}$$

Here, $H^1(\Omega)$ is the usual Hilbert space of L^2-functions with square integrable weak derivatives of order one. If the infimum in (1.2) is achieved for a function $u \in H^1(\Omega)$, then u is a second eigenfunction, i. e. u satisfies (1.1) with with $\mu = \mu_2$. On the other hand, each second eigenfunction yields the infimum in (1.2). By virtue of the condition $\int_\Omega u dx = 0$, each second eigenfunction assumes positive and negative values; hence the nodal set of u, $N(u) := \{x \in \Omega : u(x) = 0\}$, is non-empty. The connected components of $\Omega \setminus N(u)$ are called nodal domains. The Courant nodal domain theorem states that each i-th eigenfunction has at most i nodal domains.
In the case $\Omega = B^n = \{x \in \mathbb{R}^n : |x| < 1\}$, the solutions of (1.1) are explicitly known (cf. [1]). For $\nu > 0$, let J_ν be the Bessel function of index ν, and let $\mathcal{H}_k^n$ be the space

of harmonic homogeneous polynomials of degree k in n variables. Introduce spherical coordinates $(r,\xi) \in \mathbb{R}^+ \times S^{n-1}$, $x = r\xi$, in $\mathbb{R}^n$ and solve equation (1.1) by a separation of variables to get the following result: For fixed n, the functions

$$r^{1-\frac{n}{2}} J_{\frac{n}{2}-1+k}(q_{k,i}r) P_k\left(\frac{x}{r}\right),\ i = 1,2,\ldots,\ k = 1,2\ldots,\ P_k \in \mathcal{H}_k^n$$

together with the constant function form a complete system of eigenfunctions with corresponding eigenvalues $q_{k,i}^2$. The number $q_{k,i}$ is defined to be the i-th positive solution of the equation

$$qJ'_{\frac{n}{2}-1+k}(q) = (\frac{n}{2} - 1)J_{\frac{n}{2}-1+k}(q). \tag{1.3}$$

In section 2, we shall prove the estimates

$$2k + k^2 < q_{k,1}^2 < 2 + 2k^2 \quad \text{for } n = 2, \tag{1.4}$$

$$nk + k^2 < q_{k,1}^2 < nk + k^2 + \frac{2k^2}{n-2} \quad \text{for } n \geq 3. \tag{1.5}$$

Especially, each $r^{1-\frac{n}{2}} J_{\frac{n}{2}}(q_{1,1}r) P_1\left(\frac{x}{r}\right)$ is an eigenfunction to the eigenvalue $\mu_2(B^n)$. The inequalities (1.4) and (1.5) now imply that

$$\mu_2(B^n) > n + 1. \tag{1.6}$$

Relations (1.4)–(1.6) will be used in section 3 to prove the following result concerning problem (1.1) on general ellipsoids Ω_a:

Theorem 1.1 *If $a_1 > a_2 \geq a_3 \ldots \geq a_n$, then*

$$\mu_2 \textit{ is simple and } N(u) = \Omega_a \cap \{x_1 = 0\}. \tag{1.7}$$

If $a_1 = a_2 = \cdots = a_k > a_{k+1} \geq \cdots \geq a_n$ for some $k \in \{2,\ldots,n\}$, then μ_2 is of multiplicity k and each eigenfunction u to μ_2 has a representation

$$u(x) = \langle \xi, x\rangle w(r, x_{k+1}, \ldots, x_n), \tag{1.8}$$

where $r = \sqrt{x_1^2 + \ldots + x_k^2}$, $\xi \in span(e_1,\ldots,e_k) \setminus \{0\}$, $\langle\cdot,\cdot\rangle$ denotes the standard scalar product in $\mathbb{R}^n$ and w is a positive function.

2 Estimates for eigenvalues

For $\nu > 0$, consider solutions $z(r)$ to the Bessel equation

$$(rz')' = z\left(\frac{\nu^2}{r} - r\right) \text{ on } \{r > 0\} \tag{2.1}$$

under the one-sided boundary condition

$$z(0) = 0. \tag{2.2}$$

The Bessel function J_ν together with the Neumann function N_ν form a fundamental system of solutions for (2.1). Condition (2.2) now determines z to be a constant multiple of J_ν. We are looking for estimates of the first positive root $\eta = \eta(\alpha)$ of the equation

$$\eta z'(\eta) = \alpha z(\eta), \tag{2.3}$$

where

$$0 < \alpha < \nu. \tag{2.4}$$

Obviously, (1.3) is a special case of (2.3) with $\nu = \frac{n}{2} - 1 + k$, $\alpha = \frac{n}{2} - 1$. We multiply equation (2.1) by rz' and integrate from 0 to η to arrive at

$$\int_0^\eta (rz')'rz' = \int_0^\eta (\nu^2 - r^2)zz',$$

$$\left.\frac{(rz')^2}{2}\right|_0^\eta = \left.(\nu^2 - r^2)\,\frac{z^2}{2}\right|_0^\eta + \int_0^\eta rz^2.$$

Insertion of condition (2.3) on the left-hand side shows that

$$\int_0^\eta rz^2 = \frac{z^2}{2}(\eta^2 + \alpha^2 - \nu^2). \tag{2.5}$$

By means of this relation we can now prove:

Lemma 2.1 *Under the conditions* (2.1)–(2.4) *we have*

$$\eta^2 > \nu(\nu + 2) - \alpha(\alpha + 2). \tag{2.6}$$

For $\nu = \frac{n}{2} - 1 + k$, $\alpha = \frac{n}{2} - 1$, (2.6) *transforms into the left-hand side inequalities of* (1.4) *and* (1.5).

Proof. Multiplication of (2.1) by z yields

$$z''zr + zz' + z^2r = \nu^2\frac{z^2}{r}. \tag{2.7}$$

We then have

$$\begin{aligned}
0 &< \int_0^\eta (r^{-\nu}z)'^2 r^{2\nu+1} \\
&= \int_0^\eta r^{-1}(rz' - \nu z)^2 \\
&= \int_0^\eta \{rz'^2 - 2\nu zz' + \nu^2 z^2 r^{-1}\} \\
&\overset{(2.7)}{=} \int_0^\eta \{z''zr + rz'^2 + zz' - 2\nu zz' + z^2 r\} \\
&= \int_0^\eta \{(rzz')' - 2\nu zz' + z^2 r\} \\
&\overset{(2.5)}{=} rzz'|_0^\eta - 2\nu\frac{z^2}{2}\Big|_0^\eta + \frac{z^2(\eta)}{2}(\eta^2 + \alpha^2 - \nu^2) \\
&= \frac{z^2(\eta)}{2}(2\alpha - 2\nu + \eta^2 + \alpha^2 - \nu^2) \\
&= \frac{z^2(\eta)}{2}(\eta^2 - \nu(\nu+2) + \alpha(\alpha+2)),
\end{aligned}$$

from which the claim follows.

We now prove upper bounds for η.

Lemma 2.2 *Conditions* (2.1)–(2.4) *imply:*

$$\eta^2 < \nu(\nu+2) - \alpha(\alpha+2) + \frac{(\nu-\alpha)^2}{\alpha}. \tag{2.8}$$

Proof. Set $w(r) = z(\eta r)$. Then $w'(1) = \alpha w(1)$ and

$$w'' + \frac{w'}{r} + (\eta^2 - \frac{\nu^2}{r^2})w = 0. \tag{2.9}$$

A further substitution $y = r^{-\alpha}w$ yields

$$y'(1) = 0, \tag{2.10}$$

and (2.9) transforms into

$$(r^{2\alpha+1}y')' + \eta^2 y r^{2\alpha+1} - (\nu^2 - \alpha^2) y r^{2\alpha-1} = 0. \tag{2.11}$$

(2.11.) is the Euler equation of the problem

$$\frac{\int_0^1 v'^2 r^{2\alpha+1} + (\nu^2 - \alpha^2)\int_0^1 v^2 r^{2\alpha-1}}{\int_0^1 v^2 r^{2\alpha+1}} \rightsquigarrow \min \text{for } v \in \mathcal{W}\setminus\{0\}, \tag{2.12}$$

where $\mathcal{W}$ is the completion of $C_c^\infty((0,1])$ under the norm $\|\cdot\|_{\nu,\alpha}$, with

$$\| v \|_{\nu,\alpha}^2 = \int_0^1 v'^2 r^{2\alpha+1} + (\nu^2 - \alpha^2)\int_0^1 v^2 r^{2\alpha-1}.$$

Well-known compactness results in the Hilbert space $\mathcal{W}$ ensure the existence of a function $v_m \in \mathcal{W}$ for which the minimum μ in (2.12) is achieved. v_m weakly solves (2.11) with η^2 replaced by μ. Standard regularity theory guarantees that $v_m \in C^\infty((0,1])$. Resubstitution $v_m = r^{-\alpha}w_m$ and $z_m(\mu^{1/2}r) = w_m(r)$ makes z_m a solution of (2.1) and (2.3). The fact that $\int_0^1 v^2 r^{2\alpha-1}$ is finite excludes N_ν from the representation of z_m. Hence z_m is a constant multiple of J_ν and also (2.2) holds. By the requirement that η be the smallest positive solution to (2.3) and the minimum property of μ, we must have $\mu = \eta^2$. So upper bounds for η^2 can be obtained by choosing $v = r^\beta$, $\beta > 0$, in (2.12). This yields:

$$\eta^2 < \frac{\beta^2 \int_0^1 r^{2(\beta+\alpha)-1} + (\nu^2 - \alpha^2)\int_0^1 r^{2(\beta+\alpha)-1}}{\int_0^1 r^{2(\alpha+\beta)+1}} = \frac{(\alpha+\beta+1)(\beta^2+\nu^2-\alpha^2)}{\beta+\alpha}.$$

For $\beta \searrow 0$, the right-hand side converges to

$$\frac{\alpha+1}{\alpha}(\nu^2-\alpha^2) = \nu^2 - \alpha^2 + \frac{(\nu+\alpha)(\nu-\alpha)}{\alpha} = \nu^2 - \alpha^2 + 2(\nu-\alpha) + \frac{(\nu-\alpha)^2}{\alpha},$$

which is (2.8). The inequality is strict as the r^β do not converge to an eigenfunction. The upper bound in (1.5) now drops out for $\nu = \frac{n}{2} - 1 + k$, $\alpha = \frac{n}{2} - 1$. The case $\beta = 1$, $\alpha = 0$, $\nu = k$ yields $\eta^2 < 2(1+k^2)$, which is the second half of (1.4).

3 Determination of the eigenfunctions

For $1 \le i \le n$ let $S_i \in GL(n, \mathbb{R})$ be the reflection $S_i = \begin{cases} e_j & \text{for } j \neq i, \\ -e_i & \text{for } j = i. \end{cases}$

Define subspaces of $H^1(\Omega)$ according to the symmetry properties of the functions:

$$V_i^- = \{u \in H^1(\Omega) : u \circ S_i = -u\},\ V_i^+ = u \in H^1(\Omega) : u \circ S_i = u\},\ V_s = \bigcap_{i=1}^n V_i^+.$$

It is clear that $H^1(\Omega) = \bigoplus_{i=1}^n V_i^- \oplus V_s$ (this is not a direct sum). Let E be the eigenspace to μ_2, $E = \{u \in C^\infty(\overline{\Omega}) : \Delta u + \mu_2 u = 0,\ \partial_\nu u = 0 \text{ on } \partial\Omega\}$, and $E_i = E \cap V_i^-$, $E_s = E \cap V_s$. Then

$$E = \bigoplus_{i=1}^n E_i \oplus E_s.$$

This sum is direct because $u \in E_i \cap E_j$, $i \neq j$, would have at least four nodal domains, a contradiction to Courant's theorem. Furthermore, set:

$$\mu^{(i)}(\Omega) = \inf\left\{\frac{\int_\Omega |\nabla u|^2 dx}{\int_\Omega u^2 dx} : u \in V_i^-(\Omega)\setminus\{0\}\right\},\ 1 \leq i \leq n, \tag{3.1}$$

$$\mu^{(s)}(\Omega) = \inf\left\{\frac{\int_\Omega |\nabla u|^2 dx}{\int_\Omega u^2 dx} : u \in V_s(\Omega)\setminus\{0\},\ \int_\Omega u\,dx = 0\right\}, \tag{3.2}$$

$$\Omega_i^+ = \{x \in \Omega : x_i > 0\}.$$

Each eigenfunction $u \in V_i^-(\Omega)$ to the eigenvalue $\mu^{(i)}(\Omega)$ is a first eigenfunction to the problem

$$\begin{cases} \Delta u + \mu u = 0 \text{ in } \Omega_i^+, \\ \partial_\nu u = 0 \text{ on } \partial\Omega_i^+ \cap \{x_i > 0\}, \\ u = 0 \text{ on } \partial\Omega_i^+ \cap \{x_i = 0\}. \end{cases}$$

Hence each $\mu^i(\Omega)$ is simple and E_i is at most one-dimensional. The insertion of $u = x_i$ as a comparison function in (3.1) yields after a short calculation

$$\mu^{(i)}(\Omega_a) < (n+2)a_i^{-2} \text{ for all } a \in \mathcal{A}_n. \tag{3.3}$$

Lemma 3.1 *Suppose* $a_i = \max\{a_1, \ldots, a_n\}$. *Then*

$$\mu^{(s)}(\Omega_a) > a_i^{-2}(2n+4). \tag{3.4}$$

Proof. For $v \in V_s(\Omega_a)$ with $\int_{\Omega_a} v\,dx = 0$ and $A \in GL(n,\mathbb{R})$, $A = \text{diag}(a_1, \ldots, a_n)$, set $u = v \circ A$. Then $u \in V_s(B)$ and $\int_B u\,dx = 0$. Furthermore,

$$\int_{\Omega_a} v^2 dx = \int_B u^2 dx \text{ and}$$

$$\int_B |\nabla u|^2 dx = \sum_{j=1}^n \int_B a_j^2((\partial_j v) \circ A)^2 dx \leq a_i^2 \int_{\Omega_a} |\nabla v|^2 dx.$$

The function u is admissible in (3.2); hence for the Rayleigh quotient

$$\mathcal{R}_\Omega(v) = \frac{\int_\Omega |\nabla v|^2 dx}{\int_\Omega v^2 dx}$$

we have

$$\mathcal{R}_\Omega(v) \geq a_i^{-2}\mathcal{R}_B(u) \geq a_i^{-2}\mu^{(s)}(B). \tag{3.5}$$

There are two candidates for eigenfunctions for $\mu^{(s)}(B)$. The first is

$$u_1 = r^{1-\frac{n}{2}} J_{\frac{n}{2}+1}(q_{2,1}r) P_2\left(\frac{x}{r}\right),$$

where $r = |x|$ and $P_2 \in \mathcal{H}_2^n$ is invariant under S_j for each $1 \leq j \leq n$. An example is $P_2(x) = \sum_{j=1}^{n-1} x_j^2 - (n-1)x_n^2$. By invoking (1.5) we see that

$$\mathcal{R}(u_1) = q_{2,1}^2 > 2n + 4. \tag{3.6}$$

The second candidate is

$$u_2 = r^{1-\frac{n}{2}} J_{\frac{n}{2}-1}(p_n r), \text{ with } \mathcal{R}(u_2) = p_n^2.$$

p_n is defined as the first positive solution of $(r^{1-\frac{n}{2}} J_{\frac{n}{2}-1}(r))' = 0$, which is equivalent to

$$(1 - \frac{n}{2}) J_{\frac{n}{2}-1}(r) + r J_{\frac{n}{2}-1}'(r) = 0. \tag{3.7}$$

With the recursion formula (see [2], p.486)

$$r J_\nu'(r) = \nu J_\nu(r) - r J_{\nu+1}(r),$$

(3.7) becomes

$$J_{\frac{n}{2}}(r) = 0.$$

Hence p_n^2 is the first Dirichlet eigenvalue $\lambda_1(B^{n+2})$ of B^{n+2}. If we enclose B^{n+2} in an (n+2)-dimensional cube W^{n+2} of edge length 2, the domain monotonicity principle for Dirichlet eigenvalues implies

$$\mathcal{R}(u_2) = p_n^2 = \lambda_1(B^{n+2}) > \lambda_1(W^{n+2}) = (n+2)\frac{\pi^2}{4}. \tag{3.8}$$

(3.4) now follows from (3.5), (3.6) and (3.8). □

As a direct consequence of (3.4) and (3.3) we obtain $E_s(\Omega_a) = \{0\}$, and hence

$$E(\Omega_a) = \bigoplus_{j=1}^{n} E_j(\Omega_a). \tag{3.9}$$

The following lemma relies on an idea due to Hersch [4].

Lemma 3.2 *Let $u \in V_{k+1}^-$, $k \in \{1, \ldots, n-1\}$, and suppose that $a_\ell = \max\{a_{k+1}, \ldots, a_n\}$. Then*

$$\int_{\Omega_a} \sum_{j=k+1}^{n} (\partial_i u)^2 \geq \frac{n-k+1}{a_\ell^2} \int_{\Omega_a} u^2 dx. \tag{3.10}$$

Proof. We may assume that $u \in C^\infty(\Omega_a) \cap V_{k+1}^-(\Omega_a)$. Decompose $x \in \mathbb{R}^n = \mathbb{R}^k \times \mathbb{R}^{n-k}$ as $x = (x', \overline{x})$, $x' = (x_1, \ldots, x_k)$, $\overline{x} = (x_{k+1}, \ldots, x_n)$ and define

$$f(x_1, \ldots, x_k) = \sum_{j=1}^{k} \frac{x_j^2}{a_j^2}, \quad g(x_{k+1}, \ldots, x_n) = \sum_{j=k+1}^{n} \frac{x_j^2}{a_j^2}.$$

For $x' \in \{x' \in \mathbb{R}^k : f(x') < 1\}$ set $\Omega_{x'} := \{\overline{x} \in \mathbb{R}^{n-k} : g(\overline{x}) < 1 - f(x')\}$. $\Omega_{x'}$ is an $(n-k)$-dimensional ellipsoid with axes of length $a_{k+1}\alpha, \ldots, a_n\alpha$, where $\alpha = (1 - f(x'))^{1/2}$. Consider $u_{x'} = u(x', \cdot) \in V_1^-(\Omega_{x'})$. Set $A = \mathrm{diag}(a_{k+1}\alpha, \ldots, a_n\alpha)$, $v = u_{x'} \circ A$. Then $v \in V_1^-(B^{n-k})$, hence by (1.6), $n - k + 1 < \mathcal{R}(v)$. Now

$$\begin{aligned}
\int_{B^{n-k}} |\nabla v|^2 dx &= \sum_{i=k+1}^{n} \int_{B^{n-k}} a_i^2 \alpha^2 (\partial_i u_{x'})^2 \circ A d\overline{x} \\
&\leq a_\ell^2 \alpha^2 \det(A^{-1}) \sum_{i=k+1}^{n} \int_{\Omega_{x'}} (\partial_i u_{x'})^2 d\overline{x} \\
&= a_\ell^2 \alpha^2 \det(A^{-1}) \int_{\Omega_{x'}} |\nabla_{\overline{x}} u_{x'}|^2 d\overline{x},
\end{aligned}$$

$$\int_{B^{n-k}} v^2 d\overline{x} = \det(A^{-1}) \int_{\Omega_{x'}} u_{x'}^2 d\overline{x}.$$

So we obtain

$$\frac{n-k+1}{a_\ell^2} \int_{\Omega_{x'}} u^2 d\overline{x} \leq \frac{n-k+1}{a_\ell^2 \alpha^2} \int_{\Omega_{x'}} u^2 d\overline{x} < \int_{\Omega_{x'}} |\nabla_{\overline{x}} u|^2 d\overline{x}. \tag{3.11}$$

Integration of (3.11) over $\{x' \in \mathbb{R}^k : f(x') < 1\}$ yields (3.10). Let us remark that, for $k = n-1$, the sharper inequality

$$\int_{\Omega_a} (\partial_n u)^2 dx > \frac{\pi^2}{4a_n^2} \int_{\Omega_a} u^2 dx \tag{3.12}$$

holds because $\mu_2(B^1) = \mu_2((-1,1)) = \frac{\pi^2}{4}$. □

Lemma 3.3 *Let $a \in A_n$ and let $u \in V_2^-(\Omega_a)$ be an eigenfunction to the eigenvalue $\mu^{(2)}(\Omega_a)$ such that*

$$\lambda := \left(\int_{\Omega_a} (\partial_2 u)^2 dx\right) / \left(\int_{\Omega_a} (\partial_1 u)^2 dx\right) > 1.$$

Let $b_2 \in \mathbb{R}^+$ with $1 < \frac{b_2}{a_2} < \lambda^{1/2}$ and set $b = (a_1 \frac{a_2}{b_2}, b_2, a_3, \ldots, a_n)$. We then have

$$\mu^{(2)}(\Omega_b) < \mu^{(2)}(\Omega_a). \tag{3.13}$$

Proof. We first note the obvious equivalence for $A, B > 0$, $A > B$, $\Lambda > 1$:

$$\Lambda^{-1}A + \Lambda B < A + B \Leftrightarrow A < \frac{A}{B}. \tag{3.14}$$

We may assume that $\int_{\Omega_a} u^2 dx = 1$. Set $s = \frac{b_2}{a_2}$ and define $v \in V_2^-(\Omega_b)$ by $v = u \circ A$, where $A \in GL(n, \mathbb{R})$, $A = \text{diag}(s, s^{-1}, 1, \ldots, 1)$. Then $\int_{\Omega_a} v^2 dx = 1$ and

$$\begin{aligned}
\int_{\Omega_b} |\nabla v|^2 dx &= s^2 \int_{\Omega_a} (\partial_1 u)^2 dx + s^{-2} \int_{\Omega_a} (\partial_2 u)^2 dx + \sum_{i=3}^{n} \int_{\Omega_a} (\partial_i u)^2 dx \\
&\overset{(3.14)}{<} \int_{\Omega_a} [(\partial_1 u)^2 + (\partial_2 u)^2 + \sum_{i=3}^{n} (\partial_i u)^2] dx \\
&= \mu^{(2)}(\Omega_a).
\end{aligned}$$

As $\mu^{(2)}(\Omega_b) = \inf\left\{\int_{\Omega_b} |\nabla v|^2 dx : v \in V_2^-(\Omega_b),\ \int_{\Omega_b} v^2 dx = 1\right\}$, (3.13) is proved. □

Before we turn to the proof of Theorem 1.1, we investigate more closely the case $n = 3$, where a stronger result can be shown.

Theorem 3.4 *For $t \in \mathbb{R}^+$ and $a = (a_1, a_2, a_3) \in A_3$ set $a(t) = (t^{-1}a_1, ta_2, a_3)$. Then*

$$\mu^{(2)}(\Omega_{a(t)}) \text{ is strictly decreasing in } t. \tag{3.15}$$

Proof. For $\alpha > \frac{1}{2}$, define $u_\alpha \in V_2(\Omega_{a(t)})$ by $u_\alpha(x) = sign(x_2)|x_2|^\alpha$. A short calculation leads to

$$\mathcal{R}_{\Omega_{a(t)}}(u_\alpha) = \frac{\alpha^2(2\alpha + 3)}{2\alpha - 1} a_2^{-2} t^{-2}.$$

$\mathcal{R}(u_\alpha)$ achieves a minimum at the point $\alpha_0 = \frac{1}{2}\sqrt{3}$, where

$$\mathcal{R}(u_{\alpha_0}) = a_2^{-2} t^{-2} 4.84807\ldots < a_2^{-2} t^{-2} 4.85$$

So we have sharpened (3.3) to

$$\mu^{(2)}(\Omega_{a(t)}) < 4.85\, a_2^{-2}t^{-2}. \qquad (3.16)$$

Inequality (3.12) yields

$$\int_{\Omega_{a(t)}} (\partial_2 u)^2 dx \geq \frac{\pi^2}{4a_2^2t^2} \int_{\Omega_{a(t)}} u^2 dx > 2.46\, a_2^{-2}t^{-2} \int_{\Omega_{a(t)}} u^2 dx. \qquad (3.17)$$

A comparison of (3.17) and (3.18) shows that

$$\int_{\Omega_{a(t)}} (\partial_2 u)^2 > \frac{1}{2}\mu^{(2)}(\Omega_{a(t)}) \int_{\Omega_{a(t)}} u^2 dx;$$

hence

$$\int_{\Omega_{a(t)}} (\partial_2 u)^2 dx > \int_{\Omega_{a(t)}} (\partial_1 u)^2 dx.$$

With the aid of the previous lemma, we infer that for each $t > 0$ there is an $\varepsilon > 0$ such that for each $t' \in (t, t+\varepsilon)$ we have $\mu^{(2)}(\Omega_{a(t')}) < \mu^{(2)}(\Omega_{a(t)})$. A simple connectivity argument then yields (3.15). □

With the observation that $\mu^{(1)}(\Omega_{(a_1,a_2,a_3)}) = \mu^{(2)}(\Omega_{(a_2,a_1,a_3)})$, we conclude

Corollary 3.5 *If* $a_1 > a_2 > a_3$ *then* $\mu^{(1)}(\Omega_a) < \mu^{(2)}(\Omega_a) < \mu^{(3)}(\Omega_a)$. □

Proof of Theorem 1.1. Let $a = (a_1, \ldots, a_n) \in \mathcal{A}_n$, $a_1 \geq a_2 \geq \cdots \geq a_n$. We claim that

$$a_i < a_1 \text{ implies } \mu^{(1)}(\Omega_a) < \mu^{(i)}(\Omega_a). \qquad (3.18)$$

Assume by contradiction that $a_i < a_1$ and that

$$\mu^{(i)}(\Omega_a) \leq \mu^{(1)}(\Omega_a).$$

Let $u \in V_i^-(\Omega_a)$ be an eigenfunction to $\mu^{(i)}(\Omega_a)$ with $\int_{\Omega_a} u^2 dx = 1$. Recalling (3.3) we have

$$\mu^{(i)}(\Omega_a) \leq \mu^{(1)}(\Omega_a) < \frac{n+2}{a_1^2}.$$

Hence

$$\int_{\Omega_a} |\nabla u|^2 dx < \frac{n+2}{a_1^2}. \qquad (3.19)$$

Lemma 3.3 yields

$$\int_{\Omega_a} \sum_{j=2}^{n} (\partial_j u)^2 dx \geq \frac{n}{a_1^2}. \tag{3.20}$$

Subtracting (3.20) from (3.19) we arrive at

$$\int_{\Omega_a} (\partial_1 u)^2 dx < \frac{2}{a_1^2}.$$

(3.12) implies

$$\int_{\Omega_a} (\partial_i u)^2 dx \geq \frac{\pi^2}{4a_i^2};$$

hence

$$\left(\int_{\Omega_a} (\partial_i u)^2 dx\right) / \left(\int_{\Omega_a} (\partial_1 u)^2 dx\right) > a_1^2 a_i^{-2},$$

and by means of Lemma 3.3 (with 2 replaced by i) we conclude that

$$\mu^{(i)}(\Omega_{a'}) < \mu^{(i)}(\Omega_a), \text{ where } a' = (a_i, a_2, \ldots, a_{i-1}, a_1, a_{i+1}, \ldots, a_n).$$

But $\mu^{(i)}(\Omega_{a'}) = \mu^{(1)}(\Omega_a)$, as $\Omega^+_{a',i}$ and $\Omega^+_{a,1}$ are congruent. Hence $\mu^{(1)}(\Omega_a) < \mu^{(i)}(\Omega_a)$, a contradiction.

By virtue of (3.9) we have $\mu_2(\Omega_a) = \min_{1\leq j\leq n} \mu^{(j)}(\Omega_a)$. If $a_1 > a_i$ for all $2 \leq i \leq n$ it now follows that $E = E_1$, and because E_1 is at most one-dimensional, $\mu_2(\Omega_a)$ is simple.

If $a_1 = a_2 = \cdots = a_k > a_{k+1} \leq \cdots \leq a_n$, we have $\mu^{(1)}(\Omega_a) = \cdots = \mu^{(k)}(\Omega_a)$ and $\mu^{(i)}(\Omega_a) > \mu^{(1)}(\Omega_a)$ for $i > k$, hence

$$E = \bigoplus_{j=1}^{k} E_j \text{ and } \dim E = k.$$

We now prove (1.8).

For $1 \leq j \leq k$ define $A_j \in GL(k, \mathbb{R})$ by $A_j e_j = e_1$, $A_j e_1 = e_j$, $A_j e_i = e_i$ for $i \neq 1, j$. Let $u \in V_1^-(\Omega_a)$ be an eigenfunction to $\mu^{(1)}(\Omega_a)$ with $u > 0$ on $\Omega^+_{a,1}$. By the Hopf boundary lemma (see [3], Lemma 3.4), $\nabla u(0) = \alpha e_1$ with $\alpha > 0$. We may assume that $\alpha = 1$. Let us adopt the notation $x = (x', \overline{x})$ for $x \in \mathbb{R}^n = \mathbb{R}^k \times \mathbb{R}^{n-k}$ and ∇_k for the gradient in $\mathbb{R}^k$. Define functions $u_j \in V_j^-(\Omega_a)$, $1 \leq j \leq k$, by

$$u_j(x', \overline{x}) = u(A_j x', \overline{x}).$$

Then $\nabla_k u_j = e_j$, u_j is an eigenfunction to $\mu^{(j)}(\Omega_a)$ and the u_j span E.
Let now $x' \in R^k \setminus \{0\}$ be arbitrary but fixed, and set $r = |x'|$. Choose $A \in GL(k, \mathbb{R})$ such that $Ae_1 = \frac{x'}{r}$. Consider $v \in E$ with $v(y', \overline{x}) = u(Ay', \overline{x})$ for all $(y', \overline{x}) \in \mathbb{R}^n$. There exist $\alpha_1, \dots, \alpha_k \in \mathbb{R}$ such that

$$v(y', \overline{x}) = \alpha_1 u_1(y', \overline{x}) + \cdots + \alpha_k u_k(y', \overline{x}). \tag{3.21}$$

Taking derivatives at 0 we see that

$$\nabla_k v(0,0) = A^t e_1 = \sum_{i=1}^{k} \alpha_i e_i;$$

hence $\alpha_i = A_{1i}$. Set $y' = re_1$ in (3.21) to conclude

$$u(rAe_1, \overline{x}) = rA_{11} r^{-1} u(re_1, \overline{x}) =: rA_{11} w(r, \overline{x}). \tag{3.22}$$

But $A_{11} = (\frac{x'}{r}, e_1)$, where $(\cdot, \cdot)$ is the standard scalar product in $\mathbb{R}^k$. Thus (3.22) yields $u_1(x') = (x', e_1) w(|x'|, \overline{x})$, which in turn implies $u_j(x', \overline{x}) = (A_j x', e_1) w(|x'|, \overline{x}) = (x', e_j) w(|x'|, \overline{x})$. So for general $v \in E$ we obtain $v(x) = \langle x, \xi \rangle w(|x'|, \overline{x})$ for some $\xi \in span(e_1, \dots, e_k)$. □

It would be desirable to prove an analogue of Corollary 3.5 for arbitrary dimensions.

References

[1] J. Chavel: *Eigenvalues in Riemannian Geometry* , Academic Press, New York 1984.

[2] R. Courant & D. Hilbert: *Methods of Mathematical Physics*, Vol. I, Interscience Publishers, New York 1953.

[3] D. Gilbarg & N.S. Trudinger: *Elliptic Partial Differential Equations of Second Order*, Second Edition, Springer-Verlag, Berlin 1983.

[4] J. Hersch: *Sur la fréquence fondamentale d'une membrane vibrante: évaluations par défaut et principe de maximum*, ZAMP *11* (1960), 387–413.

Rolf Pütter
FB9 Universität des Saarlandes
D-6600 Saarbrücken
Germany

B RUF

Remarks on a superlinear Sturm–Liouville equation

1. Introduction

In this note we consider boundary value problems of the form

$$(1)\qquad \begin{cases} -u'' = \lambda u + g(u) + h(x), & x \in (o,1) \\ u'(0) = u'(1) = 0, \end{cases}$$

where $g \in C(\mathbb{R})$ such that $\lim_{s\to-\infty} g(s)/s = 0$, $\lim_{s\to+\infty} g(s)/s = +\infty$, and $h \in L^2(\Omega)$ is a given forcing term. As a typical and important example one may think of the equation

$$(2)\qquad \begin{cases} -u'' = \lambda u + e^u + h(x), & x \in (0,1) \\ u'(0) = u'(1) = 0 \end{cases}$$

In fact, to simplify the exposition, we will consider mostly equation (2) in what follows. For general conditions which guarantee the same results as the ones stated for (2), see [7].

It is known that for the existence of solutions for equations (1) and (2) the location of the parameter λ with respect to the spectrum of the linear operator $-\partial^2/\partial x^2$ (with Neumann boundary conditions) plays an important role. In fact, denoting by $\lambda_k = k\cdot\pi^2$, $k = 0,1,2,\ldots$, this spectrum, one has the following results:

1. If $\lambda < 0 = \lambda_o$, then equations (1) and (2) are of Ambrosetti-Prodi type [2], i.e. there exist forcing terms $h \in L^2(0,1)$ such that (2) has at least two solutions, and there exist other forcing terms in L^2 such that (1) and (2) have no solution. For results of this type see e.g. [1,6,8] .

2. Suppose that $\lambda \in (\lambda_k, \lambda_{k+1})$, and $h(x) = c$ (= constant) >> 0 . Then equations (1) and (2) have at least 2k+2 solutions. (see [11,13]).

3. Let $\lambda \in \mathbb{R}$ arbitrary; then for every $n \in \mathbb{N}$ there exists a number $C_n < 0$ such that for $h(x) = c < C_n$ equations (1) and (2) have at least n solutions. (see [12]).

By the results 2 and 3 one is lead to look for conditions such that equations (1) and (2) have solutions *for all* $h \in L^2(0,1)$. We will discuss the following result:

Theorem: Assume that $\lambda_0 = 0 < \lambda < \pi^2/4 = \lambda_1/4$. Then (1) (with some technical conditions) and (2) have a solution for every $h \in L^2(0,1)$.

In what follows we will describe *three different proofs* of this theorem. With this we try to illustrate that this simple equation has some interesting connections with other problems. In fact, the number $\pi^2/4$ will appear in three completely different ways, see (3), (6) and (8).

2. A bouncing problem

The first proof of the theorem relies on the fact that $\pi^2/4$ can be characterized as follows: Let $E_1 = \{ u \in H^1(0,1) : \int_0^1 u \, dx = 0 \}$. Then

$$(3) \qquad \pi^2/4 = \inf_{u \in E_1 \setminus \{0\}} \frac{\int |u'|^2 dx}{\int u^2 dx + |u|_\infty^2} .$$

The proof of this is given in [7]. The main difficulty of the proof lies in the non-differentiability of the L^∞-norm. To overcome this problem, methods of convex analysis and in particular the notion of the subdifferential are employed in [7]. It is interesting to note that there is no unique minimizer of (3), but that there are two functions (up to normalization) which minimize (3), namely $u_1(x) = \pm 2^{1/2}\cos(\pi x/2) \mp 2^{3/2}/\pi$, $u_2(x) = \pm$

$2^{1/2}\cos(\pi x/2) \mp 2^{3/2}/\pi$. These solutions can be interpreted as the motions of a harmonic oscillator which *bounces* against a rigid obstacle. For a related result for a Dirichlet problem, see G. Talenti [14].

Given (3), the proof of the theorem is completed quite easily. The energy functional associated to (2) is

$$I(u) = \frac{1}{2}\int_0^1 |u'|^2\, dx - \frac{\lambda}{2}\int_0^1 u^2\, dx - \int_0^1 e^u\, dx - \int_0^1 h\, u\, dx\ , \quad u \in H^1 .$$

One verifies that $I \in C^1(H^1, \mathbb{R})$ and that I satisfies the so-called Palais-Smale condition. Let **1** denote the constant function (= 1) in H^1. One notes that $I(r\cdot\mathbf{1}) \xrightarrow[r\to\pm\infty]{} -\infty$. Now, denoting by $N = \{ u \in H^1\ ;\ u \le 0 \}$ the negative cone in H^1, one proves that the boundary ∂N of N is a C^0-manifold of codimension 1 in H^1 such that $H^1\setminus \partial N = N_1 \cup N_2$ has exactly two components. Relation (3) now serves to prove that $I|_{\partial N} \ge -c$. In fact, every $u \in \partial N$ can be written in a unique way as $u = y - |y|_\infty$, with $\int_0^1 y\, dx = 0$; then we can estimate, choosing $\varepsilon = \frac{\pi^2}{8} - \frac{\lambda}{2}$

$$I(u) = \int_0^1 |y'|^2 - \lambda\int_0^1 y^2 - \lambda\, |y|_\infty^2 - \int_0^1 e^u - \int_0^1 h(y-|y|_\infty)$$

$$\ge \frac{\pi^2}{4}\Big(\int_0^1 y^2+|y|_\infty^2\Big) - \lambda\int_0^1 y^2+|y|_\infty^2\Big) - c - \varepsilon\int_0^1 (y^2+|y|^2) - c_\varepsilon$$

$$\ge -d\ , \quad \text{for every } u \in \partial N .$$

Choosing $R > 0$ such that $I(\pm r\mathbf{1}) < -d$ for all $|r| \ge R$ we are now able to apply the *mountain-pass theorem* of Ambrosetti-Rabinowitz [3] ; that is, defining the family of paths

$$\Gamma = \{ \gamma : [-1,1] \longrightarrow H^1\ ;\ \gamma(\pm 1) = \pm R \}$$

one obtains a critical point of the functional I at a level larger than $-d$ via the characterization

$$\inf_{\gamma\in\Gamma}\ \sup_{u\in\gamma}\ I(u) \quad .$$

3. The Fučik spectrum

In this proof we relate equations (1) and (2) to the following positive homogeneous equation (setting $u^+ = \max\{u,o\}$ and $u^- = u^+ - u$) :

$$(4) \qquad \begin{cases} -u'' = \alpha\, u^+ - \lambda\, u^- , & u'(0) = u'(1) = 0 , \\ \quad = \lambda\, u + (\alpha - \lambda)\, u^+ & \\ \quad =: \lambda\, u + \mu\, u^+ & \end{cases}$$

Equations of this type and related problems have been studied extensively in recent years (see the recent survey article of Lazer - McKenna [10]). S. Fučik introduced equation (4) in [8], and characterized the "resonance set" $\Sigma = \{ (\alpha,\lambda) \in \mathbb{R}^2 :$ (4) has a nontrivial solution u $\}$. We call Σ the *Fučik spectrum* of (4). Using that a solution u of (4) satisfies a linear equation on the subintervals where u is positive respectively negative, one finds that Σ consists of continuous curves Σ_i , $i = 0,1,2,\dots$, such that $(\lambda_i,\lambda_i) \in \Sigma_i$, and such that the solutions u_i corresponding to values on Σ_i have exactly i zeroes. Note that $\Sigma_0 = \mathbb{R} \times \{0\} \cup \{0\} \times \mathbb{R}$. Furthermore, characterizing Σ_1 by the third equation above, one can write $\Sigma_1 = \{ \lambda_1(\mu);\ \mu \in \mathbb{R} \}$, with

$$(5) \qquad \lambda_1(\mu) = \frac{\pi^2}{4a^2} - \mu = \frac{\pi^2}{4(1-a)^2} ,$$

where $a \in (0,1)$ denotes the unique zero of the corresponding solution $u_1(\mu)$. From (5) one now immediately deduces

$$(6) \qquad \pi^2/4 = \lim_{\mu \to +\infty} \lambda_1(\mu) .$$

To prove the theorem, it is important to have a variational characterization of the value $\lambda_1(\mu)$. This is obtained by the *mountain-pass theorem* by considering the family pf paths $G = \{ g : [-1,1] \longrightarrow H^1 \cap S_1 ,\ g(\pm 1) = \pm \mathbf{1} \}$, where S_1 denotes the unit sphere (in the L^2- norm) in H^1 ; defining

$$J(u) = \frac{1}{2}\int_0^1 |u'|^2 - \mu \int_0^1 |u^+|^2$$

and setting

(7) $$c = \inf_{g\in G} \sup_{u\in g} J(u)\ ,$$

one shows that c is a critical value, that the critical points of J are solutions of (4), and that $c = \lambda_1(\mu)$.

We remark that for the higher values $\lambda_i(\mu)$, $i \geq 2$, no variational characterization is known.

To complete the proof, consider again the functional

$$\begin{aligned} I(u) &= \frac{1}{2}\int_0^1 |u'|^2 - \frac{\lambda}{2}\int_0^1 u^2 - \int_0^1 e^u - \int_0^1 h\, u \\ &= \frac{1}{2}\int_0^1 |u'|^2 - \frac{\lambda}{2}\int_0^1 u^2 - \int_0^1 (e^u - 2) - \int_0^1 h\, u - 2\ . \end{aligned}$$

To be able to apply again the *mountain-pass theorem* of Ambrosetti-Rabinowitz, we let once more $\varepsilon = \frac{\pi^2}{8} - \frac{\lambda}{2}$ and estimate $\int_0^1 h\, u \leq \varepsilon\, \|u\|^2_{L^2} + c_\varepsilon(h)$. Let now as before $R > 0$ such that $I(\pm r\mathbf{1}) = < - c_\varepsilon - 2$, for all $|r| \geq R$ and consider the same family of paths Γ as in section 2. For every path γ let $m(\gamma) = \max_{y\in\Gamma, x\in[0,1]} y(x)$. Then we can find a number $\nu(\gamma) > 0$ such that we have $e^t - 2 \leq \frac{\nu(\gamma)}{2} t^2$, for all $0 \leq t \leq m(\gamma)$. Hence we can estimate

$$\max_{u\in\gamma} I(u) \geq \max_{u\in\gamma} \left\{\frac{1}{2}\int_0^1 |u'|^2 - \frac{\lambda+\varepsilon}{2}\int_0^1 u^2 - \frac{\nu(\gamma)}{2}\int_0^1 |u^+|^2\right\} - c_\varepsilon(h) - 2.$$

By the characterization (7) it is now easy to see that $\max_{u\in\gamma} I(u) \geq - c_\varepsilon(h) - 2$. Since this holds for every path $\gamma \in \Gamma$ we therefore get

$$\inf_{\gamma\in\Gamma} \max_{u\in\gamma} I(u) \geq - c_\varepsilon(h) - 2\ ,$$

which allows to conlude that (7) defines the critical value $\lambda_1(\mu)$.

4. An anti-maximum principle

This proof is due to Clément - deFigueiredo [4]. It is based on the following *anti-maximum principle* for the operator $L = -\partial^2/\partial x^2 - \lambda$, $u'(0) = u'(1) = 0$, see Clément - Peletier [5] :

$$(8) \qquad \begin{cases} \text{Let } 0 < \lambda \le \frac{\pi^2}{4} \text{ ; then} \\ - u'' - \lambda u \ge 0 \text{ , } u'(0) = u'(1) = 0 \text{ , } \Longrightarrow u \le 0 \text{ .} \end{cases}$$

The proof of this is based on the Green function $\Gamma(x,\xi)$ for L . In fact, one verifies that it has the form

$$\Gamma(x,\xi) = \frac{-1}{\sqrt{\lambda}} \begin{cases} \cos(\sqrt{\lambda}\xi)\sin(\sqrt{\lambda}x) + \mathrm{ctg}(\sqrt{\lambda})\cos(\sqrt{\lambda}\xi)\cos(\sqrt{\lambda}x) \text{ , } x \ge \xi \\ \mathrm{ctg}(\sqrt{\lambda})\cos(\sqrt{\lambda}\xi)\cos(\sqrt{\lambda}x) + \sin(\sqrt{\lambda}\xi)\cos(\sqrt{\lambda}x) \text{ , } x \le \xi \text{ .} \end{cases}$$

Clearly, $\Gamma(x,\xi) \le 0$ for $0 < \lambda \le \pi^2/4$, which yiels the claim.

With this the proof of the theorem (with the more restricting assumption that $h \in L^\infty$) is obtained as follows: Let u_t denote any solution of

$$(9) \qquad \begin{cases} - u_t'' - \lambda u_t = t\ (\exp(u_t) + h) \\ u'(0) = u'(1) = 0 \end{cases} \text{ , } \qquad 0 \le t \le 1$$

Now, choosing $c > 0$ such that $h + \lambda c > 0$, we obtain

$$(10) \qquad - u_t'' - \lambda\ (u_t - c) = t \exp(u_t) + th + \lambda c > 0$$

and hence $u_t - c \le 0$ by the anti-maximum principle. This implies that the right-hand side in (9) is uniformly bounded in $C^0(0,1)$, which in turn yields that

$$\|u_t\|_{H^1} \le c \text{ , for all } t \in [0,1] \text{ .}$$

The existence of a solution now follows by the Leray-Schauder principle.

4. Remarks

It is an open problem whether a similar results holds for larger parameter values λ . However, we conjecture that the following result is true:

If $\lambda \neq \lambda_k$, $k = 0,1,2,\ldots$, then equations (1) and (2) have a solution for every $h \in L_2(0,1)$.

References

[1] Amann, H., Hess, P., A multiplicity result for a class of elliptic boundary value problems, *Proc. R. Soc. Edinb.* **84-A**, 145-151 (1979).

[2] Ambrosetti, A., Prodi, G., On the inversion of some differentiable mappings between Banach spaces, *Annali Mat. pura appl.* **93**, 231-247 (1973).

[3] Ambrosetti, A., Rabinowitz, P.H., Dual variational methods in critical point theory and applications, *J. Funct. Anal.* **14** (1973), 231-247.

[4] Clément, Ph., deFigueiredo, D.J., preprint.

[5] Clément, Ph., Peletier, L.A., An Anti-Maximum Principle for Second-Order Elliptic Operators, *J. Diff. Equ.,* **34** (1979), 218-229.

[6] Dancer, E.N., On the ranges of certain weakly nonlinear elliptic differential equations, *J. Math. Pures Appl.* **57** (1978), 351-366.

[7] deFigueiredo, D.J., Ruf, B., On a superlinear Sturm-Liouville equation and a related bouncing problem, *J. reine angew. Math.,* **421** (1991), 1-22.

[8] deFigueiredo, D.J., Solimini, S., A variational approach to superlinear elliptic problems, *Comm. Partial Differential Equations,* 9 (1984), 699-717.

[9] Fucik, S., Boundary value problems with jumping nonlinearities, *Casopis Pest. Mat.*, **101** (1976), 289-307.

[10] Lazer, A., McKenna, P. J., *SIAM review,* **32,** 1990.

[11] Pádua, J.C., On a superlinear two point boundary value problem with multiple solutions, *An. Acad. Brasil. Ciênc.* **58** (1986), 177-181.

[12] Ruf, B., Solimini, P.N., On a class of superlinear Sturm-Liouville problems with arbitrary many solutions, *SIAM J. Math. Anal.*, **4**, (1986), 761-771.

[13] Ruf, B., Srikanth, P.N., Multiplicity results for ODE's with nonlinearities crossing all but a finite number of eigenvalues, *Nonlinear Analysis, TMA*, **10**, (1986), 157-163.

[14] Talenti, G., Estimates for Eigenvalues of Sturm-Liouville Problems, *Int. Ser. Num. Math.*, **71** (1984), 341-350.

Bernhard Ruf
Dip. di Matematica
Università delgi Studi
Via Saldini 50, I-20133 Milano

N SVANSTEDT AND J WYLLER

A numerical algorithm for the solution of the homogenized p-Poisson equation

1 Introduction

The present paper is devoted to the study of the following quasilinear problem

$$-\operatorname{div}(a(\frac{x}{\epsilon}, Du)) = f \text{ in } \Omega \tag{1.1}$$

with Dirichlet boundary data, where

$$a(\frac{x}{\epsilon}, Du) = \lambda(\frac{x}{\epsilon})|Du|^{p-2}Du. \tag{1.2}$$

Here Ω is an open bounded set in $\mathbf{R}^N$, $N \geq 1$, $1 < p < \infty$ and λ is a positive and bounded periodic function with period proportional to $\epsilon > 0$. For ϵ small, a numerical treatment of (1.1) is practically impossible. However, by considering $\epsilon = \epsilon_h$, $h = 1, 2, \ldots$, as a sequence, such that, $\epsilon_h \to 0^+$ as $h \to +\infty$, one can consider (1.1)-(1.2) as a sequence of quasilinear problems. Homogenization results for monotone operators then yield the existence of a corresponding homogenized problem to (1.1) on the form

$$-\operatorname{div}(b(Du)) = f \text{ in } \Omega \tag{1.3}$$

with corresponding Dirichlet boundary data. Thanks to these results and additional corrector results [4], the homogenized problem will serve as a good approximation for the original problem (1.1) as ϵ is small.

The main result in this paper consists of a construction of a numerical algorithm for determining the homogenized operator corresponding to the momentum b. This algorithm is based on a coupled system of two augmented Lagrangians, where one of them corresponds to the variational form of the local problem, the other one to the homogenized problem (1.3). The coupling yields a parameter dependence in the first one and a constraint in the latter one. The paper is organized in the following way: In section 2 some preliminaries are given concerning the p-Poisson equation (1.1) and section 3 is devoted to the construction of the numerical algorithm for problem (1.3). By $|\cdot|$, $(\cdot,\cdot)$ and $\langle\cdot,\cdot\rangle$ we denote the usual Euclidean norm and scalar product in $\mathbf{R}^N$ and the duality pairing, respectively.

2 Homogenization of the p-Poisson equation

We recall that the p-Poisson equation (1.1) can be written in the following form

$$\begin{cases} -\operatorname{div}(a(\frac{x}{\epsilon}, Du_\epsilon)) = f \text{ in } \Omega \\ u_\epsilon \in W_0^{1,p}(\Omega), \end{cases} \tag{2.1}$$

where the map a is defined as in (1.2). We collect some results concerning homogenization of the problem (2.1).

Definition 1 By $S_\Omega(\mathbf{R}^N)$ we denote the class of single-valued maps

$$a : \Omega \times \mathbf{R}^N \to \mathbf{R}^N$$

such that, for some strictly positive constants m_1 and m_2

(i) $a(\cdot, \xi)$ is Lebesgue measurable.

(ii) $a(y, \cdot)$ is maximal monotone for a.e. $y \in \Omega$.

(iii) $m_1|\xi|^p \leq (a(y, \xi, \xi))$ for a.e. $y \in \Omega$ and for all $\xi \in \mathbf{R}^N$.

(iv) $|a(y, \xi)| \leq m_2(1 + |\xi|^{p-1}$ for a.e. $y \in \Omega$ and for all $\xi \in \mathbf{R}^N$.

If the map is independent of the first variable we denote the class by $S(\mathbf{R}^N)$.

Proposition 1 *Let $\lambda : \Omega \to \mathbf{R}_+$ be a strictly positive and bounded Lebesgue measurable function. Then, the map*

$$a(y, \xi) = \lambda(y)|\xi|^{p-2}\xi$$

belongs to $S_\Omega(\mathbf{R}^N)$. Consequently, the problem (2.1) possesses a unique solution $u_\epsilon \in W_0^{1,p}(\Omega)$ for every $f \in W^{-1,p'}(\Omega)$ and each fixed $\epsilon > 0$. Here, and in the sequel, p and p' are dual exponents $(1/p + 1/p' = 1)$.

Proof. For the proof we refer to e.g. [9].

Proposition 2 *The operator $A : W_0^{1,p}(\Omega) \to W^{-1,p'}(\Omega)$ given by*

$$Au = -\text{div}(\lambda(y)|Du|^{p-2}Du)$$

is maximal monotone and is the subdifferential of the functional $\Psi : W_0^{1,p}(\Omega) \to \overline{\mathbf{R}}$ given by

$$\Psi(u) = \frac{1}{p}\int_\Omega \lambda(x)|Du|^p\, dx,$$

which is proper, lower semicontinuous and convex. Thus, the solution of (2.1) is also the unique minimum of the following problem:

$$\min_{u_\epsilon \in W_0^{1,p}(\Omega)} \{\frac{1}{p}\int_\Omega \lambda(x)|Du|^p\, dx - \langle f, u_\epsilon\rangle\}.$$

Proof. See [2] and the references given there.

Definition 2 Let $Y =]0,1[^N$ be the unit cube in $\mathbf{R}^N$. A function $u : \mathbf{R}^N \to \mathbf{R}$ is called *Y-periodic* if $u(x + e_i) = u(x)$ for every $x \in \mathbf{R}^N$ and every $i = 1, \ldots, N$, where (e_i) is the canonical basis in $\mathbf{R}^N$.

Definition 3 By $W_\sharp^{1,p}(Y)$ we denote the subset of $W^{1,p}(Y)$ of all functions with mean value zero over Y which have the same trace on the opposite faces of Y.

Proposition 3 *Every function belonging to $W_\sharp^{1,p}(Y)$ can be extended by periodicity to a function of $W_{loc}^{1,p}(\mathbf{R}^N)$.*

Proof. See [12].

Definition 4 Let $a \in S_Y(\mathbf{R}^N)$ and fix ξ in $\mathbf{R}^N$. The local problem on the unit cube, corresponding to problem (2.1), is given by

$$\begin{cases} -\mathrm{div}(a(y, Dv(y) + \xi)) = 0 \text{ on } Y \\ v \in W_\sharp^{1,p}(Y). \end{cases} \tag{2.2}$$

Definition 5 By $S_\sharp(\mathbf{R}^N)$ we denote the set of maps $a \in S_{\mathbf{R}^N}(\mathbf{R}^N)$, such that, $a(\cdot, \xi)$ is Y-periodic for all $\xi \in \mathbf{R}^N$.

Now, let $\{\epsilon_h\}$ be a sequence of positive numbers tending to zero, and consider the following sequence of Dirichlet boundary value problems

$$\begin{cases} -\mathrm{div}(a(\frac{x}{\epsilon_h}, Du_h)) = f \text{ on } \Omega \\ u_h \in W_0^{1,p}(\Omega), \end{cases} \tag{2.3}$$

where $a \in S_\sharp(\mathbf{R}^N)$ is defined as in (1.2). Moreover, let us consider the Dirichlet boundary value problem

$$\begin{cases} -\mathrm{div}(b(Du)) = f \text{ on } \Omega \\ u_h \in W_0^{1,p}(\Omega), \end{cases} \tag{2.4}$$

where $b \in S(\mathbf{R}^N)$ is given by

$$b(\xi) = \int_Y a(y, Dv(y) + \xi)\, dy, \quad \xi \in \mathbf{R}^N \tag{2.5}$$

and v is the solution of (2.2). The following homogenization result holds true:

Theorem 1 *Let u_h and u be the unique solutions of* (2.3) *and* (2.5)*, respectively. Then, for every $f \in W^{-1,p'}(\Omega)$, as $h \to +\infty$*

(i) $u_h \to u$, *weakly in* $W_0^{1,p}(\Omega)$,

(ii) $a(\frac{x}{\epsilon_h}, Du_h) \to b(Du)$, *weakly in* $(L^{p'}(\Omega))^N$.

Proof. See [3] and also [6].

3 A numerical algorithm for solving the homogenized p-Poisson equation

In this section we aim to present a numerical algorithm for the solution of the homogenized p-Poisson equation (2.4)-(2.5). This algorithm is based on augmented Lagrangian methods. A detailed description of this technique can be found in [5] and [7]. We recall the homogenized p-Poisson equation (2.4)-(2.5) and the corresponding local problem (2.2). According to Proposition 2 they are equivalent to the minimization problems

$$\min_{u\in W_0^{1,p}(\Omega)} \{\frac{1}{p}\int_\Omega b(Du), Du)\,dx - \langle f, u_\epsilon\rangle\}, \tag{3.1}$$

$$\min_{v\in W_\sharp^{1,p}(Y)} \{\frac{1}{p}\int_Y \lambda(y)|Dv(y)+\xi|^p\,dy\}, \tag{3.2}$$

where (3.1) and (3.2) are coupled via (2.5). Now let us introduce the Lagrangians

$$\mathcal{L}_\Omega : W_0^{1,p}(\Omega)\times(L^p(\Omega))^N\times(L^{p'}(\Omega))^N \to \mathbf{R}$$

and

$$\mathcal{L}_Y : W_\sharp^{1,p}(Y)\times(L^p_{loc}(\mathbf{R}^N))^N\times(L^{p'}_{loc}(\mathbf{R}^N))^N \to \mathbf{R}$$

given by

$$\mathcal{L}_\Omega(u,\xi,\mu) = \frac{1}{p}\int_\Omega (b(\xi),\xi)\,dx - \langle f,u\rangle + \int_\Omega \mu\cdot(Du-\xi)\,dx \tag{3.3}$$

and

$$\mathcal{L}_Y(v,\eta,\gamma) = \frac{1}{p}\int_Y \lambda(y)|\eta|^p\,dy + \int_Y \gamma\cdot(Dv+\xi-\eta)\,dy. \tag{3.4}$$

Moreover, for some positive constants δ_1 and δ_2, let us introduce the augmented Lagrangians $\mathcal{L}_{\Omega,\delta_1}$ and $\mathcal{L}_{Y,\delta_2}$ given by

$$\mathcal{L}_{\Omega,\delta_1}(u,\xi,\mu) = \mathcal{L}_\Omega(u,\xi,\mu) + \frac{1}{2\delta_1}\int_\Omega |Du-\xi|^2\,dx \tag{3.5}$$

and

$$\mathcal{L}_{Y,\delta_2}(v,\eta,\gamma) = \mathcal{L}_Y(v,\eta,\gamma) + \frac{1}{2\delta_2}\int_Y |Dv+\xi-\eta|^2\,dy. \tag{3.6}$$

The following result justifyes the numerical algorithms to be presented:

Proposition 4 *Suppose (u,ξ,μ) and (v,η,γ) are the unique saddle points of $\mathcal{L}_\Omega$ and $\mathcal{L}_Y$, respectively. Then, they are also the unique saddle points of $\mathcal{L}_{\Omega,\delta_1}$ and $\mathcal{L}_{Y,\delta_2}$ for all $\delta_1,\delta_2>0$, and vice versa. Moreover, u and v are the unique solutions of* (3.1) *and* (3.2) *with $Du=\xi$ and $Dv+\xi=\eta$.*

Proof. See e.g. [11].

In order to calculate saddle points of $\mathcal{L}_{\Omega,\delta_1}$ and $\mathcal{L}_{Y,\delta_2}$ we will use an iterative scheme of Uzawa type. This scheme is not immediately applicable, but it will converge for discrete analogoues of (3.5) and (3.6). Therfore, we proceed by assuming that Ω is a polygon in $\mathbf{R}^2$ with corresponding unit cube Y. Furthermore, let τ_m and τm_Y denote finite triangulations of Ω and Y, respectively, where $m>0$ and $m_Y>0$ denote the largest possible diameters of some triangles in τ_m and τ_{m_Y}. We introduce the following finite element spaces:

$$V_m = \{\tilde{u}_m \in C(\Omega) : \tilde{u}_m|_{\partial\Omega} = 0, \tilde{u}_m|_T \in P_1,\ \forall T\in\tau_m\}$$

and

$$V_{m_Y,\sharp} = \{v_{m_Y} \in C(\overline{Y}) : v_{m_Y} \text{ is } Y-\text{periodic and has mean value zero over } Y,$$
$$v_{m_Y}|_{T_Y} \in P_1,\ \forall T_Y \in \tau_{m_Y}\}.$$

It is obvious, that, $V_m \subset W_0^{1,p}(\Omega)$ and $V_{m_Y,\sharp} \subset W_\sharp^{1,p}(Y)$, with norm equivalence. We obtain the following discrete analogoues to (3.1) and (3.2):

$$\min_{\tilde{u}_m \in V_m} \{\frac{1}{p}\int_\Omega (b_{m_Y}(D\tilde{u}_m), D\tilde{u}_m)\,dx - \langle f, \tilde{u}_m\rangle\}, \tag{3.7}$$

$$\min_{v_{m_Y} \in V_{m_Y,\sharp}} \{\frac{1}{p}\int_Y \lambda(y)||Dv_{m_Y}(y) + |\xi_{m_Y}|^p\,dy\}, \tag{3.8}$$

where (3.7) and (3.8) are coupled via

$$b_{m_Y}(\xi_{m_Y}) = \int_Y \lambda(y)|Dv_{m_Y}(y) + \xi_{m_Y}|^{p-2}(Dv_{m_Y}(y) + \xi_{m_Y})\,dy,\ \xi_{m_Y} \in \mathbf{R}^2. \tag{3.9}$$

In order to formulate the discrete versions of the augmented Lagrangians we first introduce the spaces

$$L_m = \{z_m : z_m = \sum_{T\in\tau_m} z_T\chi_T,\ z_T \in \mathbf{R}^2\},$$
$$L_{m_Y} = \{z_{m_Y} : z_{m_Y} = \sum_{T_Y\in\tau_{m_Y}} z_{T_Y}\chi_{T_Y},\ z_{T_Y} \in \mathbf{R}^2\},$$

where χ_T and χ_{T_Y} denote the characteristic functions of T and T_Y. One notices, that, if $u_m \in V_m$ and $v_{m_Y} \in V_{m_Y,\sharp}$, then, $Du_m \in L_m$ and $Dv_{m_Y} \in L_{m_Y}$, respectively. Moreover, $L_m, L_{m_Y} \subset (L^p_{loc}(\mathbf{R}^2))^2$ and also $L_m, L_{m_Y} \subset (L^{p'}_{loc}(\mathbf{R}^2))^2$. Next, let us consider the augmented Lagrangians

$$\overline{\mathcal{L}}_{\Omega,\delta_1} : V_m \times L_m \times L_m \to \mathbf{R},$$
$$\overline{\mathcal{L}}_{Y,\delta_2} : V_{m_Y} \times L_{m_Y} \times L_{m_Y} \to \mathbf{R},$$

given by

$$\overline{\mathcal{L}}_{\Omega,\delta_1}(\tilde{u}_m, \xi_m, \mu_m) =$$
$$\frac{1}{p}\int_\Omega (b(\xi_m), \xi_m)\,dx - \langle f, \tilde{u}_m\rangle + \tag{3.10}$$
$$\int_\Omega \mu_m \cdot (D\tilde{u}_m - \xi_m)\,dx + \frac{1}{2\delta_1}\int_\Omega |D\tilde{u}_m - \xi_m|^2\,dx$$

and

$$\overline{\mathcal{L}}_{Y,\delta_2}(v_{m_Y}, \eta_{m_Y}, \gamma_{m_Y}) =$$
$$\frac{1}{p}\int_Y \lambda(y)|\eta_{mY}|^p\,dy + \int_Y \gamma_{m_Y} \cdot (Dv_{m_Y} + \xi_{m_Y} - \eta_{m_Y})\,dy + \tag{3.11}$$
$$\frac{1}{2\delta_2}\int_Y |Dv_{m_Y} + \xi_{m_Y} - \eta_{m_Y}|^2\,dy,\ \xi_{m_Y} \in \mathbf{R}^2.$$

From Proposition 4 we have the following corollary:

Corollary 1 *The results stated in Proposition 4 remain valid when replacing* $(\rightarrow)$

$$\{u, \xi, \mu\} \rightarrow \{\tilde{u}_m, \xi_m, \mu_m\}, \quad \{v, \eta, \gamma\} \rightarrow \{v_{mY}, \eta_{mY}, \gamma_{mY}\}$$

$$\mathcal{L}_{\Omega,\delta_1} \rightarrow \overline{\mathcal{L}}_{\Omega,\delta_1}, \quad \mathcal{L}_{Y,\delta_2} \rightarrow \overline{\mathcal{L}}_{Y,\delta_2}$$

$$Du = \xi \rightarrow D\tilde{u}_m = \xi_m, \quad Dv + \xi = \eta \rightarrow Dv_{mY} + \xi_{mY} = \eta_{mY}.$$

Now, we define u_m as the solution of

$$\min_{u_m \in V_m} \{\frac{1}{p} \int_\Omega (b(Du_m), Du_m)\, dx - \langle f, u_m \rangle\}. \tag{3.12}$$

The following convergence reult for the approximate solutions u_m and v_{mY} holds true:

Proposition 5 *Let* $\{\tau m_h\}_{h=1}^\infty$ *and* $\{\tau_{mY_h}\}_{h=1}^\infty$ *be families of triangulations of* Ω *and* Y*, respectively, such that* $m_h \rightarrow 0$ *and* $m_{Yh} \rightarrow 0$ *as* $h \rightarrow \infty$*. Suppose, that, there exist fixed angles* $\theta_\Omega > 0$ *and* $\theta_Y > 0$*, such that, all angles of triangles in* τ_{m_h} *are greater than, or equal to,* θ_Ω *and all angles of triangles in* τ_{mY_h} *are greater than, or equal to,* θ_Y*, for every* $h = 1, 2, 3, \ldots$*. Then, as* $h \rightarrow \infty$

i) $\quad u_m = u_{m_h} \rightarrow u$, *in* $W_0^{1,p}(\Omega)$,

ii) $\quad v_{mY} = v_{mY_h} \rightarrow v$, *in* $W_\sharp^{1,p}(Y)$.

Proof. We prove (i) and omit the proof of (ii) which is essentially analogous. By the assumptions made, we immediately get the apriori estimate

$$\|u_{m_h}\|_{W_0^{1,p}(\Omega)}^{p-1} \leq \|f\|_{W^{-1,p'}(\Omega)}.$$

Thus, by weak sequential compactness, there exists a subsequence, such that,

$$u_{m_{\sigma(h)}} \rightharpoonup u^*, \text{ weakly in } W_0^{1,p}(\Omega). \tag{3.13}$$

For $\varphi \in C_0^\infty(\Omega)$ we let $\pi_h\varphi$ be the interpolant of φ in V_{m_h}. Then, (see e.g. [8])

$$\pi_h\varphi \rightarrow \varphi, \text{ in } W_0^{1,p}(\Omega),$$

from which it follows, that,

$$J(\pi_h\varphi) \rightarrow J(\varphi),$$

where

$$J(z) = \frac{1}{p} \int_\Omega (b(Dz), Dz)\, dx - \langle f, z \rangle.$$

By definition

$$J(u_{m_h}) \leq J(\pi_h\varphi) \text{ for every } \varphi \in C_0^\infty(\Omega).$$

Thus,

$$\limsup_{h\rightarrow\infty} J(u_{m_h}) \leq \limsup_{h\rightarrow\infty} J(\pi_h\varphi) = J(\varphi).$$

Since $C_0^\infty(\Omega)$ is dense in $W_0^{1,p}(\Omega)$, we have

$$\limsup_{h\to\infty} J(u_{m_h}) \le \inf_{\varphi\in C_0^\infty(\Omega)} J(\varphi) = J(u)$$

and consequently,

$$J(u_{m_h}) \to J(u). \tag{3.14}$$

By Proposition 2, J is convex and lower semicontinuous and, thus, also weakly lower semicontinuous. Hence,

$$J(u^*) \le \liminf_{h\to\infty} J(u_{m_{\sigma(h)}}) = J(u).$$

But, u is the unique minimum of J, thus, $u^* = u$. Finally, by (3.13) and (3.14)

$$\|u_{m_{\sigma(h)}}\|_{W_0^{1,p}(\Omega)} \to \|u\|_{W_0^{1,p}(\Omega)}.$$

And, by the uniform convexity of the norm, we obtain

$$\|u_{m_{\sigma(h)}} - u\|_{W_0^{1,p}(\Omega)} \to 0.$$

We are now able to describe the Uzawa type algorithm for the evaluation of the saddle points of $\overline{\mathcal{L}}_{\Omega,\delta_1}$ and $\overline{\mathcal{L}}_{Y,\delta_2}$.

A Choose $\mu_m^0 \in L_m$ and $\gamma_{m_Y}^0 \in L_{m_Y}$ arbitrarily.

B Determine $\{u_m^n, \xi_m^n\} \in V_m \times L_m$ and $\{v_{m_Y}^n, \eta_{m_Y}^n\} \in V_{m_{Y,\sharp}} \times L_{m_Y}$ via

$$\overline{\mathcal{L}}_{\Omega,\delta_1}(u_m^n, \xi_m^n, \mu_m^n) \le \overline{\mathcal{L}}_{\Omega,\delta_1}(\hat{u}, \hat{\xi}, \mu_m^n) \text{ for all } \{\hat{u}, \hat{\xi}\} \in V_m \times L_m,$$

$$\overline{\mathcal{L}}_{Y,\delta_2}(v_{m_Y}^n, \eta_{m_Y}^n, \gamma_{m_Y}^n) \le \overline{\mathcal{L}}_{Y,\delta_2}(\hat{v}, \hat{\eta}, \gamma_{m_Y}^n) \text{ for all } \{\hat{v}, \hat{\eta}\} \in V_{m_{Y,\sharp}} \times L_{m_Y}.$$

C Determine μ_m^{n+1} and $\gamma_{m_Y}^{n+1}$ via

$$\mu_m^{n+1} = \mu_m^n + \rho_1^n(Du_m^n - \xi_m^n),$$

$$\gamma_{m_Y}^{n+1} = \gamma_{m_Y}^n + \rho_2^n(Dv_{m_Y} + \xi_m^n - \eta_{m_Y}^n),$$

where ρ_1^n and ρ_2^n will be specified in the following crucial result:

Proposition 6 *Consider the iterative schemes* **A**-**C**. *Suppose, that, the sequences* $\{\rho_1^n\}$ *and* $\{\rho_2^n\}$, $n = 1, 2, 3, \ldots$, *satisfy* $0 < r_1 \le \rho_1^n \le R_1 < 2/\delta_1$ *and* $0 < r_2 \le \rho_2^n \le R_2 < 2/\delta_2$, *respectively, for suitable constants* r_1, R_1, r_2 *and* R_2. *Then, for arbitrary* $\mu_m^0 \in L_m$ *and* $\gamma_{m_Y}^0 \in L_{m_Y}$, *as* $n \to \infty$

$$\begin{cases} u_m^n \to u_m, \ \text{in } V_m, \\ \xi_m^n \to Du_m, \ \text{in } L_m, \\ \mu_m^n \to |Du_m|^{p-2}Du_m, \ \text{in } L_m \end{cases}$$

and

$$\begin{cases} v^n_{mY} \to v_{mY}, \ \textit{in } V_{mY,\sharp}, \\ \eta^n_{mY} \to Dv_{mY} + \xi^n_{mY}, \ \textit{in } L_{mY}, \\ \gamma^n_{mY} \to |Dv_{mY} + \xi^n_{mY}|^{p-2}(Dv_{mY} + \xi^n_{mY}), \ \textit{in } L_{mY}, \end{cases}$$

where u_m and v_{mY} are the unique solutions of (3.12) and (3.8), respectively, and where the parameter ξ^n_{mY} is fixed in the latter convergence suite.

Proof. See [7].

By combining the Propositions 5 and 6 we obtain the following:

Corollary 2

$$\begin{cases} \lim_{h\to\infty}\lim_{n\to\infty} u^n_{m_h} = u, \\ \lim_{h\to\infty}\lim_{n\to\infty} \xi^n_{m_h} = Du, \\ \lim_{h\to\infty}\lim_{n\to\infty} \mu^n_{m_h} = |Du|^{p-2}Du \end{cases}$$

and

$$\begin{cases} \lim_{h\to\infty}\lim_{n\to\infty} v^n_{mY_h} = v, \\ \lim_{h\to\infty}\lim_{n\to\infty} \eta^n_{mY_h} = Dv + \xi^n_{mY}, \\ \lim_{h\to\infty}\lim_{n\to\infty} \gamma^n_{mY_h} = |Dv + \xi^n_{mY}|^{p-2}(Dv + \xi^n_{mY}), \end{cases}$$

where ξ^n_{mY} is a fixed parameter in the latter convergence suite.

Since ξ^n_{mY} plays the role as a parameter in the local problem and the corresponding augmented Lagrangian one can put $\xi^n_{mY} = \xi^n_m$ without loss of generality. Hence, the local and the global problems become coupled. This crucial fact enables us to apply the Uzawa schemes in the following way:

Step 1. Calculate the approximative saddle point $\{v_{mY}, \eta_{mY}, \gamma_{mY}\}$ of $\overline{\mathcal{L}}_{Y,\delta_2}$.

Step 2. Pass to the limit ($h \to \infty$) for the approximative saddle point in Step 1 and obtain the saddle point $\{v, \eta, \gamma\}$ of $\mathcal{L}_{Y,\delta_2}$.

Step 3. Insert $\{v, \eta, \gamma\}$ into the global problem, i.e. into the augmented Lagrangian $\overline{\mathcal{L}}_{\Omega,\delta_1}$, and calculate the saddle point $\{u_m, \xi_m, \mu_m\}$.

Step 4. Pass to the limit ($h \to \infty$) for the approximative saddle point in Step 3 and obtain the saddle point $\{u, \xi, \mu\}$ of $\mathcal{L}_{\Omega,\delta_1}$, where $\xi = Du$ and where $\mu = |Du|^{p-2}Du$.

We can now describe a practical performance for the numerical algorithm.

Suppose $\mu_m \in L_m$ and $\gamma_{m_Y} \in L_{m_Y}$ are known and that $\{u_m^n, \xi_m^n\} \in V_m \times L_m$ and $\{v_{m_Y}^n, \eta_{m_Y}^n\} \in V_{m_{Y,\sharp}} \times L_{m_Y}$ are solutions of

$$\overline{\mathcal{L}}_{\Omega,\delta_1}(u_m^n, \xi_m^n, \mu_m^n) \leq \overline{\mathcal{L}}_{\Omega,\delta_1}(\hat{u}, \hat{\xi}, \mu_m^n) \text{ for all } \{\hat{u}, \hat{\xi}\} \in V_m \times L_m,$$

$$\overline{\mathcal{L}}_{Y,\delta_2}(v_{m_Y}^n, \eta_{m_Y}^n, \gamma_{m_Y}^n) \leq \overline{\mathcal{L}}_{Y,\delta_2}(\hat{v}, \hat{\eta}, \gamma_{m_Y}^n) \text{ for all } \{\hat{v}, \hat{\eta}\} \in V_{m_{Y,\sharp}} \times L_{m_Y}.$$

By using the convexity of $\overline{\mathcal{L}}_{\Omega,\delta_1}$ and $\overline{\mathcal{L}}_{Y,\delta_2}$ the above system can be transformed into a system of four variational equations.

$$\begin{cases} \frac{\partial}{\partial s}\overline{\mathcal{L}}_s^v|_{s=0} = 0 \\ \frac{\partial}{\partial s}\overline{\mathcal{L}}_s^\eta|_{s=0} = 0 \end{cases} \tag{3.15}$$

and

$$\begin{cases} \frac{\partial}{\partial s}\overline{\mathcal{L}}_s^u|_{s=0} = 0 \\ \frac{\partial}{\partial s}\overline{\mathcal{L}}_s^\xi|_{s=0} = 0 \end{cases} \tag{3.16}$$

where

$$\begin{cases} \overline{\mathcal{L}}_s^v = \overline{\mathcal{L}}_{Y,\delta_2}(v_{m_Y} + s\hat{v}, \eta_{m_Y}, \gamma_{m_Y}) \\ \overline{\mathcal{L}}_s^\eta = \overline{\mathcal{L}}_{Y,\delta_2}(v_{m_Y}, \eta_{m_Y} + s\hat{\eta}, \gamma_{m_Y}) \end{cases} \tag{3.17}$$

and

$$\begin{cases} \overline{\mathcal{L}}_s^u = \overline{\mathcal{L}}_{\Omega,\delta_1}(u_m + s\hat{u}, \xi_m, \mu_m) \\ \overline{\mathcal{L}}_s^\xi = \overline{\mathcal{L}}_{\Omega,\delta_1}(u_m, \xi_m + s\hat{\xi}, \mu_m) \end{cases} \tag{3.18}$$

By inserting (3.17)-(3.18) into (3.15)-(3.16), they can be written as

$$\begin{gathered} \int_Y (Dv_{m_Y} + \xi_{m_Y} - \eta_{m_Y}) \cdot D\hat{v}\, dy = -\delta_2 \int_Y \gamma_{m_Y} \cdot D\hat{v}\, dy, \\ \int_Y \lambda(y)|\eta_{m_Y}|^{p-2}\eta_{m_Y} \cdot \hat{\eta}\, dy - \frac{1}{\delta_2}\int_Y (Dv_{m_Y} + \xi_{m_Y} - \eta_{m_Y}) \cdot \hat{\eta}\, dy = \\ \int_Y \gamma_{m_Y} \cdot \hat{\eta}\, dy, \\ \text{for all } \{\hat{v}, \hat{\eta}\} \in V_{m_{Y,\sharp}} \times L_{m_Y}, \end{gathered} \tag{3.19}$$

and

$$\int_\Omega (Du_m - \xi_m)\cdot D\hat{u}\,dx = \delta_1(\langle f,\hat{u}\rangle - \int_\Omega \mu_m \cdot D\hat{u}\,dx),$$

$$\frac{p-1}{p}\int_\Omega \left(\int_Y \lambda(y)|Dv_{m_Y}+\xi_m|^{p-2}\hat{\xi}\,dy\cdot\xi_m\right)dx+$$

$$\frac{1}{p}\int_\Omega \left(\int_Y \lambda(y)|Dv_{m_Y}+\xi_m|^{p-2}(Dv_{m_Y}+\xi_m)\,dy\cdot\hat{\xi}\right)dx- \tag{3.20}$$

$$\frac{1}{\delta_1}\int_\Omega (Du_m-\xi_m)\cdot\hat{\xi}\,dx = \int_\Omega \mu_m\cdot\hat{\xi}\,dx,$$

$$\text{for all } \{\hat{u},\hat{\xi}\}\in V_m\times L_m.$$

The system (3.19) can be solved numerically by a slight modification of the FORTRAN-code PPO (see [10] and also [1]). Recall, that, by keeping $\xi_{m_Y}=\xi_m$ fixed in (3.19) we get the natural coupling between the systems (3.19) and (3.20). The system (3.20) is solved in an analogous way as (3.19), In a forthcoming paper we will make a detailed numerical study of the homogenized p-Poisson equation based upon this algorithm.

Acknowledgements. The present work was mainly carried out in May 1991 at Narvik Institute of Technology. One of the authors (N.S.) would like to express his sincere appreciation of the kind hospitality during his stay at this institute. Both authors will express their deep gratitude to Professor Gunnar Aronsson for many helpful and inspiring discussions in the preparation of this paper. We will also thank Professor Anneliese Defranceschi for many inspiring discussions during her visit in Luleå, April 1991. Finally we are grateful to Professor Lars Erik Persson for suggesting us this research and for his constant interest and support.

References

[1] T. Andersson, *FEM-solution to the p-Poisson equation in 2D*, Report no:LiTH-Mat-EX-90-03, Linköping, (1990).

[2] V. Chiadò Piat, G. Dal Maso and A. Defranceschi, *G-convergence of monotone operators*, Ann. Inst. H. Poincare. Anal. Non Lineare, 7 , (1990), 123-160.

[3] V. Chiadò Piat and A. Defranceschi, *Homogenization of monotone operators*, Nonlinear Anal., 14, (1990), 717-732.

[4] G. Dal Maso and A. Defranceschi, *Correctors for the homogenization of monotone operators*, Differential and Integral Equations 3, (1990), 1137-1152.

[5] M. Fortin and R. Glowinski, *Augmented Lagrangian methods*, North-Holland Publ., Amsterdam, 1983.

[6] N. Fusco and G. Moscariello, *On the homogenization of quasilinear divergence structure operators*, Ann. Mat. Pura Appl., 146, (1987), 1-13.

[7] R. Glowinski and A. Marrocco, *Sur l'approximation par elements finis d'ordre un, et la resolution par penalisation-dualite, d'une classe de problemes de Dirichlet non lineares*, Rapport de Recherche no:115, Iria, Rocqencort, (1975).

[8] C. Johnson, *Numerical solutions of partial differential equations by the finite element method*, Studentlitteratur Publ., Lund, (1987).

[9] J.L. Lions, *Quelques methodes de resolution des problemes aux limites non lineares*, Dunod, Gautier-Villars, Paris, 1967.

[10] L. Petterson, *FEM-lösning av p-Poissons ekvation i planet*, Report no:LiTH-MAT-EX-1989-01, Linköping, (1989).

[11] T. Rockafellar, *Convex Analysis*, Princton, 1970.

[12] P, Suquet, *Plasticite et homogeneisation*, These, Paris VI, (1982).

Authors adresses

Nils Svanstedt, Department of Applied Mathematics, LuleåUniversity of Technology, S-95187 Luleå, Sweden.

John Wyller, Narvik Institute of Technology, P.O. Box 385, N-8501 Narvik, Norway.

G SWEERS*

A sign-changing global minimizer on a convex domain

Introduction: Recently one has established the existence of stable sign–changing solutions for the elliptic problem

$$(1)\qquad \begin{cases} -\Delta u = f(u) & \text{in } \Omega , \\ u = 0 & \text{on } \partial\Omega . \end{cases}$$

In [5] there is an example of a sign changing stable solution on a convex domain with $f(0) \neq 0$. Matano [2] shows the existence of a sign–changing stable solution even with $f(0) = 0$. A next question will be: does a global minimizer have a fixed sign? It has been guessed that the answer is positive if the domain is convex.

In this note we will recall a proof for the ball and give a counterexample for a triangle.

We will assume that $f \in C^{0,1}$, Ω is bounded with $\partial\Omega \in C^{0,1}$ and that (1) has a solution u that minimizes the energy functional J. This functional $J : H_0^1(\Omega) \to \mathbb{R}$ is defined by

$$(2)\qquad J(v) = \tfrac{1}{2}\int_\Omega |\nabla v|^2 dx - \int_\Omega F(v)\, dx ,$$

where $F(s) = \int_0^s f(t)\, dt$. It is classical that if $J(u) = \min \{ J(v);\ v \in H_0^1(\Omega) \}$, then u is a $C^2(\Omega) \cap C(\bar\Omega)$ – solution,

$$(3)\qquad J'(u)v := \int_\Omega \nabla u.\nabla v\, dx - \int_\Omega f(u)\, v\, dx = 0 \quad \text{for all } v \in H_0^1(\Omega)$$

and

$$(4)\qquad J''(u)(v) := \int_\Omega |\nabla v|^2\, dx - \int_\Omega f'(u)v^2\, dx \geq 0 \quad \text{for all } v \in H_0^1(\Omega).$$

* Supported by the Netherlands Organization for Scientific Research N.W.O.

Proposition 1: *If* f *is antisymmetric or* Ω *is a ball in* $\mathbb{R}^n$, *then the global minimizer* u *has a fixed sign.*

Remark 1: The result that a (local) minimizer cannot change sign on a ball is due to Lin and Ni, [1]. In their unpublished preprint they also prove the result for Ω being the difference of two balls with the same center. We will sketch their proof.

Proof: i) If f is antisymmetric, $f(s) = -f(-s)$, then $J(|u|) = J(u)$. Hence $|u|$ is a minimizing solution and $|u| \in C^2(\Omega)$. It follows from $x \in \Omega$ and $u(x) = 0$, that $\nabla u(x) = 0$. Then the strong maximum principle shows that $u \equiv 0$. Hence u has a fixed sign.

ii) Suppose Ω is a ball with center 0. Then differentiate the solution u in a tangential direction, that is, apply $\frac{d}{d\theta} = x_i \frac{d}{dx_j} - x_j \frac{d}{dx_i}$. Since $\frac{d}{d\theta}$ and Δ commute, the function $\varphi = \frac{d}{d\theta} u$ satisfies $-\Delta\varphi = f'(u)\varphi$ in Ω. Moreover $\varphi = 0$ on $\partial\Omega$. Then either $\varphi = 0$ or φ is an eigenfunction (with eigenvalue 0) of

$$(5)\qquad \begin{cases} -\Delta v - f'(u)v = \lambda v & \text{in } \Omega\,, \\ v = 0 & \text{on } \partial\Omega\,. \end{cases}$$

From (4) one finds that all eigenvalues, except maybe the first, are strictly positive. Hence φ is a multiple of the first eigenfunction. If φ is nonzero this shows φ has a fixed sign, which contradicts $\int_{\theta=0}^{2\pi} \varphi \, d\theta = 0$.

Since this holds for all i and j, u is radially symmetric. Now suppose $u = u(r)$ changes sign; then there is a positive number r_0 such that $u_r(r_0) = 0$. Set

$$v(r) = \begin{cases} u_r(r) & \text{for } r < r_0\,, \\ 0 & \text{for } r \geq r_0\,. \end{cases}$$

Then $v \in H_0^1(\Omega)$ and

$$0 \leq J''(u)(v) = \int_{|x|<r_0} \left(|\nabla u_r|^2 - f'(u)\, u_r^2 \right) dx =$$

$$= \int_{|x|<r_0} u_r(-\Delta u_r - f'(u)u_r)\, dx = -(n-1) \int_{|x|<r_0} r^{-2} u_r^2 dx\,,$$

which gives a contradiction for nonconstant u. □

Proposition 2: *There is* $f \in C^{0,1}(\mathbb{R})$, *with* $f(0) = 0$, *and* $\Omega \subset \mathbb{R}^2$, *bounded and convex, such that the global minimizer changes sign.*

Remark 2: In this note we will construct just one example. A forthcoming paper of Matano will certainly have a more rigorous approach to sign–changing stable solutions. However, it is not clear if this considers global minimizers.

Remark 3: Without the condition $f(0) = 0$ one can modify the example in [5] to obtain the result of proposition 2.

Proof: Set $\Omega = \{(x_1,x_2) \in \mathbb{R}^2;\ 2|x_2| < x_1 < 1\}$ and define the Lipschitz–continuous functions f_ε for $\varepsilon > 0$ by

$$
\begin{aligned}
f_\varepsilon(s) &= 0 && \text{on } (-\infty,-2\varepsilon],\\
f_\varepsilon(s) &= -\varepsilon^{-2}(s+2\varepsilon) && \text{on } (-2\varepsilon,-\varepsilon],\\
f_\varepsilon(s) &= \varepsilon^{-2}s && \text{on } (-\varepsilon,0],\\
f_\varepsilon(s) &= s && \text{on } (0,2],\\
f_\varepsilon(s) &= 4 - s && \text{on } (2,4],\\
f_\varepsilon(s) &= 0 && \text{on } (4,\infty].
\end{aligned}
$$

Ω:

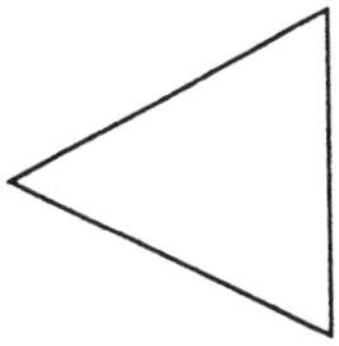

f_ε:

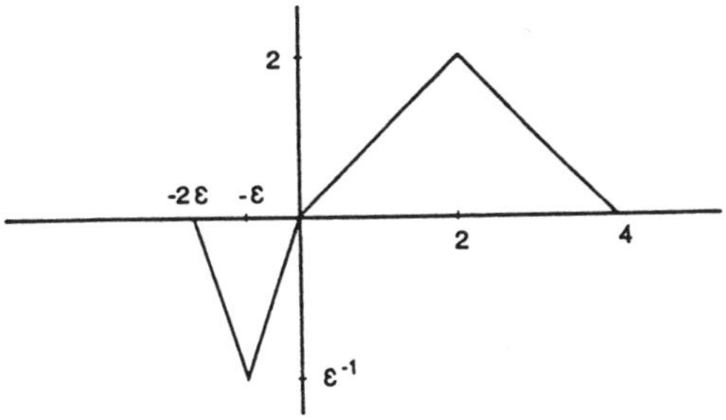

Note that $f(s) = -2\varepsilon\, f(-\tfrac{1}{2}\varepsilon s)$ for $s > 0$.

Let λ_0 denote the first eigenvalue of

$$(6)\qquad \begin{cases} -\Delta\varphi = \lambda\varphi & \text{in } \Omega, \\ \varphi = 0 & \text{on } \partial\Omega, \end{cases}$$

then the bifurcation picture for solutions with fixed sign of

$$(7)\qquad \begin{cases} -\Delta u = \lambda f_\varepsilon(u) & \text{in } \Omega, \\ u = 0 & \text{on } \partial\Omega, \end{cases}$$

looks as follows.

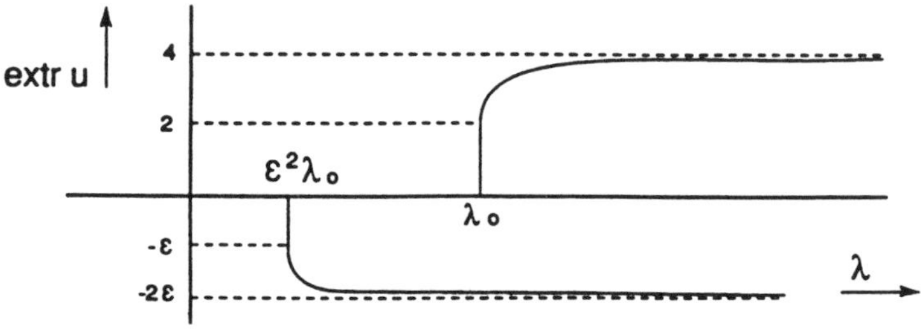

Since $s^{-1}f_\varepsilon(s)$ is decreasing on [0,4] (and strictly on [2,4]) it is known that there is a unique positive solution for every $\lambda > \lambda_0$. See [4].
There is no positive solution for $\lambda < \lambda_0$. Similar arguments hold for negative solutions. Let U_λ and V_λ^ε denote the positive, respectively the negative solution of (7) for $\lambda > \lambda_0$.
Let $J_\varepsilon(\lambda,u)$ denote the energy functional for (7), that is

$$J_\varepsilon(\lambda,u) = \int_\Omega \Big(\tfrac{1}{2}|\nabla u|^2 - \lambda \int_0^u f_\varepsilon(s)\, ds \Big)\, dx.$$

Lemma 3: $\lim\limits_{\lambda\to\infty} \lambda^{-1} J_\varepsilon(\lambda,U_\lambda) = -4|\Omega|$ *and* $\lim\limits_{\lambda\to\infty} \lambda^{-1} J_\varepsilon(\lambda,V_\lambda^\varepsilon) = -|\Omega|$, *uniformly for* $\varepsilon \in [0,1]$, *where* $|\Omega|$ *is the Lebesgue measure of* Ω.

Proof: We will show the second statement. Since V_λ^ε is the only stable solution of

$$\begin{cases} -\Delta u = \lambda \min\big(f_\varepsilon(u),0\big) & \text{in } \Omega, \\ u = 0 & \text{on } \partial\Omega, \end{cases}$$

the function minimizes

$$J_\varepsilon^-(\lambda,u) = \int_\Omega \Big(\tfrac{1}{2}|\nabla u|^2 - \lambda \int_0^u \min(f_\varepsilon(s),0)\, ds \Big)\, dx \quad \text{for } \lambda > \lambda_0.$$

Since we can estimate $J_\varepsilon^-(\lambda,u)$ from below by $-\lambda|\Omega|$:

$$J_\varepsilon^-(\lambda,u) \geq -\lambda \int_\Omega \int_0^u \min(f_\varepsilon(s),0)\, ds\, dx \geq -\lambda \int_\Omega 1\, dx\,;$$

it is sufficient to show that for all $\sigma > 0$ there is $\varphi_\varepsilon \in H_0^1(\Omega)$ such that uniformly for $\varepsilon \in [0,1]$

$$\lim_{\lambda\to\infty} \lambda^{-1} J_\varepsilon^-(\lambda,\varphi_\varepsilon) < -|\Omega| + \sigma.$$

Take $\varphi \in C_0^\infty(\Omega)$ with $\varphi = -2$ in a closed subset of Ω with measure larger than $|\Omega| - \frac{1}{2}\sigma$. The result follows for λ large since

$$\lambda^{-1}J_\varepsilon^-(\lambda,\varepsilon\varphi) < \lambda^{-1}\varepsilon^2 \int_\Omega \tfrac{1}{2}|\nabla\varphi|^2 dx - |\Omega| + \tfrac{1}{2}\sigma \leq$$
$$\leq \lambda^{-1} \int_\Omega \tfrac{1}{2}|\nabla\varphi|^2 dx - |\Omega| + \tfrac{1}{2}\sigma\,. \qquad \square$$

Because of Lemma 3 there is $\lambda_1 > \lambda_0$ such that

$$J_\varepsilon(\lambda,U_\lambda) < J_\varepsilon(\lambda,V_\lambda^\varepsilon) < -\tfrac{1}{2}|\Omega| \qquad \text{for all } \lambda \geq \lambda_1 \text{ and } \varepsilon \in [0,1].$$

Lemma 4: $U_{\lambda_1}(x_1,x_2) < \frac{1}{3}\lambda_1(x_1^2 - 4x_2^2)$.

Proof: For t large enough

(8) $$U_{\lambda_1}(x_1,x_2) < \tfrac{1}{3}\lambda_1((x_1+t)^2 - 4x_2^2) \quad \text{in } \Omega.$$

Since $-\Delta\lambda_1((x_1+t)^2 - 4x_2^2) = 6\lambda_1 > \lambda_1 \max f_\varepsilon$ and since $\lambda_1((x_1+t)^2 - 4x_2^2) > 0$ in $\bar\Omega$ for $t > 0$, this function is a supersolution for $t \geq 0$.
By the Sweeping Principle [3, Theorem 9] one finds (8) for all $t \geq 0$. $\square$

Finally we will show, for $\varepsilon > 0$ but small enough, that U_{λ_1} does not minimize $J_\varepsilon(\lambda_1,\cdot)$. We will modify U_{λ_1} near (0,0) to obtain a $H_0^1(\Omega)$–function with lower energy. Hence the solution of (7) for $\lambda = \lambda_1$ that minimizes $J_\varepsilon(\lambda_1,\cdot)$ is not U_{λ_1} or $V^\varepsilon_{\lambda_1}$, which are the only stable solutions with fixed sign.

Set
$$\Omega^1_\delta = \{(x_1,x_2) \in \Omega \,;\, x_1 < \delta\}\,,$$
$$\Omega^2_\delta = \{(x_1,x_2) \in \Omega \,;\, \delta < x_1 < 2\delta\} \text{ and}$$
$$\Omega^3_\delta = \{(x_1,x_2) \in \Omega \,;\, x_1 < 2\delta\}\,.$$

Then $|\Omega^3_\delta| = 2\delta^2$.

Moreover define $z \in C^{0,1}(\mathbb{R})$ by
$$\begin{cases} z(s) = 0 & \text{for } s \le 1\,, \\ z(s) = s-1 & \text{for } 1 < s \le 2\,, \\ z(s) = 1 & \text{for } 2 < s\,, \end{cases}$$

and set
$$u_\delta(x_1,x_2) = z(\delta^{-1}x_1)\, U_{\lambda_1}(x_1,x_2).$$

Then $u_\delta \in H^1_0(\Omega)$ and
$$\nabla u_\delta(x_1,x_2) = \delta^{-1} U_{\lambda_1}(x_1,x_2)(1,0) + z(\delta^{-1}x_1)\, \nabla U_{\lambda_1}(x_1,x_2) \qquad \text{in } \Omega^2_\delta.$$

By using lemma 4 we can estimate the difference in energy as follows:

$$(9) \qquad J_\varepsilon(\lambda_1,u_\delta) - J_\varepsilon(\lambda_1,U_{\lambda_1}) \le \tfrac12 \int_{\Omega^3_\delta} \left(|\nabla u_\delta|^2 - |\nabla U_{\lambda_1}|^2\right) dx + \lambda_1 \int_{\Omega^3_\delta} \tfrac12 U^2_{\lambda_1}\, dx \le$$

$$\le \int_{\Omega^2_\delta} \left(\tfrac12 \delta^{-2}\, U^2_{\lambda_1} dx + \delta^{-1}\, U_{\lambda_1} z(\delta^{-1}x_1) \tfrac{d}{dx_1} U_{\lambda_1}\right) dx + \lambda_1 \int_{\Omega^3_\delta} \tfrac12 U^2_{\lambda_1}\, dx \le$$

$$\le |\Omega^3_\delta| \left(\tfrac12 \delta^{-2} (\tfrac13 \lambda_1 4\delta^2)^2 + \delta^{-1} (\tfrac13 \lambda_1 4\delta^2)\, \|\nabla U_{\lambda_1}\|_\infty + \tfrac12 \lambda_1 (\tfrac13 \lambda_1 4\delta^2)^2 \right) \le$$

$$\le 2\delta^2 \left(\tfrac89 \lambda_1^2 \delta^2 + \tfrac43 \lambda_1 \delta\, \|\nabla U_{\lambda_1}\|_\infty + \tfrac89 \lambda_1^3 \delta^4 \right) \le C(\lambda_1)\, \delta^3 \qquad \text{for } 2\delta < 1.$$

The function v_δ defined by
$$v_\delta(x_1,x_2) = -\tfrac12 \delta\, U_{\lambda_1}(\delta^{-1}x_1,\, \delta^{-1}x_2)$$

satisfies:

$$
\begin{aligned}
-\Delta v_\delta(x) &= \tfrac{1}{2}\delta^{-1}(\Delta U_{\lambda_1})(\delta^{-1}x) = \\
&= -\tfrac{1}{2}\delta^{-1}\lambda_1 f_\delta\big(U_{\lambda_1}(\delta^{-1}x)\big) = \\
&= \tfrac{1}{2}\delta^{-1}\lambda_1 2\delta\, f_\delta\big(-\tfrac{1}{2}\delta\, U_{\lambda_1}(\delta^{-1}x)\big) = \\
&= \lambda_1 f_\delta(v_\delta(x)) .
\end{aligned}
$$

Hence v_δ is a solution of (7) with $\varepsilon = \delta$ and Ω replaced by Ω_δ^1.

After extending v_δ by 0 outside of Ω_δ^1 we obtain:

$$
\begin{aligned}
(10) \qquad J_\delta(\lambda_1, v_\delta) &= \int_{\Omega_\delta^1} \Big(\tfrac{1}{2}|\nabla v_\delta|^2 - \lambda_1 \int_0^{v_\delta} f_\delta(s)\, ds\Big)\, dx = \\
&= \tfrac{1}{4}\int_{\Omega_\delta^1} \Big(\tfrac{1}{2}|(\nabla U_{\lambda_1})(\delta^{-1}x)|^2 - \lambda_1 \int_0^{U_{\lambda_1}(\delta^{-1}x)} f_\delta(t)dt \Big)\, dx = \\
&= \tfrac{1}{4}\delta^2 J_\delta(\lambda_1, U_{\lambda_1}) .
\end{aligned}
$$

Finally, we set $w_\delta = u_\delta + v_\delta$ and we find, since $\operatorname{supp} u_\delta \in \bar\Omega\backslash\Omega_\delta^1$ and $\operatorname{supp} v_\delta \in \bar\Omega_\delta^1$ that by (9) and (10) for δ sufficiently small:

$$
\begin{aligned}
J_\delta(\lambda_1, w_\delta) &= J_\delta(\lambda_1, u_\delta) + J_\delta(\lambda_1, v_\delta) \le \\
&\le (1+\tfrac{1}{4}\delta^2)\, J_\delta(\lambda_1, U_{\lambda_1}) + C(\lambda_1)\,\delta^3 < J_\delta(\lambda_1, U_{\lambda_1}) . \qquad \square
\end{aligned}
$$

The example uses a triangle for a domain and a piecewise linear right hand side. One can modify both Ω and f to have the same result on a smooth, strictly convex domain with a C^∞–function f.

Acknowledgement:

I thank Wei Ming Ni for showing me the main part of Proposition 1.

References:

[1] Chang–Shou Lin and Wei–Ming Ni, On stable steady states of semilinear diffusion equations, preprint 1986.

[2] H. Matano, in preparation.

[3] A. McNabb, Strong comparison theorems for elliptic equations of second order, Journal of Math. Mech. 10, (1961), 431–440.

[4] D.H. Sattinger, Topics in stability and bifurcation theory. Lecture Notes in Math. 309, Springer Verlag, Berlin, 1973.

[5] G. Sweers, Semilinear elliptic problems on domains with corners, Commun. in Partial Differential Equations 14 (1989), 1229–1247.

Department of Pure Mathematics, Delft University of Technology,

POBox 5031, 2600 GA Delft, The Netherlands

M A VIVALDI

Oscillation and energy decay of solutions to obstacle problems involving quasi-linear, degenerate-elliptic operators

Introduction

This paper concerns *obstacle problems* relative to *quasi-linear degenerate elliptic* operators of the form: $Lu + H(u)$

$$\begin{cases} Lu = -\sum \frac{\partial}{\partial x_j}\left(a_{ij}\frac{\partial u}{\partial x_i}\right) + a_0 u \\ \lambda|\xi|^2 w(x) \le \sum a_{ij}(x)\xi_i\xi_j \le \Lambda|\xi|^2 w(x) \quad \forall x \in \Omega\,, \quad \forall \xi \in \mathbb{R}^N\,. \end{cases} \tag{0.1}$$

The 'weight' $w(x)$ is a nonnegative, measurable function satisfying the Muckenhoupt's condition A_2.

The nonlinear part is a Nemytsky operator associated with a Caratheodory function with quadratic growth in the gradient variable.

$$\begin{cases} H(u) = H(x,u,\nabla u) \\ |H(x,\eta,\xi)| \le (k_1 + k_2(|\eta|)|\xi|^2)w(x) \quad \forall x \in \Omega \quad , \forall(\eta,\xi) \in \mathbb{R}\times\mathbb{R}^N\,. \end{cases} \tag{0.2}$$

Linear degenerate elliptic equations have been previously studied by D.E. Edmunds–L.A. Peletier, S.N. Kruskov, M.K.V. Murty, G. Stampacchia and N. Trudinger in [7], [11], [15], [17] assuming the weight in a restrictive class; a larger '*sharp*' class of weights has been more recently introduced by E. Fabes, D. Jerison, C. Kenig and R. Serapioni. A Harnack inequality, the Hölder continuity of the solutions and a Wiener test for regular points in the Dirichlet problem are established in [7] and [8]. Estimates of the oscillation of the weak solution in terms of barrier functions and of the smoothness of the boundary have been established by M. Biroli and S. Marchi in [2] and by S. Marchi in [12], for bounded solutions of quasi-linear operators. The existence of such bounded solutions is supposed.

Section II of this paper is devoted to show the existence of bounded weak solutions for obstacle problems involving quasi-linear, degenerate-elliptic operators $Lu + H(u)$. The proof is done by an approximation method. A convenient map in the weighted Sobolev space $\overset{0}{H}{}^1_2(\Omega, w)$ is introduced and evalued in the L_∞-norm and in the weighted-Sobolev norm.

By the Schauder fixed point theorem the existence of approximated solutions u_m is proved. Owing to the quadratic nonlinearity of the term H, the *standard* monotonicity arguments dont seem work but still comparesons with suitable constants, independent of the approximation index m, are established by using suitable test functions of exponential type. Uniform estimates and strong convergences in $H_2^1(\Omega, w)$–norm are then proved; the ***natural*** passage to the limit in the approximation procedure conclude the proof (see Theorem 2.1).

Obstacle problems involving *quasi-linear uniformly* elliptic operators, with quadratic growth in the gradient (i.e. $w \equiv 1$ in the previous notations (0.1), (0.2)), have been studied by many authors and existence results have been established by different methods and with different assumptions. I only mention [1] and [5] and i refer to the extensive literature there quoted for more details.

The pointwise regularity of local solutions for obstacle problems, in the *linear uniformly elliptic* case, has been recently studied by U. Mosco [14], by G. Dal Maso–U. Mosco–M.A. Vivaldi [6] and, in a more general contest and by different methods, by J.H. Michael–W.P. Ziemer [13] and T. Kipelainen–W.P. Ziemer [10]. I refer to the above quoted papers [14] and [6] for "*historical*" remarks and more complete reference. In a recent paper [3] M. Biroli–U. Mosco studied the *linear* degenerate case.

Let me recall that in the theory developed in [9], [8], [12], [4], [3] an intrinsic notion of w–capacity associated to the weighted Sobolev space $H_2^1(\Omega, w)$ plays an important role. To point out the difference between the *uniformly* elliptic and the *degenerate* case we remark that a single point of $\mathbf{R}^N$ may have, for particular weights w, a positive capacity so the energy can be so sharply concentrated around a single point that the total energy in a ball can not be estimated in terms of the total energy in an anulus surrounding the same ball (see [4], [6], [14]).

Section III of this paper is devoted to evaluate the oscillation and the energy decay of a bounded weak solution of obstacle problems, involving quasi-linear, degenerate elliptic operators; the existence of such solutions being proved in section II. The first step is to obtain inequalities of Caccioppoli–De–Giorgi type. Let me remark that owing to the quadratic nonlinearity of H the 'linear' test functions used in [4], [3], [6], [14]... do not seem work; this difficulty is here overcome by means of '*particular*' test functions of '*hyperbolic cosinus*' type. Finally a *refinement* of the Poincaré inequality is used and at this point the choise of the weight in the Muckenhoupt's class A_2 seems to be crucial (see [9], [8]). The proof is concluded by the hole filling technique (see Theorem 3.1).

I Notations and preliminaries

In this part we fix the notations and we recall some preliminary results.

I.1 Functions spaces

Here and in the following $w(x)$ denote the *weight* i.e. a nonnegative function belonging to $L^1_{\text{loc}}(\mathbf{R}^N)$ and satisfying the *Muckenhoupt's condition* A_2:

$$\sup_B \{ \fint_B w \, dx \cdot \fint_B w^{-1} \, dx \} \leq k_0 \ . \tag{1.1}$$

The supremum is taken over all euclidean balls B and the symbol $\oint_B g\,dx$ denotes the mean value i.e. $\int_B g\,dx \cdot (\int_B dx)^{-1}$. We shall use also the notations for any measurable set E:

$$w(E) = \int_E w\,dx\ , \quad |E| = \int_E dx\ .$$

Let us recall three well-known properties of the weights w (see [9] and [8]).

Rermak 1.1. Let us consider the weight w satisfying the Muckenhoupt's condition A_2 then:

$$\exists q > 1 \quad \text{s.t} \quad \left(\oint_B w^q dx\right)^{1/q} \le C \oint_B w\,dx \quad \forall \text{ ball } B \tag{1.2}$$

$$\exists \eta_1 > 0 \quad \text{s.t} \quad \forall \text{ measurable set } E \subset B \text{ and any ball } B \tag{1.3}$$

$$\frac{w(E)}{w(B)} \le C\left(\frac{|E|}{|B|}\right)^{\eta_1}$$

$$\exists \eta_2 > 0 \quad \text{s.t} \quad \forall \text{ measurable set } E \subset B \text{ and any ball } B \tag{1.4}$$

$$\frac{w(E)}{w(B)} \ge C\left(\frac{|E|}{|B|}\right)^{\eta_2}\ .$$

Let Ω be always a bounded, open connected subset of $\mathbf{R}^N$. Denote by $L^p(\Omega, w)$ the Lebesgue class with norm

$$\|f\|_{0,p} = \left(\int_\Omega |f(x)|^p w\,dx\right)^{1/p} \qquad p \in [1, +\infty)\ ;$$

by $H^1_p(\Omega, w)$ the completion of Lip $(\overline{\Omega})$ for the norm

$$\|f\|_{1,p} = \{\|f\|^p_{0,p} + \|\nabla f\|^p_{0,p}\}^{1/p}\ ,$$

and by $\overset{0}{H}{}^1_p(\Omega, w)$ the closure of $C_0^\infty(\Omega)$ in $H^1_p(\Omega, w)$. The dual space of $\overset{0}{H}{}^1_p(\Omega, w)$ is the space $H^{-1}_p, (\Omega, w)$, $\frac{1}{p} + \frac{1}{p'} = 1$, $p \ge 2$.

$$H^{-1}_p(\Omega, w) = \{f_0 - \text{div}\, \vec{f}\colon \vec{f} = (f_1, \cdots f_N) \text{ and } f_i w^{-1} \in L^{p'}(\Omega, w), \quad i = 0, 1, \cdots, N\}\ .$$

Finally the following Poincaré inequality, consequence of a stronger inequality, holds for any $f \in C_0^\infty(\Omega)$:

$$\int_\Omega |f|^2 w\,dx \le C(\operatorname{diam}\Omega)^2 \int_\Omega |\nabla f|^2 w\,dx\ . \tag{1.5}$$

I refer to [7], [8] for proofs and details.

I.2 Operators and Green's functions

We denote by L a linear bounded operator from $\overset{0}{H}{}_2^1(\Omega, w)$ to $H_2^{-1}(\Omega, w)$:

$$\begin{cases} \text{i)} & Lu = -\sum_{ij=1}^{N} \frac{\partial}{\partial x_j}\left(a_{ij}\frac{\partial u}{\partial x_i}\right) + a_0 u \equiv L_0 u + a_0 u a_{ij} = a_{ji} \ , \\ \text{ii)} & \lambda w(x)|\xi|^2 \le \sum_{i,j=1}^{N} a_{ij}(x)\xi_i\xi_j \le \Lambda w(x)|\xi|^2 \ , \ \Lambda, \ \lambda \in \mathbf{R}, \ \lambda > 0 \ , \\ \text{iii)} & \underline{a}_0 w(x) \le a_0(x) \le \overline{a_0} w(x) \ , \ \overline{a_0}, \ \underline{a}_0 \in \mathbf{R} \ , \ \underline{a}_0 > 0 \ ; \end{cases} \tag{1.6}$$

by $H(x, u, \nabla u)$ the Nemytsky operator associated to the function $H(x, \eta, \xi)$:

$$\begin{cases} H(x,\eta,\xi) \text{ is a Carathéodory function} \\ |H(x,\eta,\xi)| \le (k_1 + k_2(|\eta|)|\xi|^2)w(x) \\ \text{of course } k_2(|\eta|) \text{ can be always supposed not decreasing in } |\eta| \ . \end{cases} \tag{1.7}$$

For every $y \in \Omega$ there exists the *Green's function* $G_\Omega^y(x)$ for the Dirichlet problem relative to L_0 in Ω with singularity at y, which is defined as solution of

$$a(\varphi, G_\Omega^y) = \int_\Omega a_{ij} \frac{\partial \varphi}{\partial x_j} \frac{\partial G_\Omega^y}{\partial x_j} \, dx = \varphi(y) \quad \forall \varphi \in C_0^\infty(\Omega) \ .$$

Here and in the following the symbol "$\sum_{ij=1}^{N}$" is understood. Such a Green function is symmetric in x, y and satisfies the following growth conditions

$$\gamma_1 \int_r^{R_0} (w(B_s(y)))^{-1} \quad s\, ds \le G_\Omega^y(x) \le \gamma_2 \int_r^{R_1} (w(B_s(y)))^{-1} \quad s\, ds \tag{1.8}$$

where γ_1, γ_2 are positive constants depending on N, Λ, λ and k_0 in (1.1), $B_{R_0}(y) \subset \Omega \subset B_{R_1}(y)$, and $|x - y| = r < R_0/2$. Here and in the following $B_\rho(y)$ denotes the ball of radious ρ centered at y.

Let us consider also the *regularized* Green function $G_{\Omega,\ \rho}^y, \rho > 0$ that is the unique solution in $\overset{0}{H}{}_2^1(\Omega, w)$ of the problem:

$$a(f, G_{\Omega,\rho}^y) = (w(B_\rho(y)))^{-1} \int_{B_\rho(y)} f(x) w(x)\, dx \qquad \forall f \in C_0^\infty(\Omega) \tag{1.9}$$

Remark 1.2. The following facts hold for $G_{\Omega,\rho}^y$ (see [14], [9], [8], [4])

$$\int_\Omega |\nabla G_{\Omega,\rho}^y|^2 v^2 w \, dx \le 4\Lambda^2 \lambda^{-2} \int_\Omega |G_{\Omega,\rho}^y|^2 |\nabla v|^2 w \, dx \tag{1.10}$$

$\forall v \in L_\infty(\Omega) \cap \overset{0}{H}{}_2^1(\Omega, w)$, $v \equiv 0$ in $B_r(y)$, $\rho \in (0, r]$.

$$\begin{cases} G_{\Omega,\rho}^y \to G_\Omega^y \text{ uniformly in any compact set of } \Omega \backslash \{y\} \\ \quad \text{and weakly in } L^p(\Omega, w) \ , \ p \in \left(1, \frac{2N}{2N-1}\right) \ . \end{cases} \tag{1.11}$$

I.3 The w–capacity notion

Let K be a compact subset of Ω: we define

$$w-\text{cap }(K,\Omega)=\inf\left\{\int_\Omega |\nabla v|^2 w\,\mathrm{d}x\ ,\ \forall v\in C_0^1(\Omega)\ ,\ v\geq 1 \text{ on } K\right\}$$

and we extend the definition to any open A and any set E by the usual passage to the supremum and to the infimum (respectively).

We say that a set $E\subseteq \mathbf{R}^N$ has w–capacity zero if $w-\text{cap }(E,D)=0$ for some bounded open D s.t. $\overline{E}\subset D$.

If a property depending on $x\in E\subset\Omega$ holds for every $x\in E$ except a subset N of w–capacity zero then we say that this property holds (w–q–e) w–quasi–everywhere in E.

A function $u:\Omega\to[-\infty,+\infty]$ is said to be w–quasi–continuous if for every $\varepsilon>0$ there exists a set $A\subset\Omega$ with w–cap $(A,\Omega)<\varepsilon$ such that the restriction of u to $\Omega\backslash A$ is continuous on $\Omega\backslash A$.

For every $v\in H_2^1(\Omega,w)$ there exists a w–quasi continuous representative (unique up to w.q.e. equivalence)

$$\tilde v(x)=\liminf_{\rho\to 0}(w(B_\rho(x)))^{-1}\int_{B_\rho(x)} vw\,\mathrm{d}x$$

and the 'lim inf' can be replaced by 'lim' for w–q–e $x\in\Omega$.

For any function $v:\Omega\to[-\infty,+\infty]$ and any set $F\subset\Omega$ we denote by $w-\inf_F(w-\sup_F)$ the essential infimum (supremum) with respect to the w–capacity on F. If $v\in H^1(\Omega,w)$ and if A is open $A\subset\Omega$ then the conditions $\tilde v\geq 0$ w–q–e on A and $v\geq 0$ a.e. in A are equivalent; so the $\sup\limits_A$ and $\inf\limits_A$ are defined unambigously (see [4]).

By $v_E=v_{E,\Omega}$ where $\overline{E}\subset\Omega$, we denote the *$w$-capacitary potential* of E in Ω with respect to the operator $L_0: v_E$ is the unique solution of the problem

$$\begin{cases} v_E\in \overset{0}{H}{}_2^1(\Omega,w)\ \ v_E\geq 1\quad w\text{–q–e}\quad \text{in } E \\ a(v_E,v_E-f)\leq 0\quad \forall f\in \overset{0}{H}{}_2^1(\Omega,w)\ ,\ f\geq 1\ w\text{–q–e in } E\ . \end{cases}\tag{1.12}$$

L_0v_E belongs to $H_2^{-1}(\Omega,w)$ and is a *positive Radon measure $\mu_E=\mu_{E,L_0}$, supported on ∂E: the w-capacitary measure* of E in Ω (with respect to L_0) and satisfies the equality:

$$\langle L_0v_E,f\rangle=a(v_E,f)=\int_\Omega \tilde f\,\mathrm{d}\mu_E\quad \forall f\in \overset{0}{H}{}_2^1(\Omega,w)\ .\tag{1.13}$$

See [4], [9], [8] for some others important properties, proofs and details.

We shall use in section III the following estimates $\forall r\in(0,\ R/2)$

$$\begin{cases} w-\text{cap }(B_r(x_0),B_R(x_0))\simeq\left(\int_r^R s(w(B_s(x_0))^{-1}\,\mathrm{d}s\right)^{-1}\ , \\ w-\text{cap }(B_{qR}(x_0),B_R(x_0))\simeq\dfrac{w(B_R(x_0))}{R^2}\ ,\quad \forall q\in(0,\ \tfrac12)\ . \end{cases}\tag{1.14}$$

The symbol $\simeq$ denotes, here and in the following, that each side can be estimated by the other with constants dependent only on N, λ, Λ q and k_0 in (1.1).

I.4 Wiener modulus and Wiener integral

Let the obstacle $\psi : \mathbf{R}^N \to [-\infty, +\infty)$ be given up to set of w–capacity zero, and let x_0 be an arbitrarily fixed point.

The pointwise value $\overline{v}(x_0) \in [-\infty, +\infty]$ of an arbitrary function $v : \Omega \to [-\infty, +\infty]$ is defined by:

$$\overline{v}(x_0) = \inf_{\rho>0} \left(w - \sup_{B\rho(x_0)} v \right) . \tag{1.15}$$

Let us now consider the level sets of ψ, $\varepsilon > 0$

$$E_\rho(\varepsilon) = E(\psi, x_0, \varepsilon, \rho) = \left\{ x \in B_\rho(x_0) : \psi(x) \geq w - \sup_{B_\rho(x_0)} \psi - \varepsilon \right\} \tag{1.16}$$

and their relative w–capacities

$$\delta(\rho) = \delta(\varepsilon, \rho) = \frac{w - \text{cap }\ (E_\rho(\varepsilon), B_{2\rho}(x_0))}{w - \text{cap }\ (B_\rho(x_0), B_{2\rho}(x_0))} . \tag{1.17}$$

DEFINITION 1.1. We call *Wiener integral* of ψ the following:

$$\int_0^1 \delta(\varepsilon, \rho) \frac{\mathrm{d}\rho}{\rho} \tag{1.18}$$

and we call *Wiener modulus* of ψ the following:

(1.19) $$\omega_\sigma(r, R) = \inf \left\{ \omega > 0 \colon \omega \exp \left(\int_r^R \delta(\sigma\omega, \rho) \frac{\mathrm{d}\rho}{\rho} \right) \geq 1 \right\} ,$$

$$\sigma > 0 \text{ and } r \in (0, R) .$$

DEFINITION 1.2. We say that x_0 is a *Wiener* point of ψ if $\overline{\psi}(x_0) = -\infty$ or if the Wiener integral defined in (1.18) is divergent.

We shall say that a function $v : \Omega \to (-\infty, +\infty)$, defined up to set of w–capacity zero, is continuous at a given point $x_0 \in \Omega$ if

$$\lim_{\rho \to 0} w - \underset{B_\rho(x_0)}{osc}\ v = 0 \tag{1.20}$$

where $\forall E \subset \Omega$ we put

$$\begin{cases} w - \underset{E}{osc}\, v = (w - \sup_E v) - (w - \inf_E) v \\ \text{if } w - \text{cap}(E, \Omega) > 0 \text{ and } w - \sup_E v > -\infty \\ w - \underset{E}{osc}\, v = 0 \text{ otherwise} . \end{cases} \tag{1.21}$$

We conclude this section by the following remarks. See [4] and [14] for more details.

Remark 1.3. Let $\overline{\psi}(x_0) > -\infty$, then the following properties are equivalent:

$$x_0 \text{ is a } \textit{Wiener point of } \psi \tag{1.22}$$

$$\begin{cases} \forall \varepsilon > 0 \ \exists \ R > 0 : \lim_{r\to 0} \omega_\sigma(r,R) = 0 \\ \text{for suitable } \sigma = \sigma(\varepsilon, r, R) : \sigma\omega_\sigma(r,R) \le \varepsilon, \ r \in (0,R) \ . \end{cases} \tag{1.23}$$

Remark 1.4. The notion of Wiener point depends on the *weight* w but do not depend on the particular choice of the operator L_0 satisfying (1.6) ((i), (ii)).

II Existence of solutions

In this section obstacle problems involving quasi–linear degenerate elliptic operators are considered and an existence result of solution is proved.

Consider, according to notations of section I, the problem

$$\begin{cases} u \in \mathbf{K} \cap L_\infty(\Omega) : \\ a(u, v-u) + \int_\Omega (a_0 u + H(u))(v-u)\,\mathrm{d}x \ge 0 \quad \forall v \in \mathbf{K} \cap L_\infty(\Omega) \end{cases} \tag{2.1}$$

where

$$\mathbf{K} = \{v \in \overset{0}{H}{}^1_2(\Omega, w) : v \ge \psi \quad w\text{–q–e}\} \ .$$

Suppose

$$\mathbf{K} \cap L_\infty(\Omega) \ne \emptyset \tag{2.2}$$

then the following theorem holds.

THEOREM 2.1. *Assume notations and hypotheses (1.6), (1.7), and (2.2) then problem (2.1) admits solution.*

To prove Theorem 2.1 an approximation procedure is used: existence of approximate solutions, uniform bounds in the L_∞–norm and in the $H^1_2(\Omega, w)$–norm are then obtained, finally the strong convergence in the $H^1_2(\Omega, w)$–norm and the 'natural' passage to the limit is shown. Let me here point out only the two principal steps and let me refer to [2] for the proofs.

Let me start by establishing some intermediary results.

Replace the *quadratic* term $H(x,\eta,\xi)$ by $H_m(x,\eta,\xi)$

$$H_m(x,\eta,\xi) = \frac{H(x,\eta,\xi)}{1 + m^{-1}w^{-1}(x)|H(x,\eta,\xi)|} \tag{2.3}$$

of course we have: (see [8])

$$\begin{cases} \text{i) } H_m w^{-1} \in L^2(\Omega, w) \text{ and so } H_m \in H^{-1}(\Omega, w) \\ \text{ii) } |H_m| \le mw \\ \text{iii) } |H_m| \le |H| \ . \end{cases} \tag{2.4}$$

Consider the approximate problems

$$\begin{cases} u_m \in \mathbf{K} : \forall v \in \mathbf{K} \\ a(u_m, v - u_m) + \int_\Omega (a_0 u_m + H_m(u_m))(v - u_m)\,\mathrm{d}x \geq 0 \ . \end{cases} \tag{2.1$_m$}$$

The following propositions hold:

PROPOSITION 2.1. *Assume previous notations and hypotheses then problem $(2.1)_m$ admits solution.*

PROPOSITION 2.2. *Assume previous notations and hypotheses then the following estimates hold for any solution u_m of $(2.1)_m$*

$$\|u_m\|_\infty \leq k_3 \tag{2.5}$$

$$\|u_m\|_{1,2} \leq k_4 \tag{2.6}$$

where $k_3 = w - \sup_\Omega \psi \vee (k_1 \underline{a}_0^{-1})$ and k_4 is a constant dependent on $\lambda, \Lambda, k_0, k_1, k_2, k_3, a_0 \ldots$ but independent of m.

III Oscillation and energy estimates

The main result of this section is the estimate of the oscillation and of the energy decay for an arbitrary (local) bounded weak solution of the obstacle problem: the existence of such a solution being proved in section II.

Let us remark that in this part we do not need the assumption $\underline{a}_0 > 0$ in the condition (1.6) iii) of I.2 so, for simplicity, we assume $a_0 \equiv 0$ i.e. $L = L_0$, the general case being analogous.

Consider the problem:

$$\begin{cases} u \in H_2^1(\Omega, w) \cap L_\infty(\Omega),\ u \geq \psi \quad w\text{–q–e} \\ a(u, v - u) + \int_\Omega H(u)(v - u)\,\mathrm{d}x \geq 0 \ \ \forall v \in H_2^1(\Omega, w) \cap L_\infty(\Omega)\ , \\ v \geq \psi \quad w\text{–q–e and } v - u \in \overset{0}{H}{}_2^1(\Omega, w)\ . \end{cases} \tag{3.1}$$

For every $r > 0$ such that $B_{2r/q}(x_0) \subset \Omega, q \in (0, 1/5)$ 'parameter', that we shall fix after, we define the seminorm $\mathcal{V}(r) = \mathcal{V}(r, u, x_0)$

$$\mathcal{V}(r) = \left\{ \underset{B_r(x_0)}{osc}{}^2\, u + \int_{B_r(x_0)} |\nabla u|^2 G^{x_0}_{B_{2rq^{-1}}(x_0)}\, w\,\mathrm{d}x \right\}^{1/2} \tag{3.2}$$

where $G^{(x_0)}_{B_{2rq^{-1}}(x_0)}$ is the Green function associated to L and x_0 in the ball $B_{2rq^{-1}}(x_0)$ see I.2.

According to notations of section I and II we have:

THEOREM 3.1. *Assume notations and hypotheses (1.6), (1.7) and (1.16)–(1.19) then for any solution u of the previous problem (3.1) the following estimate holds.*

$$\mathcal{V}(r) \leq C\{\mathcal{V}(R)\omega_\sigma^\beta(r, R) + \sigma\omega_\sigma(r, R) + R\} \tag{3.3}$$

$\forall \sigma > 0$ *and* $r \in (0, qR)$, $B_{2R}(x_0) \subset \Omega$. *The constant* $\beta > 0$ *depends on* $N, \lambda, \Lambda, w, q$, *but is independent of* r *and* R. *Moreover if* $\overline{B}_{2R_0}(x_0) \subset \Omega, R_0 \geq R$ *we have:*

$$\mathcal{V}(R) \leq C \left\{ \|u\|_{L^2(B_{2R0}(x_0),w)} \cdot (w(B_{R_0}(x_0)))^{-1/2} + \left(w - \sup_{B_{2R_0}(x_0)} \psi \right)^+ \right\} . \tag{3.4}$$

Here and in the following the 'symbol' $\underline{C}$ denotes (possibly) different constants depending on $N, \lambda, \Lambda, w, q \dots$ but independent of r and R.

Remark 3.1. From Theorem 3.1 follows, in particular, that if x_0 is a Wiener point of ψ then any solution u of problem (3.1) is continuous at x_0 (see (1.20), (1.21) and Remark 1.3). Moreover the energy and the oscillation decay can be evalued in terms of the Wiener modulus of ψ.

In order to prove Theorem 3.1 we shall establish some preliminary inequalities of Caccioppoli–De Giorgi type and potential estimates.

PROPOSITION 3.1. *Assume previous notations and hypotheses then for every* z *such that* $\overline{B}_R(z) \subset \Omega$, *for every* $d \geq w - \sup_{B_R(z)} \psi$, $s \in (0, t)$ *and* $t \in (0, 1)$ *we have*

$$\begin{cases} ((u(z) - d)^{\pm})^2 + \lambda \int_{B_{sR}(z)} |\nabla (u-d)^{\pm}|^2 G^z_{B_{2R}(z)} w \, \mathrm{d}x \leq \\ \leq C \left\{ (w(B_{tR}(z)))^{-1} \int_{B_{tR}(z) - B_{sR}(z)} |(u-d)^{\pm}|^2 w dx + R^2 \right\} \end{cases} \tag{3.5}$$

where $C = C(N, \lambda, \Lambda, w, s, t)$ *but is independent of* R.

Proof. Choose as test function in (3.1)

$$v = u - \varepsilon \tau^2 G_p (u-d)^+ \cdot \cosh M((u-d)^+)^2$$

where

$$\begin{cases} G_p \text{ is the regularized Green function relative to } G^z_{B_{2R}(z)} (\text{see I.2}), \\ \rho \leq \frac{sR}{2}, \tau \in \operatorname{Lip}(\Omega) \text{ a cut off function, } 0 \leq \tau \leq 1, \tau \equiv 1 \text{ on } B_{sR}(z) , \\ \operatorname{supp} \tau \subset B_{tR}(z),\ |\nabla \tau| \leq \frac{C(s,t)}{R}; M \text{ and } \varepsilon \text{ are positive constants} \\ M \text{ 'big' enough and } \varepsilon \text{ 'small' enough} . \end{cases} \tag{3.6}$$

Set $X_1 = \cosh M((u-d)^+)^2$, $X_2 = \sinh M((u-d)^+)^2$ and $B_r = B_r(2)$, $r > 0$ we obtain:

$$\begin{cases} 0 \geq \lambda \int_{B_{tR}} |\nabla (u-d)^+|^2 \{X_1 + 2M |(u-d)^+|^2 X_2\} G_\rho \tau^2 w \, \mathrm{d}x + \\ + \int_{B_{tR}} \tau^2 a_{ij} \frac{\partial u}{\partial x_i} (u-d)^+ \frac{\partial G_\rho}{\partial x_j} \eta^2 X_1 \, \mathrm{d}x + \\ +2 \int_{B_{tR} - B_{SR}} a_{ij} \frac{\partial u}{\partial x_i} (u-d)^+ \tau \frac{\partial \tau}{\partial x_j} X_1 \mathrm{d}x - k_1 \int_{B_{tR}} \tau^2 G_\rho (u-d)^+ X_1 w \, \mathrm{d}x \\ -k_2(\|u\|_\infty) \int_{B_{tR}} |\nabla (u-d)^+|^2 \tau^2 G_\rho (u-d)^+ X_1 w \, \mathrm{d}x \equiv \\ \equiv I + II + III + IV + V . \end{cases} \tag{3.7}$$

It is easy to see that there exists a constant M such that:

$$I - V \geq \frac{5}{6}\lambda \int_{B_{tR}} G_\rho |\nabla(u-d)^+|^2 X_1 \tau^2 w \, dx \ . \tag{3.8}$$

On the other hand

$$\begin{cases} |III| \leq \frac{\lambda}{6} \int_{B_{tR}-B_{sR}} |\nabla(u-d)^+|^2 X_1 G_\rho \tau^2 w \, dx + \\ 6\lambda^{-1}\Lambda^2 \int_{B_{tR}-B_{SR}} |\nabla\tau|^2 G_\rho X_1((u-d)^+)^2 w \, dx \ . \end{cases} \tag{3.9}$$

Let us use definitions and properties of I.2 (in particular (1.10)). Denote by σ a cut off function, $\sigma \in \mathrm{Lip}\,(\overline{\Omega})$, $\leq \sigma \leq 1$, $\sigma \equiv 1$ on $B_R - B_{tR}$ $\sigma \equiv 0$ on B_{sR}, $|\nabla\sigma| \leq \frac{C(s,t)}{R}$, η, γ small positive constants (that we shall choose after) we obtain:

$$\begin{cases} II \geq \frac{1}{2}(w(B_\rho))^{-1} \int_{B_\rho}((u-d)^+)^2 w \, dx - \\ -\frac{8\gamma\eta\Lambda^3}{M\lambda^2} \int_{B_{tR}-B_{sR}} G_\rho^2 \cdot \{X_2|\nabla(\tau\sigma)|^2 + 4M((u-d)^+)^2 X_1^2 X_2^{-2} \cdot \\ \cdot |\nabla(u-d)^+|^2\tau^2\sigma^2\} w \, dx - \frac{\Lambda}{M\gamma\eta} \int_{B_{tR}-B_{sR}} |\nabla\tau|^2 X_2 w \, dx \ . \end{cases} \tag{3.10}$$

Let us pass to the limit as $\rho \to 0^+$ and choose $\gamma \sim \left(\int_{sR}^{tR} \omega(B_r)^{-1} r \, dr\right)^{-1}$. From (1.4), (1.8), (3.6), (3,7), (3.8), (3.9) and (3.10) we derive:

$$\begin{cases} \frac{2\lambda}{3} \int_{B_{tR}} |\nabla(u-d)^+|^2 G^z_{B_{2R}} X_1 \tau^2 w \, dx + \frac{1}{2}((u-d)^+)^2(z) \\ \leq C\left\{\frac{\eta+\eta^{-1}+6\Lambda^2\lambda^{-1}}{w(B_{tR})}\right\} \int_{B_{tR}-B_{sR}}((u-d)^+)^2 w \, dx + \\ +\eta \int_{B_{tR}-B_{sR}} |\nabla(u-d)^+|^2 G^z_{2R} X_1 \tau^2 w \, dx\} + \\ +k_1 \int_{B_{tR}} \tau^2 G^z_{2R}(u-d)^+ X_1 w \, dx \ . \end{cases} \tag{3.11}$$

From (1.8), (1.3) and (1.4) we derive:

$$\int_{B_{tR}} G_{2R}^2 w \, dx \sim \int_{|x-z|<tR} \left(\int_{|x-z|}^{2R} \frac{s^2}{w(B_s)} \frac{ds}{s}\right) w(x) \, dx \leq$$

$$\leq C \sum_{j=0}^{+\infty} \sum_{k=0}^{j} 2^{-k\varepsilon} 2R^2 \cdot \frac{w(B_{tR2^{-j}})}{w(B_{R2^{-j-1}})} \cdot 2^{(-j+k)2} \leq CR^2 \ . \tag{3.12}$$

From (3.11), (3.12) and by choosing η small ($\eta < \lambda/6C$) we deduce inequality (3.5) for the positive part $(u-d)^+$.

If we proceed in an analogous way and we choose as test function in (3.1) $v = u + \varepsilon\tau^2 G_\rho(u-d)^- \cosh M((u-d)^-)^2$ we prove inequality (3.5) for the negative part $(u-d)^-$.

From Proposition 3.1 follows:

COROLLARY 3.1. *Assume notations and hypotheses of Proposition 3.1 then* $\forall q < 1/3$ *we have (if* $B_{3R}(x_0) \subset \Omega$*):*

$$\begin{cases} \sup_{B_{qR}(x_0)}((u-d)^{\pm})^2 + \lambda \int_{B_{qR}(x_0)} |\nabla(u-d)^{\pm}|^2 G^{x_0}_{B_{2R}(x_0)} w \, \mathrm{d}x \leq \\ C\left\{(w(B_R(x_0)))^{-1} \int_{B_R(x_0)-B_{qR}(x_0)} |(u-d)^{\pm}|^2 w \, \mathrm{d}x + R^2\right\} \ . \end{cases} \tag{3.13}$$

Proof. If we choose in (3.5) $s \geq 2q$ and $t+q \leq 1$, and we use (1.4), then (3.13) follows direct by (3.5).

Before proving Theorem 3.1 we establish another intermediary result.

PROPOSITION 3.2. *Assume notations and hypotheses of previous Proposition 3.1 then* $\forall \gamma > 0$, $s \in (0,\ t/2)$, $t \in (0,\ 1/2)$ *we have:*

$$\begin{cases} \lambda \int_{B_{sR}(z)} |\nabla(u-d)^{\pm}|^2 G^z_{B_{tR}(z)} w \, \mathrm{d}x + (|u-d)^{\pm})^2(z) \leq \\ \leq (C+\gamma) \sup\limits_{B_{tR}(z)} |(u-d)^{\pm}|^2 + \\ +\frac{C}{\gamma} \int_{B_{tR}(z)-B_{sR}(z)} \sup_{B_{tR}(z)} |\nabla(u-d)^{\pm}|^2 G^z_{tR} w \, \mathrm{d}x + CR^2 \ . \end{cases} \tag{3.14}$$

Proof. We now choose as test function in (3.1)

$$v = u - \varepsilon\varphi G_\rho (u-d)^+ \cdot \cosh M(u-d)^+)^2$$

where $\varphi = \varphi_{sR}$ is the potential of $B_{sR}(z)$ in $B_{tR}(z)$ with respect to L. Let be $M, G_\rho, \varepsilon, X_1, X_2$ and B_r as in the proof of Proposition 3.1 we obtain:

$$\begin{cases} 0 \geq \lambda \int_{B_{tR}} |\nabla(u-d)^+|^2 \{X_1 + 2M((u-d)^+)^2 X_2\} G_\rho \varphi w \, \mathrm{d}x + \\ + \int_{B_{tR}} a_{ij} \frac{\partial u}{\partial x_i}(u-d)^+ \frac{\partial G_\rho}{\partial x_j} \varphi X_1 \, \mathrm{d}x + \int_{B_{tR}} a_{ij} \frac{\partial u}{\partial x_i}(u-d)^+ \frac{\partial \varphi}{\partial x_j} G_\rho X_1 \, \mathrm{d}x - \\ -k_1 \int_{B_{tR}} G_\rho (u-d)^+ X_1 w \, \mathrm{d}x - \\ -k_2(\|u\|_\infty) \int_{B_{tR}} |\nabla(u-d)^+|^2 \varphi G_\rho (u-d)^+ X_1 w \, \mathrm{d}x \equiv \\ \equiv I + II + III + IV + V \ . \end{cases} \tag{3.15}$$

As previously:

$$I - V \geq \frac{\lambda}{2} \int_{B_{tR}} G_\rho |\nabla(u-d)^+|^2 X_1 \varphi w \, \mathrm{d}x \ . \tag{3.16}$$

On the other hand:

$$II \geq \frac{1}{2}(w(B_\rho))^{-1} \int_{B_\rho} ((u-d)^+)^2 \varphi w \, \mathrm{d}x - \frac{1}{2M} a(\varphi, G_\rho X_2) + III \ . \tag{3.17}$$

Moreover by (1.13) and relative properties we deduce:

$$\left|\frac{1}{2M} a(\varphi, G_\rho X_2)\right| \leq C \sup_{B_{tR}} |(u-d)^+|^2 \ . \tag{3.18}$$

Let us pass to the limit $\rho \to 0^+$. We obtain form (3.15), (3.16), (3.17) (3.18) and (3.12)

$$\frac{\lambda}{2}\int_{B_{tR}} G^z_{B_{2R}}|\nabla(u-d)^+|^2 X_1\varphi w\,dx + \frac{1}{2}|(u-d)^+|^2(z) \le$$
$$\le C(\sup_{B_{tR}}|(u-d)^+|^2 + R^2) + 2\int_{B_{tR}-B_{sR}} a_{ij}\frac{\partial u}{\partial x_j}(u-d)^+\frac{\partial \varphi}{\partial x_j}G^z_{B_{2R}}X_1\,dx \quad .(3.19)$$

From the properties of $G^z_{B_{2R}}$ and φ (see section 1.2) we derive $\forall \eta > 0$

$$2\int_{B_{tR}-B_{sR}} a_{ij}\frac{\partial u}{\partial x_i}(u-d)^+\frac{\partial \varphi}{\partial x_j}G^z_{B_{2R}}X_1\,dx \le$$
$$\le \frac{\Lambda}{\eta}\int_{B_{tR}-B_{sR}} G^z_{B_{2R}}|\nabla(u-d)^+|^2 X_1 w\,dx +$$
$$+ c\eta\sup_{B_{tR}}|(u-d)^+|^2\cdot\sup_{B_{sR}}G^z_{B_{2R}}w\!-\!\mathrm{cap}\,(B_{sR},B_{tR}) \ . \tag{3.20}$$

Now by using (3.19), (3.20), (1.8), (1.12) and choosing η small 'enough' ($\eta < \gamma C^{-1}$) we obtain the estimate in (3.14) for the positive part.

If we proceed in an analogous way, choosing as test function in (3.1) $v = u + \varepsilon\varphi G_\rho(u-d)^- \cosh M((u-d)^-)^2$ we prove the estimate (3.14) for the negative part $(u-d)^-$.

To show Theorem 3.1 we will use also the following refinement of the Poincaré inequality (1.5) (see [10] and [4]).

PROPOSITION 3.3. *Let $v \in H^1(B_R(x_0), w)$ and $N(v) = \{x : x \in B_R(x_0), v = 0\}$ then:*

$$\int_{B_R(x_0)} v^2 w\,dx \le \frac{Cw(B_R(x_0))}{w-\mathrm{cap}(N(v), B_{2R}(x_0))}\int_{B_R(x_0)}|\nabla v|^2 w\,dx \ . \tag{3.21}$$

Proof of Theorem 3.1. Choose $t = 1-q$, $s = 2q$, $q \in \left(0, \frac{1}{5}\right)$. By proceeding as in the proof of Corollary 3.1, and by using (1.8) and (1.4) we deduce from (3.14)

$$\begin{cases} \sup_{B_{qR}(x_0)}|(u-d)^\pm|^2 \le (C_1+\gamma)\sup_{B_R(x_0)}|(u-d)^\pm|^2 + \\ +\frac{C_2}{\gamma}\int_{B_R(x_0)-B_{qR}(x_0)}|\nabla(u-d)^\pm|^2 G^{x_0}_{B_{2R}(x_0)}w\,dx + C_3R^2 \ . \end{cases} \tag{3.22}$$

From Corollary 3.1 we derive in particular

$$\lambda\int_{B_{qR}(x_0)}|\nabla(u-d)^\pm|^2 G^{x_0}_{B_{2R}(x_0)}w\,dx \le C_4\{\sup_{B_R(x_0)}|(u-d)^\pm|^2 + R^2\} \tag{3.23}$$

From (3.22) and (3.23) we have:

$$\begin{cases} \lambda\int_{B_{qR}(x_0)}|\nabla(u-d)^\pm|^2 G^{x_0}_{B_{2R}(x_0)}w\,dx + \sup_{B_{qR}(x_0)}|(u-d)^\pm|^2 \le \\ \le (C_1+C_4+\gamma)\sup_{B_R(x_0)}|(u-d)^\pm|^2 + \\ +\frac{C_2}{\gamma}\int_{B_R(x_0)-B_{qR}(x_0)}|\nabla(u-d)^\pm|^2 G^{x_0}_{B_{2R}(x_0)}w\,dx + (C_3+C_4)R^2 \ . \end{cases} \tag{3.24}$$

If we now multiply (both the sides) of (3.24) by γ and if we add the term $C_2 \int_{B_{qR}(x_0)} |\nabla(u-d)^{\pm}|^2 G^{x_0}_{B_{2R}(x_0)} w\, dx$ then we obtain:

$$\begin{cases} (\gamma\lambda + C_2) \int_{B_{qR}(x_0)} |\nabla(u-d)^{\pm}|^2 G^{x_0}_{B_{2R}(x_0)} w\, dx + \\ +\gamma \sup_{B_{qR}(x_0)} |(u-d)^{\pm}|^2 \le \gamma(C_1 + C_4 + \gamma) \sup_{B_R(x_0)} |(u-d)^{\pm}|^2 + \\ +C_2 \int_{B_R(x_0)} |\nabla(u-d)^{\pm}|^2 G^{x_0}_{B_{2R}(x_0)} w\, dx + (C_3 + C_4)\gamma R^2 \ . \end{cases} \tag{3.25}$$

Let us split the second term in the right side in two parts and still use (1.8) and (1.4) we obtain

$$\begin{cases} \int_{B_R(x_0)} |\nabla(u-d)^{\pm}|^2 G^{x_0}_{B_{2R}(x_0)} w\, dx = \\ = \int_{B_R(x_0)} |\nabla(u-d)^{\pm}|^2 G^{x_0}_{B_{2Rq^{-1}}(x_0)} w\, dx + \\ + \int_{B_R(x_0)} |\nabla(u-d)^{\pm}|^2 \left(G^{x_0}_{B_{2R}(x_0)} - G^{x_0}_{B_{2Rq^{-1}}}(x_0) \right) w\, dx \le \\ \le \int_{B_R(x_0)} |\nabla(u-d)^{\pm}|^2 G^{x_0}_{B_{2Rq^{-1}}(x_0)} w\, dx - \\ - \frac{C(s,q)R^2}{w(B_R(x_0))} \int_{B_R(x_0)} |\nabla(u-d)^{\pm}|^2 w\, dx \ . \end{cases} \tag{3.26}$$

Moreover as $\forall r > 0$:

$$\sup_{B_r(x_0)} |(u-d)^+|^2 + \inf_{B_r(x_0)} |(u-d)^-|^2 = \operatorname{osc}^2_{B_r(x_0)} u - 2 \sup_{B_r(x_0).} |(u-d)^+| \cdot \sup_{B_r(x_0)} |(u-d)^-|$$

form (3.25) and (3.26) we deduce:

$$\begin{aligned} (3.27)\quad (\lambda\gamma + C_2) \int_{B_{qR}(x_0)} |\nabla u|^2 G^{x_0}_{B_{2R}(x_0)} w\, dx + \gamma \operatorname{osc}^2_{B_{qR}(x_0)} u \le \\ \le \gamma(C_5 + \gamma) \operatorname{osc}^2_{B_R(x_0)} u + C_2 \int_{B_R(x_0)} |\nabla u|^2 G^{x_0}_{B_{2q^{-1}R}(x_0)} w\, dx \\ - \frac{C_6 R^2}{w(B_R(x_0))} \int_{B_R(x_0)} |\nabla u|^2 w\, dx + C\gamma R^2 \ . \end{aligned}$$

Now to evaluate the last integral in (3.27) we apply the Poincaré inequality (3.21) to the functions $(u-\hat{d})^{\pm}$, where $\hat{d} \in \left(w-\inf_{E_R(\varepsilon)} u,\ w-\sup_{E_R(\varepsilon)} u \right)$, $E_R(\varepsilon)$ is the level set of the obstacle ψ defined in (1.16).

As in [14] we can prove that there exists $\hat{d} \in (w-\inf_{E_R(\varepsilon)} u,\ w-\sup_{E_R(\varepsilon)} u)$ such that $w-\operatorname{cap}(E_R(\varepsilon),\ B_{2R}(x_0)) \le 4w-\operatorname{cap}(x \in B_R(x_0), (u-d)^{\pm} = 0,\ B_{2R}(x_0))$.

Hence using (3.21), (1.17), (1.14), (1.4) we obtain:

$$\begin{cases} \frac{R^2}{w(B_R(x_0))} \int_{B_R(x_0)} |\nabla(u-\hat{d})^{\pm}|^2 w\, dx \ge \\ \ge \frac{CR^2 w\text{-cap}(E_R(\varepsilon), B_{2R}(x_0))}{w^2(B_R(x_0))} \int_{B_R(x_0)} (|u-\hat{d}|^{\pm})^2 w\, dx \ge \\ \ge \frac{C\delta(R)}{w(B_R(x_0))} \int_{B_R(x_0)} |(u-\hat{d}-\varepsilon)^{\pm}|^2 w\, dx + C\,\varepsilon^2 \ . \end{cases} \tag{3.28}$$

On the other hand $d = \hat{d} + \varepsilon$ is admissible in Corollary 3.1, by the definition of the level set $E_R(\varepsilon)$ (see (1.16)). Hence by (3.13), (3.28) we find:

$$\frac{R^2}{w(B_R(x_0))}\int_{B_R(x_0)} |\nabla u|^2 w\, dx \geq C\frac{\delta(R)}{2}\operatorname{osc}^2_{B_{qR}(x_0)} u - C(R^2+\varepsilon^2) \ . \tag{3.29}$$

Let us now add on both sides of (3.27) the term $C_2 h \operatorname{osc}^2_{B_{qR}(x_0)} u$ where h is a positive 'big' constant that we shall choose below, and use (3.29). We obtain:

$$\begin{cases} (\lambda\gamma + C_2)\int_{B_{qR}(x_0)} |\nabla u|^2 G^{x_0}_{B_{2R}(x_0)} w\, dx + \\ +h(C_2 + (\gamma + C_7, \delta(R))h^{-1}) \underset{B_{qR}(x_0)}{osc}{}^2 u \leq \\ \leq h(C_2 + (\gamma(C_5+\gamma))h^{-1}) \operatorname{osc}^2_{B_R(x_0)} u + \\ +C_2 \int_{B_R(x_0)} |\nabla u|^2 G^{x_0}_{B_{2R_{q^{-1}}}(x_0)} w\, dx + C(\gamma+1)R^2 + C\varepsilon^2 \ . \end{cases} \tag{3.30}$$

Now we choose $\gamma = C_7\delta(R)(\lambda h - 1)^{-1}$, $h \geq 2(C_5 + C_7) \vee 2\lambda^{-1}$, and we devide both the sides of (3.30) by $C_2 + C_7\delta(R)(\lambda h-1)^{-1}$, if we denote by

$$k = \frac{C_7(\lambda h - 1)^{-1}}{2C_2 + C_7(\lambda h-1)^{-1}} \text{ and by } \mathcal{V}^2_h(r) = \int_{B_r(x_0)} |\nabla u|^2 G^{x_0}_{B_{2rq^{-1}}(x_0)} w\, dx + h \underset{B_r(x_0)}{osc^2}\, u$$

then we obtain

$$\mathcal{V}^2_h(qR) \leq \frac{1}{1+k\delta(R)}\mathcal{V}^2_h(R) + CR^2 + C\varepsilon^2 \ .$$

By an integration lemma (see [3], [14]) we prove that there exists $\beta > 0$ such that:

$$\mathcal{V}_h(r) \leq C\left\{\mathcal{V}_h(R)\exp\left(-\beta\int_r^R \delta(\rho)\frac{d\rho}{\rho}\right) + R + \varepsilon\right\}$$

and hence:

$$\mathcal{V}(r) \leq C\left\{\mathcal{V}(R)\exp\left(-\beta\int_r^R \delta(\rho)\frac{d\rho}{\rho}\right) + R + \varepsilon\right\} \ . \tag{3.31}$$

If we choose $\varepsilon = \sigma w_\sigma(r, R)$ (see Definition 1.1 and also Remark 1.3) then estimate (3.3) follows immediately by (3.31).

Finally we remark that (3.4) can be deduce by estimate (3.13) of the Corollary 3.1. The proof of Theorem 3.1 is now complete.

References

[1] A. Bensoussan — J. Frehse — U. Mosco: *A stochastic impulse control problem with quadratic growth Hamiltonian and the corresponding quasi-variational inequality*, J. Reine Ang. Math. **331** (1982), 124–145.

[2] I. Birindelli — M.A. Vivaldi: *Pointwise regularity for two-ubstacle problems involving degenerate-elliptic operators and quadratic Hamiltonian.* In preparation.

[3] M. Biroli — S. Marchi: *Wiener estimates for degenerate elliptic equation II*, to appear in Diff. Int. Eq.

[4] M. Biroli — U. Mosco: *Wiener criterion and potential estimates for obstacle problems relative to degenerate elliptic operators*, to appear in Ann. Mat. Pura Appl.

[5] L. Boccardo — F. Murat — J.P. Puel: *Existence de solutions faibles pour des équations elliptiques quasi-lineares à croissance quadratique.* Non linear partial differential equations and their applications, Collège de France Seminar, **IV** ed. by J.L. Lions and H. Brézis. Research Notes in Mathematics **84**, Pitman, London (1983), 19–73.

[6] G. Dal Maso — U. Mosco — M.A. Vivaldi: *A pointwise regularity theory for the two-obstacle problems*, Acta Math. **163** (1989), 57–107.

[7] D.E. Edmunds — L.A. Peletier: *A Harnack inequality for weak solutions of degenerate quasilinear elliptic equations*, J. London Math. Soc. **5** (1972), 21–31.

[8] E.B. Fabes — D.S. Jerison — C.E. Kenig: *The Wiener test for degenerate elliptic equations*, Ann. Inst. Fourier **32** (1982), 151–182.

[9] E.B. Fabes — C.E. Kenig — R.P. Serapioni: *The local regularity of solutions of degenerate elliptic equations*, Comm. P.D.E. 7 (1982), 77–116.

[10] T. Kipeläinen — W.P. Ziemer: *Pointwise regularity of solutions to nonlinear double obstacle problems*, preprint.

[11] S.N. Kruskov: *Certain properties of solutions to elliptic equations*, Soviet Mathematics **4** (1963), 686–695.

[12] S. Marchi: *Wiener estimates at boundary points for degenerate quasi-linear elliptic equations*, Istituto Lombardo (Rend. Sc.) **120** (1986), 17–33.

[13] J.H. Michael — W.P. Ziemer: *Interior regularity for solutions to obstacle problems*, Nonlinear Anal. T.M.A. **10** (1986), 1427–1448.

[14] U. Mosco: *Wiener criterion and potential estimates for the obstacle problem*, Indiana Un. Math. **3** (1987), 455–494.

[15] M.K.V. Murthy — G. Stampacchia: *Boundary value problems for some degenerate elliptic operators*, Ann. Mat. Pura. Appl. **80** (1968), 1–122.

[16] M. Struwe — M.A. Vivaldi: *On the Hölder-continuity of bounded weak solutions of quasi-linear parabolic inequalities*, Ann. Mat. Pura. Appl. **139** (1984), 175–190.

[17] N. Trudinger: *On the regularity of generalized solutions of linear non-uniformly elliptic equations*, Arch. Rat. Mech. Anal. **42** (1971), 51–62.

Acknowledgment: the author wishes to thank professors E. Fabes, C. Kenig and U. Mosco for many useful, interesting conversations.

M.A. Vivaldi
Dipartimento di Metodi e Modelli
Matematici per le Scienze Applicate
Università di Roma "La Sapienza"
00185 Roma - Italia

J VON BELOW

An existence result for semilinear parabolic network equations with dynamical node conditions

In this paper we investigate the classical global solvability for a class of semilinear parabolic equations (see (2)) on ramified networks under the following conditions at the vertices. At the ramification nodes we require continuity. Moreover, we assume the vertices to be partitioned into classes E_1, E_2 and E_3 corresponding to different transition conditions. In the first class we consider inhomogeneous Dirichlet conditions, while at the vertices in E_2 we impose a weighted Kirchhoff condition. At each vertex $e_i \in E_3$ we prescribe a dynamical node condition of the form

$$u_t(e_i,t) = \rho_i u(e_i,t) - \sum_j d_{ij} c_{ij} u_{jx_j}(e_i,t).$$

The properties of the nonlinearities in the differential equations will be such that they admit an a priori estimate of the time-derivative of a solution at the vertices in $E_2 \cup E_3$. In this way the application of classical estimates for domains established in [9] becomes possible. We then establish the classical solvability in the class $C^{2+\alpha,1+\frac{\alpha}{2}}$ with the aid of the Leray-Schauder-Principle.

1. Networks and Vertex Conditions

Let G denote a c^ν-network ($\nu \geq 3$) with finite sets of vertices $E = \{e_i | 1 \leq i \leq n\}$ and edges $K = \{k_j | 1 \leq j \leq N\}$ as defined in [2, Chap.2]. Thus G is the union of Jordan curves k_j in $\mathbb{R}^m$ with arc length parametrizations $\pi_j \in C^\nu([0,l_j], \mathbb{R}^m)$. The arc length parameter of an edge k_j is denoted by x_j. The topological graph Γ belonging to G is assumed to be simple and connected. Thus, by definition, $\Gamma = (E,K)$ consists in a collection of N Jordan curves k_j with the following properties: Each k_j has its endpoints in the set E, any two vertices in E can be connected by a path with arcs in K, and any two edges $k_j \neq k_h$ satisfy $k_j \cap k_h \subset E$ and $|k_j \cap k_h| \leq 1$. Endowed with the induced topology G is a connected and compact space in $\mathbb{R}^m$. The valency of each vertex is denoted by $\gamma_i = \gamma(e_i)$. We distinguish the ramification nodes $E_r = \{e_i \in E | \gamma_i > 1\}$ from the boundary vertices $E_b = \{e_i \in E | \gamma_i = 1\}$. The orientation of G is given by the incidence matrix $D = (d_{ij})_{n \times N}$ with

$$d_{ij} = \begin{cases} 1 & \text{if } \pi_j(l_j) = e_i, \\ -1 & \text{if } \pi_j(0) = e_i, \\ 0 & \text{otherwise.} \end{cases}$$

We introduce t as the time variable and for $T > 0$

$$\Omega = G \times [0,T], \quad \Omega_j = [0,l_j] \times [0,T], \quad \Omega_{jp} = (0,l_j) \times (0,T],$$

and for $u : \Omega \to \mathbb{R}$ we define $u_j = u \circ (\pi_j, id) : \Omega_j \to \mathbb{R}$ and use the abbreviations $u_j(e_i,t) := u_j(\pi_j^{-1}(e_i),t)$, $u_{x_j}(e_i,t) := \frac{\partial}{\partial x_j} u_j(\pi_j^{-1}(e_i),t)$ etc. The supremum norm for any set c is denoted by $|\cdot|_c^{(0)}$.

As special subspaces of $C(\Omega)$ we introduce for $\mu \in \mathbb{N}$, $\alpha \in [0,1), \mu + \alpha \leq \nu$

$$C^{\mu+\alpha,\frac{\mu+\alpha}{2}}(\Omega) = \{u \in C(\Omega) | \forall j \in \{1,...,N\} : u_j \in C^{\mu+\alpha,\frac{\mu+\alpha}{2}}(\Omega_j)\},$$

where $C^{\mu+\alpha,\frac{\mu+\alpha}{2}}(\Omega_j)$ with a Hölder norm $|\cdot|_{\Omega_j}^{(\mu+\alpha)}$ denotes the Banach space of functions u on Ω_j that have continuous derivatives $\frac{\partial^{r+s} u}{\partial t^r \partial x_j^s}$ for $2r+s \leq \mu$ and finite Hölder constants

$$\sum_{2r+s=\mu} H^{\alpha}_{x_j \Omega_j}\left(\frac{\partial^{r+s}}{\partial t^r \partial x_j^s} u\right) + \sum_{0<\mu+\alpha-2r-s<2} H_{t\Omega_j}^{\frac{\mu+\alpha-2r-s}{2}}\left(\frac{\partial^{r+s}}{\partial t^r \partial x_j^s} u\right)$$

in the case $\alpha > 0$. Again, $C^{\mu+a,\frac{\mu+\alpha}{2}}(\Omega)$ is a Banach space endowed with the norm

$$|u|_{\Omega}^{(\mu+\alpha)} = \sum_{j=1}^{N} |u_j|_{\Omega_j}^{(\mu+\alpha)}.$$

At each ramification node e_i we impose the continuity condition $k_j \cap k_s = \{e_i\} \Rightarrow u_j(e_i,t) = u_s(e_i,t)$, that clearly is contained in the condition $u(\cdot,t) \in C(G)$. We decompose the vertex set E into three disjoint parts

$$E = E_1 \uplus E_2 \uplus E_3 \qquad n_i := |E_i|$$

with respect to different transition conditions. At the vertices in E_1 we consider Dirichlet conditions of the form

$$u(e_i,t) = \psi(e_i,t) \qquad \text{in} \quad E_1 \times [0,T],$$

while at the vertices in E_2 we impose classical Kirchhoff laws

$$\sum_{j=1}^{N} d_{ij} c_{ij} u_{jx_j}(e_i,t) = 0 \qquad \text{in} \quad E_2 \times [0,T], \tag{1a}$$

with positive conductivity factors c_{ij}. At the vertices in E_3 we prescribe dynamical transition conditions

$$u_t(e_i,t) = \rho_i u(e_i,t) - \sum_{j=1}^{N} d_{ij} c_{ij} u_{jx_j}(e_i,t) \qquad \text{in} \quad E_3 \times [0,T] \tag{1b}$$

with c_{ij} as above and arbitrary constants ρ_i. If a function u satisfies (1a) and (1b), it is said to fulfill a generalized Kirchhoff condition, (GK) for short. For function spaces on Ω let the subscript GK indicate the validity of (GK) in $(E_2 \uplus E_3) \times [0,T]$. We may assume that

$$E_r \subseteq E_2 \uplus E_3 \quad \text{and} \quad E_1 \subseteq E_b,$$

since a Dirichlet condition at a ramification node e_i corresponds to the same Dirichlet condition at γ_i boundary vertices. Note that for vertices in $E_b \cap E_2$ condition (1a) is just the Neumann boundary condition. We define the parabolic network interior Ω_p and the parabolic network boundary ω_p with respect to the decomposition $E = E_1 \uplus E_2 \uplus E_3$ as

$$\Omega_p = (G \backslash E_1) \times (0,T] \qquad \omega_p = (G \times \{0\}) \cup (E_1 \times (0,T]).$$

Note that E_1 can be empty. An existence result for semilinear equations in the case $E_2 = E_r$ has been obtained in [5]. An example arising in neurobiology including a vertex condition (1b) in a tree as in Fig.1 is given in [3], further remarks on the (GK) condition can be found in [4] and [6]. Other types of nondynamical transition conditons are considered in [1].

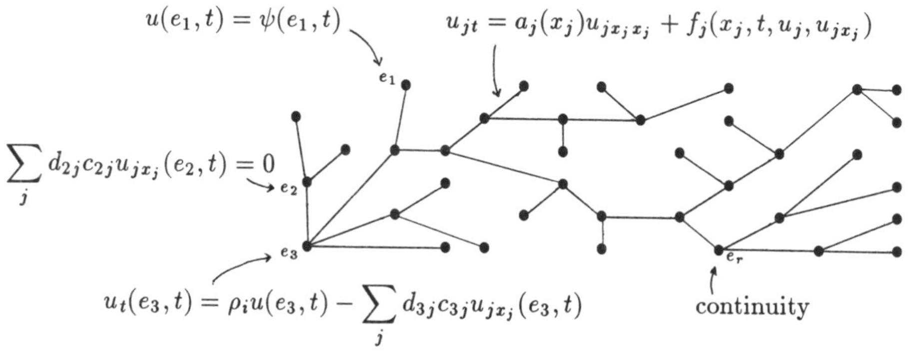

Fig.1

2. The Initial Boundary Value Problem

For a given function $\psi \in C^{2,1}(\Omega)$ we want to solve the IBVP (2) on Ω

$$(2) \quad \begin{cases} u \in C^{2,1}_{GK}(\Omega) & (2a) \\ u_{jt} = a_j(x_j)u_{jx_jx_j} + f_j(x_j,t,u_j,u_{jx_j}) \text{ in } \Omega_{jp} \quad \text{for } 1 \le j \le N & (2b) \\ u|_{\omega_p} = \psi|_{\omega_p} & (2c) \end{cases}$$

under the following conditions (3) - (8): There are positive constants $\mu_1, \mu_2, K_1, L_1, K_2$ and L_2 such that for all $j \in \{1,...,N\}$

$$0 < \mu_1 \le a_j(x) \le \mu_2 \qquad \text{in} \quad [0,l_j], \tag{3}$$

$$a_j \in C^1([0,l_j]), \tag{4}$$

$$zf_j(x,t,z,0) \le K_1 z^2 + L_1 \qquad \text{in} \quad \Omega_j \times \mathbb{R}, \tag{5}$$

$$f_j \in C(\Omega_j \times [-M_0, M_0] \times \mathbb{R}), \tag{6}$$

$$|f_j(x,t,z,p)| \le K_2 p^2 + L_2 \qquad \text{in} \quad \Omega_j \times [-M_0, M_0] \times \mathbb{R}, \tag{7}$$

where

$$b := \max_{e_i \in E_3} \rho_i \quad \text{and} \quad M_0 := \inf_{\lambda > K_1, \lambda \ge b} \Big(e^{\lambda T} \max\Big\{ |\psi|_\Omega^{(0)}, \sqrt{\frac{L_1}{\lambda - K_1}} \Big\} \Big).$$

(8) All f_j are differentiable with respect to $z = u_j$ and $p = u_{jx_j}$ and locally Lipschitz continuous with respect to t in $\Omega_j \times [-M_0, M_0] \times \mathbb{R}$, and $\left| \frac{f_j(x,t+h,z,p) - f_j(x,t,z,p)}{h} \right|$ for $0 \le t+h \le T, h \ne 0$ and $\frac{\partial f_i}{\partial z}$ and $\frac{\partial f_j}{\partial p}$ remain bounded on bounded subsets of $\Omega_j \times [-M_0, M_0] \times \mathbb{R}$.

We may assume $\max_{1 \le j \le N} |a_{jx_j}|_{[0,l_j]}^{(0)} \le L_2$. Note that by [4, Thm.2] any solution u of (2) satisfies

$$|u|_\Omega^{(0)} \le M_0.$$

The conditions (3) - (7) allow the application of the interior estimate of $|u_x|$ given in [9, V. Thm. (3.1)]. For each $d \in (0, \frac{1}{2} \min_j l_j)$ an estimate

$$\max_{1 \le j \le N} |u_{jx_j}|_{[d, l_j - d] \times [0,T]}^{(0)} \le M_1(d) \tag{9}$$

holds with a constant $M_1(d) = M_1(d; \mu_1, \mu_2, M_0, K_2, L_2, \Omega, |\psi|_\Omega^{(1)})$. Furthermore, set

$$M_2 = (\mu_2 + K_2 + 1)|\psi|_\Omega^{(2)} + L_2.$$

Finding a solution of (2) consists mainly of establishing an a priori estimate of $|u|_\Omega^{(1+\beta)}$ in order to apply the Leray-Schauder-Principle. Under the conditions imposed at $E_2 \uplus E_3$ the crucial step is to find an estimate of $|u_t|$ in

$$R = \big(E_2 \uplus E_3\big) \times [0,T].$$

Lemma 1: *Suppose conditions (3) - (8) are fulfilled. Let M^* be a constant such that any solution of (2) satisfies the estimate*

$$\max_R |u_t| \le M^*. \tag{10}$$

Then there is a constant M_3 depending only on $\mu_1, \mu_2, K_2, L_2, M_0, M^*, \Omega, |\psi|_\Omega^{(2)}$, and $\max_j |a_{jx_j}|_{\Omega_j}^{(0)}$, such that

$$|u|_\Omega^{(2)} \le M_3. \tag{11}$$

Proof: By (3) - (7) and [9, V. Thm.(4.2)] there is a constant $M_4 = M_4(\mu_1, \mu_2, K_2, L_2, M_0, M^*, \Omega, |\psi|_\Omega^{(2)})$ such that

$$\max_{1\le j\le N} |u_{jx_j}|_{\Omega_j}^{(0)} \le M_4.$$

By (8), (10) and [9, V. Thm.(5.1)] the norm $|u_t|_\Omega^{(0)}$ can be estimated from above by a constant $M_5 = M_5\left(\mu_1, \mu_2, M_0, M_4, M^*, K_4, |\psi_t|_\Omega^{(0)}\right)$, where K_4 is an upper bound for $|\frac{\partial f_j}{\partial z}|$, $|\frac{\partial f_j}{\partial \rho}|$ and

$$\sup_{0\le t+h\le T, h\ne 0}\left|\frac{f_j(x,t+h,z,p)-f_j(x,t,z,p)}{h}\right| \quad \text{in} \quad \Omega_j\times[-M_0,M_0]\times[-M_4,M_4]$$

for all f_j. Finally, the differential equations (2b) and condition (4), (6), and (7) yield a constant $M_3 = M_3(\mu_1, \mu_2, K_2, L_2, M_0, M^*, M_4, \Omega)$ such that $|u|_\Omega^{(2)} \le M_3$. ◇

3. An a priori Bound for $|\ u_t\ |$ on R

For the sake of simplicity we establish the existence of an a priori bound M^* only under the restriction (12a) that we assume to hold in a neighbourhood of the vertices:

(12) There is a number $\varepsilon > 0$ and a constant $K_3 \ge b$ such that for each f_j restricted to $([0,\varepsilon]\cup[l_j-\varepsilon, l_j])\times[0,T]\times[-M_0,M_0]\times\mathbb{R}$:

$$\frac{\partial}{\partial z} f_j(x_j,t,z,p), \quad \sup_{0\le t+h\le T, h\ne 0}\left|\frac{f_j(x_j,t+h,z,p)-f_j(x_j,t,z,p)}{h}\right| \le K_3. \tag{12a}$$

With (8) we can assume K_3 be chosen such that for each f_j restricted to $[\varepsilon, l_j-\varepsilon]\times[0,T]\times[-M_0,M_0]\times[-M_1(\varepsilon),M_1(\varepsilon)]$:

$$\left|\frac{\partial}{\partial z} f_j(x_j,t,z,p)\right|, \left|\frac{\partial}{\partial p} f_j(x_j,t,z,p)\right|,$$

$$\sup_{0\le t+h\le T, h\ne 0}\left|\frac{f_j(x_j,t+h,z,p)-f_j(x_j,t,z,p)}{h}\right| \le K_3. \tag{12b}$$

Clearly, (12a) is not needed in E_1. Due to (9), $\varepsilon > 0$ can be chosen arbitrarily small. While conditions (12b) and (3) - (8) are natural, (12a) is of course restrictive. Anyhow, it includes nonlinearities of the form $f_j(x,t,z,p) = g_j(x,t,z) + h_j(x,p) + m_j(z)|d_j(x,p)|$ subject to $\frac{\partial}{\partial z} m_j(z) \le 0$ and to the natural conditions (3) - (8).

Lemma 2: *Suppose conditons (3) - (8) and (12) are satisfied. Let u be a solution of (2). Then $|u_t|_R|$ can be estimated from above by a constant M^* that depends only on $\mu_1, \mu_2, \varepsilon, \Omega, M_0, M_1(\varepsilon/4), M_2, L_2, K_2$ and K_3.*

Proof: By (2) - (4) and (6) we have $|u_t(\cdot,0)|_G^{(0)} \leq M_2$. By (8), (9) and the local estimate of $|u_t|$ in [9, V. Thm. (5.1)] we find a constant $M_8 = M_8(\mu_1, K_3, M_2, \varepsilon/4, M_1(\varepsilon/4), L_2, K_2, M_0)$ such that

$$\max_j |u_{jt}|^{(0)}_{[\varepsilon/2, l_j - \varepsilon/2]\times[0,T]} \leq M_8.$$

The global estimate in [9, V. Thm.(5.1)] yields the assertion at vertices in E_1. Therefore we may assume without restriction that Γ is a star graph with

$$E_r \subseteq E_2 \uplus E_3 = \{e_i\},\quad N = \gamma_i \geq 1,\quad l_j = \varepsilon,\quad \pi_j(0) = e_i \quad \text{for all} \quad j \in \{1, ..., \gamma_i\}.$$

For $h > 0$ sufficiently small set $\Omega_{T-h} = G \times [0, T-h]$, and define $v^h \in C(\Omega_{T-h})$ by the divided differences

$$v_j^h(x_j, t) = \frac{u_j(x_j, t+h) - u_j(x_j, t)}{h}.$$

Furthermore, set

$$\Xi = \Big(x_j, t+h, \tau u_j(x_j, t+h) + (1-\tau) u_j(x_j,t), \tau u_{jx_j}(x_j, t+h) + (1-\tau) u_{jx_j}(x_j, t)\Big),$$

$$B_j^h(x_j,t) = \int_0^1 \frac{\partial}{\partial p} f_j(\Xi) d\tau, \qquad C_j^h(x_j,t) = \int_0^1 \frac{\partial}{\partial z} f_j(\Xi) d\tau,$$

$$F_j^h(x_j,t) = \frac{1}{h} f_j\Big(x_j, t+h, u_j(x_j,t), u_{jx_j}(x_j,t)\Big) - \frac{1}{h} f_j\Big(x_j, t, u_j(x_j,t), u_{jx_j}(x_j,t)\Big).$$

Then v^h belongs to $C^{2,1}_{GK}(\Omega_{T-h})$ and satisfies the linear equation

$$\frac{\partial}{\partial t} v_j^h = a_j v^h_{jx_jx_j} + B_j^h v^h_{jx_j} + C_j^h v_j^h + F_j^h \quad \text{in} \quad \Omega_{j,T-h}.$$

By (12a) we have for any $(x_j, t, z) \in \Omega_{j,T-h} \times \mathbb{R}$

$$z^2 C_j^h(x_j,t) + z F_j^h(x_j,t) \leq 2K_3 z^2 + K_3.$$

Thus the estimate [4, Thm.2] can be applied to v^h in Ω_{T-h}. With $b \leq K_3$ we conclude

$$|v^h|^{(0)}_{\Omega_{T-h}} \leq e^{3K_3T} \max\Big\{1, \max_{t \leq T-h, (x,t)\in\omega_p} |v^h(x,t)|\Big\}.$$

As h tends to zero, the *r.h.s.* goes to an expression that is bounded from above by $M^* = e^{3K_3T}\max\{1, M_2, M_8\}$

Thus $|u_t|^{(0)}_\Omega \leq M^*$. $\diamondsuit$

4. The Linear Case

In [7,III] the classical solvability in $C_{GK}^{2+\alpha,1+\frac{\alpha}{2}}(\Omega)$ is established for general linear parabolic equations on networks with general inhomogeneous dynamical and nondynamical linear vertex conditions. The techniques are the same as the ones used in [2]. These problems include as a special case the IBVP

$$(13)\quad \begin{cases} u \in C_{GK}^{2,1}(\Omega) & (13a)\\ u_{jt} = a_j(x_j)u_{jx_jx_j} + g_j(x_j,t) \text{ in } \Omega_{jp} \quad \text{for } 1 \le j \le N & (13b)\\ u|_{\omega_p} = \psi|_{\omega_p}. & (13c)\end{cases}$$

Besides the compatibility conditions contained in the condition $\psi \in C_{GK}^{2,1}(\Omega)$ (instead of (GK), (1a) would be sufficient), we need the following compatibility condition (14):

$$\psi_t(e_i,0) = a_j(e_i)\psi_{jx_jx_j}(e_i,0) + g_j(e_i,0) \quad \text{for} \quad e_i \in E_1,\ d_{ij} \ne 0 \qquad (14a)$$

$$a_j(e_i)\psi_{jx_jx_j}(e_i,0) + g_j(e_i,0) = a_s(e_i)\psi_{sx_sx_s}(e_i,0) + g_s(e_i,0) \text{ for } k_j \cap k_s = \{e_i\} \qquad (14b)$$

$$a_s(e_i)\psi_{jx_sx_s}(e_i,0) + g_s(e_i,0) = \rho_i\psi(e_i,0) - \sum_{j=1}^{N} d_{ij}c_{ij}\psi_{jx_j}(e_i,0)$$

$$\text{for } e_i \in E_3,\ d_{is} \ne 0 \qquad (14c)$$

Theorem 1 ([7]): *Assume all a_j satisfy (3) and (4) and all g_j belong to $C^{\alpha,\frac{\alpha}{2}}(\Omega_j)$. Suppose ψ belongs to $C_{GK}^{2+\alpha,1+\frac{\alpha}{2}}(\Omega)$ and fulfills the compatibility condition (14). Then problem (13) has a unique solution $u \in C_{GK}^{2+\alpha,1+\frac{\alpha}{2}}(\Omega)$. It satisfies the estimate*

$$|u|_\Omega^{(2+\alpha)} \le K_0\Big(\sum_{j=1}^{N} |g_j|_\Omega^{(\alpha)} + |\psi|_\Omega^{(2+\alpha)}\Big), \qquad (15)$$

where the constant K_0 depends only on all a_j, c_{ij}, ρ_i and on Ω.

Remark: Using the eigenvalue approach in [3], the dissipativity result in [10], the perturbation results for analytic semigroups [8,I.6] and the variation of constants formula [8,II.1], we can also achieve the solvability of the special linear problem (13) in the class $C_{GK}^{2,1}(\Omega) \cap \{u \mid u|_{E_1} \equiv 0\}$. We then use standard techniques in order to handle the general inhomogeneous Dirichlet condition in E_1 too.

5. The Existence Theorem in the Semilinear Case

Before stating the main result, we introduce the semilinear compatibility condition:

$$\text{Condition (14) holds for } \psi \text{ and } g_j(e_i,0) := f_j\Big(e_i, 0, \psi_j(e_i,0), \psi_{jx_j}(e_i,0)\Big) \qquad (16)$$

As noted above, it is no restriction to assume that ψ satisfies (GK), though only condition (1a) is needed. Let the constant M_4 be defined as in the proof of Lemma 1.

Theorem 2: *Let the diffusion coefficients a_j and the nonlinearities f_j satisfy conditions (3) - (8) and (12). Let the compatibility conditions (16) be satisfied. Suppose that for some $\alpha \in (0,1)$ and all $j \in \{1, ..., N\}$*

$$f_j \in C^{\alpha,\frac{\alpha}{2},\alpha,\alpha}(\Omega_j \times [-M_0, M_0] \times [-M_4, M_4])$$

and

$$\psi \in C_{GK}^{2+\alpha,1+\frac{\alpha}{2}}(\Omega).$$

Then problem (2) has a unique solution $u \in C_{GK}^{2+\alpha,1+\frac{\alpha}{2}}(\Omega)$.

Proof: Extend each f_j such that $f_j \in C^{\alpha,\frac{\alpha}{2},\alpha,\alpha}(\Omega_j \times \mathbb{R}^2)$. Let B denote the Banach space

$$B = \left(C^{2,1}(\Omega), |\cdot|_\Omega^{(2)}\right).$$

For $u \in B$ there exists by Theorem 1 a unique solution

$$v = \theta(u) \in C_{GK}^{2+\alpha,1+\frac{\alpha}{2}}(\Omega)$$

of the linear problem

$$(17) \qquad \begin{cases} v \in C_{GK}^{2,1}(\Omega), \\ v_{jt} = a_j v_{jx_jx_j} + f_j(x_j, t, u_j, u_{jx_j}) \\ \qquad -f_j\big(x_j, 0, u_j(x_j, 0), u_{jx_j}(x_j, 0)\big) \\ \qquad +f_j\big(x_j, 0, \psi_j(x_j, 0), \psi_{jx_j}(x_j, 0)\big) \quad \text{in } \Omega_{jp} \text{ for } 1 \le j \le N, \\ v\big|_{\omega_p} = \psi\big|_{\omega_p}. \end{cases}$$

In this way a nonlinear operator θ in the Banach space B is defined. Since the modified nonlinearities

$$f_j(x_j, t, z, p) - f_j\big(x_j, 0, \lambda\psi_j(x_j, 0), \lambda\psi_{jx_j}(x_j, 0)\big) + f_j\big(x_j, 0, \psi_j(x_j, 0), \psi_{jx_j}(x_j, 0)\big)$$

satisfy the same conditions as f_j up to adding a constant depending only on K_2, L_2 and $|\psi|_\Omega^{(2)}$ to K_1, L_1 and L_2, we find by Lemmata 1 and 2 that the set

$$\left\{u \in B | \exists \lambda \in [0,1] : u = \lambda\theta(u)\right\}$$

is bounded in the norm $|\cdot|_\Omega^{(2)}$ by a constant M not depending on λ.

For the mappings

$$\rho_j(u) = \Big((x_j, t) \longmapsto f_j\big(x_j, t, u_j(x_j, t), u_{jx_j}(x_j, t)\big)\Big)$$

we find constants H_1 and H_2 depending only on K_2, L_2 and the Hölder constants of all f_j such that

$$\sum_{j=1}^N |\rho_j(u)|_\Omega^{(\alpha)} \le H_1 |u|_\Omega^{(2)} + H_2.$$

By the $(2+\alpha)$-estimate (15) for the equation (17) we have for $u \in B$

$$|\theta(u)|_{\Omega}^{(2+\alpha)} \leq K_0\Big(|\psi|_{\Omega}^{(2+\alpha)} + \sum_{j=1}^{N} |\rho_j(u)|_{\Omega_j}^{(\alpha)} + \sum_{j=1}^{N}\big(|\rho_j(u)|_{t=0}|_{[0,l_j]}^{(\alpha)} + |\psi_j|_{t=0}|_{[0,l_j]}^{(\alpha)}\big)\Big) \tag{18}$$
$$\leq K_5(1 + |u|_{\Omega}^{(2)}),$$

where the constant K_5 depends only on K_0, H_1, H_2 and $|\psi|_{\Omega}^{(2+\alpha)}$. This shows that $\theta : B \to B$ is a compact operator, since the embeddings

$$C^{\lambda+\beta,\frac{\lambda+\beta}{2}}(\Omega) \hookrightarrow C^{\mu+\alpha,\frac{\mu+\alpha}{2}}(\Omega)$$

are compact for $\mu + \alpha < \lambda + \beta \leq \nu$. We note in passing that the embeddings remain compact with the subscripts (1a), GK for $1 \leq \mu + \alpha$, $2 \leq \mu + \alpha$ respectively.

As for the continuity of θ, we observe that for a sequence $U = \{u_k | k \in \mathbb{N}\}$ with $|\cdot|_{\Omega}^{(2)} - \lim_{k\to\infty} u_k = u \in B$ the sequence $V = \theta(U)$ contains by (18) a convergent subsequence in B. Any $\{v_m | m \in \mathbb{N}\} \subseteq V$ with $|\cdot|_{\Omega}^{(2)} - \lim_{m\to\infty} v_m = v \in B$ contains a $|\cdot|_{\Omega}^{(2+\delta)}$ - convergent subsequence for $0 \leq \delta < \alpha$, whose limit has to be v. Therefore v belongs to $C_{GK}^{2+\delta,1+\delta/2}(\Omega)$. By the continuity assumption (6) we have for $v_m = \theta(u_{k_m})$

$$\lim_{m\to\infty} \rho_j(u_{k_m})(x_j, t) = \rho_j(u)(x_j, t) \quad \text{for all} \quad (x_j, t) \in \Omega_j.$$

The uniqueness result of [6] applied to (17), v and $\theta(u)$ shows that $v = \theta(u)$ and that $|\cdot|_{\Omega}^{(2)} - \lim_{k\to\infty} \theta(u_k) = \theta(u)$. Thus $\theta : B \to B$ is shown to be continuous.

By the Leray-Schauder Theorem in the special case [11] there exists a solution $u = \theta(u)$ of problem (2) with

$$u \in C_{GK}^{2+\alpha,1+\frac{\alpha}{2}}(\Omega).$$

Since we have the inequality

$$\frac{\partial}{\partial z} f_j(x_j, t, z, p) \leq K_4 \quad \text{in} \quad \Omega_j \times [-M_0, M_0] \times [-M_4, M_4],$$

we obtain a one-sided Lipschitz condition in z by

$$f_j(x, t, w, p) - f_j(x, t, z, p) = (w - z) \int_0^1 \frac{\partial}{\partial z} f_j(x, t, \zeta w + (1-\zeta)z, p) d\zeta.$$

Thus u is the unique solution of (2) in the class $C_{GK}^{2,1}(\Omega)$ by [6, Cor.]. $\diamond$

References:

[1] F. Ali Mehmeti and S. Nicaise, Nonlinear interaction problems, *Pub. IRMA, Lille* Vol.23, No.V, 1991.

[2] J. v. Below, Classical solvability of linear parabolic equations on networks, *J. Differential Equ.* **72** (1988) 316-337.

[3] J. v. Below, Sturm-Liouville eigenvalue problems on networks, *Math. Meth. Applied Sciences* **10** (1988) 383-395.

[4] J. v. Below, A maximum principle for semilinear parabolic network equations, in: J. A. Goldstein, F. Kappel, and W. Schappacher (eds.): Differential equations with applications in biology, physics, and engineering, *Lect. Not. Pure and Appl.Math.* Vol. 133, M. Dekker Inc. New York 1991, pp. 37-45.

[5] J. v. Below, An existence result for semilinear parabolic network equations, *Semesterbericht Funktionalanalysis Tübingen* **15** (1988/89) 33-41.

[6] J. v. Below, Comparison theorems for parabolic network equations with dynamical node conditions, *Semesterbericht Funktionalanalysis Tübingen*, **18**(1990)17-22.

[7] J. v. Below, Parabolic network equations, *to appear.*

[8] J. A. Goldstein, Semigroups of linear operators and applications, Oxford University Press New York 1985.

[9] O. A. Ladyženskaja, V. A. Solonnikov, and N. N. Ural'ceva, Linear and quasilinear equations of parabolic type, *Amer. Math. Soc. Providence RI*, 1968.

[10] G. Lumer, Connecting of local operators and evolution equations on networks, in: Potential Theory Copenhagen 1979, (Proceedings), *Lect. Not. Math.* Vol. 787, Springer - Verlag Berlin 1980, pp. 219-234.

[11] H. H. Schaefer, Über die Methode der a priori- Schranken, *Math. Ann.* **129** (1955) 415-416.

Joachim von Below
Lehrstuhl für Biomathematik, Universität Tübingen
Auf der Morgenstelle 10
D - W 7400 Tübingen 1
Germany

HONG-MING YIN

Remarks on regularity of the interface in the heat equation with strong absorption

1. Introduction: Consider the following heat equation with strong absorption:

$$u_t = \Delta u - \lambda u^p, \qquad (x,t) \in G = \Omega \times (0,\infty), \tag{1.1}$$

$$u(x,t) = 1, \qquad (x,t) \in S = \partial\Omega \times (0,\infty), \tag{1.2}$$

$$u(x,0) = 1, \qquad x \in \Omega, \tag{1.3}$$

where Ω is a bounded domain in R^n with smooth boundary $\partial\Omega$ while $p \in (0,1)$ and $\lambda > 0$ are given constants.

It is known [7] that the problem admits a unique classical solution $u(x,t)$. Moreover, for larege λ, there exists a region D, called dead core, such that the solution becomes identically zero on D. The boundary of the dead core is called free boundary or the interface. We are interested in the regularity of the interface. The standard parabolic equation theory implies that the best regularity of $u(x,t)$ is

$$u(x,t) \in C^{2+\frac{2p}{1-p},1+\frac{2p}{1-p}}(G).$$

For $\alpha > 0$ small, assume that the level set $\{(x,t) : u(x,t) = \alpha\}$ can be represented by a differentiable function $t = f(x)$. Formally, we differentiate the equation $u(x, f(x)) = \alpha$ to obtain

$$f_{x_i}(x) = -\frac{u_{x_i}(x, f(x))}{u_t(x, f(x))}.$$

This indicates that the best regularity of the interface is in $C^{1+\frac{p}{1-p}}(\Omega)$. At the same time, the above intuitive observation suggests us the way to prove the Lipschitz continuity of the interface.

There are some previous results on the subject. Kennington in [6] showed that the dead core is convex if the domain Ω is convex. In his proof, the convexity condition plays a crucial role. Moreover, the positivity on the fixed boundary is also necessary since with the zero value on the boundary, the dead core is no longer convex. His argument seems difficult to be applied for the Cauchy problem. The continuity of the interface for the on-dimensional Cauchy problem was obtained by Chen, Matano and Mimura [2]. Recently, the author of [9] proved that the interface for the problem (1.1)-(1.3) is Lipschitz continuous without convexity assumption on the domain. However, the positivity condition is still needed. In the present note we shall indicate that how our method can be used to study the regularity of the interfaces for the Cauchy problem and some other type of problems. We show that the interface in the n-dimensional Cauchy problem is Lipschitz continuous. For the one dimensional problem with the mixed Dirichlet-Neumann boundary conditions, we have a partial answer (see Theorem 2 in Section 2 for details).

2. The Main Results: We begin the following Cauchy problem:

$$u_t = \Delta u - u^p, \qquad in\ R^n \times (0,\infty), \tag{2.1}$$

$$u(x,0) = u_0(x), \qquad x \in R^n. \tag{2.2}$$

Under some mild conditions on the initial value $u_0(x)$ (cf. [3]), the problem (2.1)-(2.2) has a classical solution which becomes identically zero after a finite time T. T is called the extinction time. To study the regularity of the interface, we need the following assumption.

Hypothesis (A): $u_0(x) \in C(R^n)\bigcap C^2(\Omega_s)$ and $u_0(x) \geq 0$ is uniformly bounded. Moreover, there exists a constant $k_0 > 0$ such that

$$\Delta u_0(x) - u_0(x)^p \leq -k_0, \text{ and } u_{0x_i}(x)\ (0 \leq i \leq n) \text{ is uniformly bounded} \tag{2.3}$$

for $x \in \Omega_s = \{x \in R^n : u_0(x) \neq 0\}$.

It is clear that at least $u_0(x) = constant$ satisfies the condition (2.3). Note that the above condition allows that $u_0(x)$ has a compact support. Our first result is

Theorem 1: Under the above assumption, the interface is Lipschitz continuous.

Next result is concerned with the following mixed boundary conditions:

$$u_t = u_{xx} - u^p, \quad (x,t) \in G = \{0 < x < 1, 0 < t < \infty\} \tag{2.4}$$

$$u_x(0,t) = 0,\ u(1,t) = g(t), \qquad t \in [0,\infty), \tag{2.5}$$

$$u(x,0) = u_0(x), \qquad x \in [0,1]. \tag{2.6}$$

Again under the proper conditions on known data, the problem (2.4)-(2.6) admits a unique classical solution. Moreover, there exists a finite time $T > 0$ such that $u(x,t)$ becomes identically zero for $t \geq T$ (cf.[4]).

Hypothesis (B): $u_0(x) \in C^2[0,1]$ and $u_0 \geq 0$; there exists a positive constant $k_1 > 0$ such that $u_0''(x) - u_0(x)^p < -k_1$. $g(t) \in C^{1+\frac{p}{1-p}}[0,\infty)$ and $g(t) \geq 0,\ g'(t) \leq 0$ and $g(t) \equiv 0$ for $t \geq T$. The consistency conditions hold: $u_0'(0) = 0,\ u_0(1) = g(0)$.

With the above assumption, we have:

Theorem 2 : The interface in $[0, x^*)$ is Lipschitz continuous, where x^* is any position such that (x^*, T) is the extinction point.

3. Proofs: Consider the following regularized problem $(P)_\varepsilon$:

$$u_t = u_{xx} - u^p + \varepsilon^p, \qquad (x,t) \in G = R^n \times (0,\infty), \tag{3.1}$$

$$u(x,0) = u_{0\varepsilon}(x), \qquad x \in R^n, \tag{3.2}$$

where $u_{0\varepsilon}(x)$ is the smooth approximation of $u_0(x)$ with

$$u_{0\varepsilon}(x) \geq \varepsilon, \ \Delta u_{0\varepsilon}(x) - u_{0\varepsilon}(x) < 0 \text{ and } \frac{u_{0\varepsilon x_i}(x)}{\Delta u_{0\varepsilon}(x) - u_{0\varepsilon}(x)^p}$$

is uniformly bounded.

Note that the problem $(P)_\varepsilon$ admits a lower solution ε and an upper solution $M = ||u_0||_0 + 1$. We have

$$\varepsilon \leq u_\varepsilon(x,t) \leq M.$$

By Friedman [5], $u_\varepsilon(x,t)$ converges to $u(x,t)$ uniformly over G as $\varepsilon \to 0$.

Lemma 3.1: $u_{\varepsilon t}(x,t) < 0$ in G.

This can be shown by differentiating the equation (3.1) with respect to t and using the positivity of the fundamental solution.

Now for any fixed $\alpha > 0$ small ($\alpha < ||u_0||_0$), we consider the level set

$$\Gamma_\varepsilon^\alpha = \{(x,t) \in G : u_\varepsilon(x,t) = \alpha.\}$$

It is clear that Γ_ε is non-empty. As $u_{\varepsilon t}(x,t) < 0$, the implicit function theorem implies that there exists a function, denoted by $f_\varepsilon^\alpha(x)$, such that $\Gamma_\varepsilon^\alpha$ is the graph of the function $t = f_\varepsilon^\alpha(x)$. On G, we define

$$w_i(x,t) = \frac{u_{\varepsilon x_i}(x,t)}{u_{\varepsilon t}(x,t)}, \qquad (1 \leq i \leq n).$$

By a direct calculation, $w_i(x,t)$ satisfies

$$w_{it} - \Delta w_i - \frac{2\nabla v}{v}\nabla w_i = 0,$$

$$w_i(x,0) = \frac{u_{0\varepsilon x_i}(x)}{\Delta u_{0\varepsilon}(x) - u_{0\varepsilon}(x)^p},$$

where $v(x,t) = u_{\varepsilon t}(x,t)$.

By the hypothesis (A), $w_i(x,0)$ is uniformly bounded. Hence the solution $w_i(x,t)$ is also uniformly bounded. We differentiate with respect to x_i for $u_\varepsilon(x, f_\varepsilon(x)) = \alpha$ to obtain

$$u_{\varepsilon x_i} + u_{\varepsilon t} f_{\varepsilon x_i}^\alpha(x) = 0.$$

Hence,
$$f^{\alpha}_{\varepsilon x_i}(x) = -\frac{u_{\varepsilon x_i}(x, f_\varepsilon(x))}{u_{\varepsilon t}(x, f_\varepsilon(x))},$$
which is uniformly bounded for any $x \in R^n$ and any i .

For any fixed $\alpha \in (0,1)$ we first extract a subsequence which converges to $f^{\alpha}(x)$. We then take a subsequence $f^{\alpha_k}(x)$ of $\{f^{\alpha}(x)\}$, which converges to a function (denoted by $f(x)$) as α_k tends to 0. Note that $u(x, f^{\alpha}(x)) = \alpha$. The uniqueness indicates that $t = f(x)$ is the boundary of the dead core. The proof of Theorem 1 is completed.

Proof of Theorem 2: To prove Theorem 2, we need the following Lemma which is similar to the result obtained by Stakgold in [7]:

Lemma 3.2: The solution is monotonically decreasing in t. Moreover, if $u(x^*, t^*) > 0$ for some $t^* > 0$ and $x^* \in R$, then there exists a positive constant δ which depends on (x^*, t^*) such that $u_t(x^*, t) \le -\delta$ for $t \in [0, t^*]$.

The proof is the same to that of [7]. We do not repeat here.

Let (x^*, T) be a extinction point. We may assume that $x^* > 0$ (otherwise it is trivial). Let $T^* < T$ be a number very close to T. By Lemma 3.2, we know that
$$u_t(x^*, t) \le -\delta, \qquad \text{for all } t \in [0, T^*].$$

We consider the regularized problem:
$$u_t = u_{xx} - u^p + \varepsilon^p, \quad (x,t) \in G = \{0 < x < 1, 0 < t < \infty\} \tag{3.3}$$
$$u_x(0,t) = 0,\ u(1,t) = g(t) + \varepsilon, \qquad t \in (0, \infty), \tag{3.4}$$
$$u(x, 0) = u_0(x) + \varepsilon, \qquad 0 < x < 1. \tag{3.5}$$

It is easy to see that $\varepsilon \le u_\varepsilon(x,t) \le M = ||u_0||_0 + 1$.

Moreover, the Schauder theory implies
$$||u - u_\varepsilon||_{C^{2,1}(0,1)\times(0,T]} \to 0, \qquad as\ \varepsilon \to 0.$$

Let $Q^{x^*}_{T^*} = \{(x,t) : 0 < x < x^*,\ 0 < t < T^*\}$.

Lemma 3.3: $u_{\varepsilon t}(x,t) < 0$ on $\bar{Q}^{x^*}_{T^*}$.

Proof: Let $v(x,t) = u_{\varepsilon t}(x,t)$. Then $v(x,t)$ satisfies
$$v_t = v_{xx} - pu_\varepsilon^{p-1} v,$$
$$v_x(0,t) = 0, v(x^*, t) < 0,$$
$$v(x, 0) = u_0''(x) - u_0^p(x) + \varepsilon^p$$

By the hypothesis (B), $v(x,0) < 0$ if ε is small enough. The maximum principle implies the desired result.

Let $H_{T^*} = (-x^*, x^*) \times (0, T^*]$. On $\bar{H}_{T^*}$, we define a new function

$$h(x,t) = \begin{cases} u_\varepsilon(x,t), & \text{for } (x,t) \in H_1 = [-x^*, 0] \times [0, T^*], \\ u_\varepsilon(-x,t), & \text{for } (x,t) \in H_2 = [0, x^*] \times [0, T^*]. \end{cases}$$

It is easy to see that $h(x,t)$ satisfies the equation (3.3) both in H_1 and H_2. Moreover, $h(0-,t) = h(0+,t)$ and $h_x(0-,t) = h_x(0+,t)$ since $u_{\varepsilon x}(0,t) = 0$, where we denote by $g(a+)$ ($g(a-)$) the right (left) limit of $g(x)$ as $x \to a$.

On H_{T^*}, we define

$$w(x,t) = \frac{h_{\varepsilon x}(x,t)}{h_{\varepsilon t}(x,t)}.$$

Then $w(x,t)$ satisfies

$$w_t - w_{xx} - \frac{2h_{tx}}{h_x} w_x = 0, \ (x,t) \in H_1 \bigcup H_2.$$

Moreover, on the boundary $x = 0$: we find

$$w(0-,t) = w(0+,t), \ w_x(0-,t) = w_x(0+,t).$$

Therefore we can apply the strong maximum principle to obtain

$$\max_{(x,t)\in \bar{H}_{T^*}} |w(x,t)| \le \max_{(x,t)\in \partial_p \bar{H}_{T^*}} |w(x,t)|,$$

where ∂_p means the parabolic boundary of the domain H_{T^*}.

To show the uniform boundness of $w(x,t)$, we only need to show that $w(x,t)$ is uniformly bounded on $\partial_p H_{T^*}$. Since $u_\varepsilon(x,t)$ converges to $u(x,t)$ uniformly in the norm of $C^{2,1}(0,1) \times (0,T)$, we have seen that $u_t(x^*,t) < -\delta$ for $t \in [0,T^*]$, it follows that

$$v(x^*,t) = u_{\varepsilon t}(x^*,t) < -\frac{\delta}{2} \qquad \text{for } t \in [0,T^*].$$

Moreover, by the assumption H(B), $v(x,0) \le \frac{\delta}{2}$, $x \in [0,1]$. By the definition of $h(x,t)$, we see that $w(-x^*,t)$, $w(x^*,t)$ on $[0,T^*]$ and $w(x,0)$ on $[-x^*,x^*]$ are uniformly bounded. It follows that $w(x,t)$ is uniformly bounded in H_{T^*}.

The rest of the proof is the same as that of Theorem 1.

References

1. J. R. Cannon and H. M. Yin, A periodic free boundary problem arising from chemical reaction-diffusion processes, Nonlinear Analysis, theory, meth. and appl., 15(1990), 639-648.

2. X.-Y. Chen, H. Matano and M. Mimura, Finite-point extinction and continuity of interface in a nonlinear diffusion equation with strong absorbtion, preprint.

3. L. C. Evans and B. F. Knerr, Instantaneous shrinking of the support of nonnegative solutions to certain nonlinear parabolic equations and variational inequalities, Illinois J. of Math. 23(1979), 153-166.

4. A. Friedman and M. A. Herrero, Extinction properties of semilinear heat equations with strong absorption, J. Math. Anal. and Appl., 124(1987), 530-546.

5. A. Friedman, Partial Differential Equations of Parabolic Type, Prentice-Hall, Inc, 1964.

6. A. V. Kennington, Convexity of level curves for an initial value problem, J. Math. Anal. and Appl., 133(1988), 324-330.

7. I. Stakgold, partial extinction in reaction-diffusion, Conferenze del Seminario Di Matematica, Dell'Universita Di Bari, 224, 1987.

8. H. M. Yin, Regularity of interfaces in the Stefan problem with mushy regions, Canadian Mathematical Bulletin, to appear.

9. H. M. Yin, The Lipschitz continuous of the interface in the heat equation with strong absorption, Preprint, University of Toronto.

Hong-Ming Yin
Department of Mathematics
University of Toronto, Toronto, Ontario M5S 1A1, Canada.

This work is supported partially by Natural Science and Engineering Research Council of Canada